3rd International Conference on Particle Physics and Astrophysics (ICPPA 2017)

Journal of Physics: Conference Series Volume 934

Moscow, Russia
2 – 5 October 2017

ISBN: 978-1-5108-5403-1
ISSN: 1742-6588

Printed from e-media with permission by:

Curran Associates, Inc.
57 Morehouse Lane
Red Hook, NY 12571

CURRAN ASSOCIATES INC. proceedings.com

Some format issues inherent in the e-media version may also appear in this print version.

Printed by Curran Associates, Inc. (2018)

For permission requests, please contact the Institute of Physics
at the address below.

Institute of Physics
Dirac House, Temple Back
Bristol BS1 6BE UK

Phone: 44 1 17 929 7481
Fax: 44 1 17 920 0979

techtracking@iop.org

Additional copies of this publication are available from:

Curran Associates, Inc.
57 Morehouse Lane
Red Hook, NY 12571 USA
Phone: 845-758-0400
Fax: 845-758-2633
Email: curran@proceedings.com
Web: www.proceedings.com

Table of contents

Volume 934

3rd International Conference on Particle Physics and Astrophysics (ICPPA 2017)
2–5 October 2017, Moscow, Russian Federation

Accepted papers received: 6 December 2017
Published online: 20 December 2017

Preface

Papers

Neutrino and astroparticle physics

Gravitation and cosmology

High energy physics

IOP Conf. Series: Journal of Physics: Conf. Series **934** (2017) 011001 doi:10.1088/1742-6596/934/1/011001

Editorial for the 3rd International Conference on Particle Physics and Astrophysics (ICPPA-2017)

1. Preface

The 3rd International Conference on Particle Physics and Astrophysics (ICPPA-2017) was held in Moscow, Russia, from October 2 to 5, 2017. The conference is organized by the National Research Nuclear University "MEPhI". The aim of the Conference is to promote contacts between scientists and to develop new ideas in fundamental research. We bring together experts and young scientists working on experimental and theoretical aspects of particle physics on the Earth and in the Space and cosmology (figure 1). The conference covers a wide range of topics such as accelerator physics, (astro) particle physics, cosmic rays, cosmology and methods of experimental physics: detectors and instruments. These directions are unified by development of the Standard Model which is evidently not complete. There are deviations from the Standard Model: neutrino oscillations, the dark matter existence, which are main subjects of the Conference. Various aspects of Standard Model testing and search for new phenomena are main subjects of the conference. Among them: flavor physics (LHCb, Belle, BESIII and other experiments), precision multi-boson production measurements (ATLAS, CMS and other experiments), dark matter searches. Electroweak interaction was discussed in the talks given by participants of neutrino physics experiments (Borexino, Neutrino-4, SOX, T2K and others).

Enigmatic properties of neutrinos such as their tiny masses, oscillations between different neutrino types, cannot be explained in frame of the modern theory and require new approaches. Properties of neutrinos influenced the formation of the large-scale structure of the Universe and may be neutrinos are partially responsible for the excess of matter over anti-matter. Also our current challenge is the nature of the Dark matter. Many opportunities arise with the development of neutrino detection technologies comprising: measurements of neutrinos from the Sun, measurements of geo-neutrino from the deep interior of the Earth, detecting neutrino fluxes from supernovae and different astrophysical sources. They were the subject of discussion which have been held at the conference. Neutrino physics experiments make huge demands to technologies like ultra-high precision measurements, huge detector size, excellent reliability and ultra-low radioactive background.

Several plenary talks were devoted to the status of gravitational waves experiment and further development in cosmology and physics of black holes – their formation in the early Universe and internal structure.

Physics of cosmic rays as well as cosmology made the giant step ahead with the obtained last year results. The main contribution has been done by PAMELA experiment which continued during nearly 10 years and provided us with absolutely new results in all fields of space physics, Solar physics and dark matter physics as well.

Figure 1. 3rd International Conference on Particle Physics and Astrophysics (ICPPA-2017).

2. Lists of committees

International Program Committee:

- V. Matveev, JINR, Dubna, Russia – chair
- M. Skorokhvatov, NRC Kurchatov Institute, Moscow, Russia – co-chair
- M. Strikhanov, MEPhI, Moscow, Russia – co-chair
- I. Arsene, University of Oslo, Oslo, Norway
- A. Bolozdynya, MEPhI, Moscow, Russia
- I. Dremin, LPI RAS, Moscow, Russia
- J. Fabris, UFES, Vitória, Brazil
- D. Froidevaux, CERN, Geneva, Switzerland
- H. Gutbrod , FRRC, Darmstadt, Germany
- M. Khlopov, APC, Paris, France
- R. Konoplich, NYU, New York, USA
- Yu. Kudenko, INR RAS, Moscow, Russia
- V. Obraztsov, NRC Kurchatov Institute, IHEP, Protvino, Russia
- A. Ogloblin, NRC Kurchatov Institute, Moscow, Russia
- A. Romaniouk, MEPhI, Moscow, Russia
- P. Senger, GSI, Darmstadt, Germany
- V. Shevchenko, NRC Kurchatov Institute, Moscow, Russia
- C. Spiering, DESY, Zeuthen, Germany
- A. Starobinsky, Landau ITP RAS, Moscow, Russia
- Yu. Suvorov, UCLA, Los Angeles, USA
- A. Vasiliev, NRC Kurchatov IHEP, Protvino, Russia
- R. Veenhof, Uludag University, Bursa, Turkey
- A. Vorobyev, NRC Kurchatov Institute, PNPI, Gatchina, Russia
- M. Vysotsky, NRC Kurchatov Institute, ITEP, Moscow, Russia

Organizing Committee:

- M. Strikhanov, MEPhI, Moscow, Russia – chair
- O. Nagornov, MEPhI, Moscow, Russia – co-chair
- S. Rubin, MEPhI, Moscow, Russia – co-chair

- T. Antonova, MEPhI, Moscow, Russia
- N. Barbashina, MEPhI, Moscow, Russia
- N. Belyaev, MEPhI, Moscow, Russia
- A. Grobov, MEPhI, Moscow, Russia
- M. Karpychev, MEPhI, Moscow, Russia
- A. Kirillov, MEPhI, Moscow, Russia
- S. Koldobskiy, MEPhI, Moscow, Russia
- D. Krasnopevtsev, MEPhI, Moscow, Russia - website support
- S. Smirnov, MEPhI, Moscow, Russia
- E. Soldatov, MEPhI, Moscow, Russia – academic secretary
- A. Taranenko, MEPhI, Moscow, Russia
- P. Teterin, MEPhI, Moscow, Russia
- N. Voznyuk, MEPhI, Moscow, Russia

IOP Conf. Series: Journal of Physics: Conf. Series **934** (2017) 011002 doi:10.1088/1742-6596/934/1/011002

Peer review statement

All papers published in this volume of *Journal of Physics: Conference Series* have been peer reviewed through processes administered by the proceedings Editors. Reviews were conducted by expert referees to the professional and scientific standards expected of a proceedings journal published by IOP Publishing.

IOP Conf. Series: Journal of Physics: Conf. Series **934** (2017) 012001 doi:10.1088/1742-6596/934/1/012001

Neutrino oscillations: status and prospects of accelerator and reactor experiments

Yury Kudenko[1,2,3] **and David Wark**[4,5]

[1] Institute for Nuclear Research of the Russian Academy of Sciences, Moscow, Russia
[2] Moscow Institute of Physics and Technology, Moscow Region, Russia
[3] National Research Nuclear University "MEPhI" (Moscow Engineering Physics Institute) Moscow, Russia
[4] Oxford University, Department of Physics, Oxford, United Kingdom
[5] STFC, Rutherford Appleton Laboratory, Harwell Oxford, and Daresbury Laboratory, Warrington, United Kingdom

E-mail: kudenko@inr.ru, david.wark@stfc.ac.uk

Abstract. A brief overview of recent results from accelerator and reactor neutrino experiments and future prospects are presented. The experimental status of sterile neutrinos is also discussed.

1. Introduction

The discovery of neutrino oscillations [1, 2] has provided convincing evidence for non-zero neutrino masses and leptonic mixing. This phenomenon is the first clear example of the new physics beyond the Standard Model (SM). Oscillation data obtained by now in atmospheric, solar, accelerator and reactor experiments are well described in the framework of three active massive neutrinos which flavour eigenstates ν_e, ν_μ, ν_τ and mass eigenstates ν_1, ν_2, ν_3 with masses m_1, m_2, m_3 are related by the Pontecorvo-Maki-Nakagawa-Sakata (PMNS) 3×3 unitary mixing matrix. This matrix is parametrized by three mixing angles $\theta_{12} \sim 34°$, $\theta_{23} \sim 45°$, $\theta_{13} \sim 9°$ and a CP-violating phase $\delta_{\rm CP}$. Non-zero masses appear in oscillations as two independent squared-mass differences between mass eigenstates $\Delta m^2_{21} = m^2_2 - m^2_1 \simeq 7.5 \times 10^{-5}$ eV^2 and $|\Delta m^2_{32}| = |m^2_3 - m^2_2| \simeq 2.4 \times 10^{-3}$ eV^2.

The number of light neutrinos constrained by LEP experiments is $N_\nu = 2.9840 \pm 0.0082$. However, this result does not exclude existence of sterile neutrinos which could mix with active neutrinos and therefore could modify the 3×3 mixing matrix. There are three exceptions - short baseline neutrino oscillation anomalies - which are not described by the PMNS matrix. 1.The observation of a $\bar{\nu}_e$ excess in a $\bar{\nu}_\mu$ beam by the LSND experiment [3]. 2.The Gallium anomaly - a deficit of ν_e found in measurements with a radioactive neutrino source by the SAGE [4] and GALLEX [5] experiments. 3.The reactor antineutrino anomaly - an observed in several short baseline experiments deficit of reactor antineutrinos relative to the calculated flux [6]. These results, if correct, could be explained by introducing at least one additional squared-mass difference $\Delta m^2 \geq 1$ eV^2. This requires the existence of at least one extra massive neutrino ν_4 which corresponds to a sterile neutrino in flavour basis.

This brief review covers the recent progress in study of neutrino oscillations with an emphasis on a search for CP violation. The new results on sterile neutrinos obtained in various disappearance experiments and expectations from running and future experiments are discussed. Next decade projects aiming at the discovery of CP violation and measurement of $\delta_{\rm CP}$ and determination of the neutrino mass hierarchy are also outlined.

2. Active neutrinos

2.1. New T2K result

The T2K (Tokai-to-Kamioka) experiment [7] collects data since 2010 and released a new preliminary result reported at several conferences this Summer. The result is based on 14.7×10^{20} protons on target (POT) for neutrino mode and 7.6×10^{20} POT for antineutrino mode. For details, one can see Ref. [8]. Statistics accumulated in T2K is summarized in Table 1. As seen

Table 1. Number of electron and muon neutrino and antineutrino detected in T2K and predicted for several values of $\delta_{\rm CP}$. The normal mass hierarchy is assumed.

	T2K data	$\delta = -\pi/2$	$\delta = 0$	$\delta = \pi/2$	$\delta = \pi$
ν_e	74	73.5	61.5	49.9	62.0
$\nu_e + 1\pi$	15	6.92	6.01	4.87	5.78
$\bar{\nu}_e$	7	7.93	9.04	10.04	8.93
ν_μ	240	267.8	267.4	267.7	268.2
$\bar{\nu}_\mu$	68	63.1	62.9	63.1	63.1

from Table 1, T2K observed 89 ν_e and 7 $\bar{\nu}_e$ events while 67 ν_e and 9 $\bar{\nu}_e$ are expected in case of CP conservation. The experimental data show some excess of ν_e events and deficit for $\bar{\nu}_e$ events relative to Monte Carlo data for all values of $\delta_{\rm CP}$. T2K already reached the sensitivity to $\delta_{\rm CP}$ without using reactor information on θ_{13}. The results of the oscillation analysis obtained combining data taken with neutrino and antineutrino beams are shown in Fig. 1.

T2K obtained that the CP conservation hypothesis ($\sin\delta_{\ rmCP} = 0$ or π) is excluded at the confidence level of 2σ. The best fit value $\delta_{\rm CP} = -1.83^{+0.60}_{-0.66}$ rad and a 90% CL interval for allowed $\delta_{\rm CP}$ values is $[-161^\circ,\ -48^\circ]$. T2K data favor the maximal CP violation ($\delta_{\rm CP} \sim -\pi/2$) and the normal mass hierarchy.

It is expected that T2K will accumulate 7.8×10^{21} POT according to the plan to reach the J-PARC beam power about 1 MW by 2021 and then icrease up to 1.3 MW by 2026. An upgrade of near detectors is also needed to reduce the systematic uncertainties from the current level of $\sim 6\%$ to 4%. T2K collaboration proposed to extend the experiment to the second phase T2K-II with the goal to accumulate 20×10^{21} POT by 2026 and to reach a 3σ sensitivity to CP violation in case of $\delta_{\rm CP} \sim -\pi/2$.

2.2. Results from NOνA

The NOνA experiment has measured a ν_μ disappearance using 6.05×10^{20} POT [10] and selecting μ-like candidates at the far detector observed 78 events, while 473 ± 30 were expected without oscillations. NOνA found an indication on a non-maximal value of θ_{23} [9], with two best fit points, both for normal mass hierarchy: $\sin^2\theta_{23} = 0.404$, $\delta_{\rm CP} = 1.48\pi$ and $\sin^2\theta_{23} = 0.623$, $\delta_{\rm CP} = 0.72\pi$, as shown in Fig. 2. The inverted mass hierarchy in the lower octant ($\theta_{23} < 45^\circ$) is disfavored at greater than 93% C.L. for all values of $\delta_{\rm CP}$, and excluded at $> 3\sigma$ significance outside the range $0.97\pi < \delta_{\rm CP} < 1.94\pi$. In general, NO$\nu$A results are in agreement with T2K data, but more statistics is needed to conclude if θ_{23} is maximal or not. It should be also noted

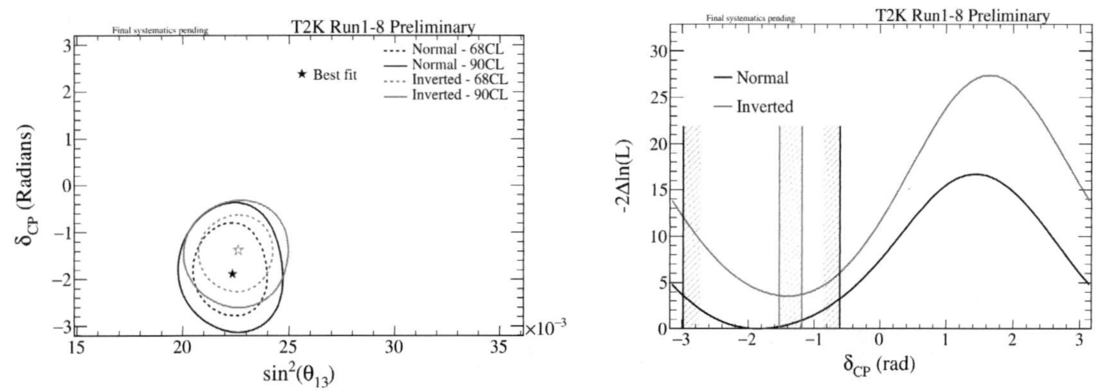

Figure 1. Left: best fit points (stars) and allowed 68% and 90% CL regions of $\sin^2\theta_{13}$ and δ_{CP} for the normal (black) and inverted (red) mass hierarchy obtained in the T2K experiment. Reactor constraint $\pm 1\sigma$ on θ_{13} is used. Right: black and red curves show $2\Delta\ln(L)$ significance at which each value of δ_{CP} is disfavored for normal and inverted mass hierarchy, respectively. Vertical lines limit the allowed regions for δ_{CP} values by the 2σ confidence level. The best fit point is $\delta_{CP} = -1.83$ rad for the normal mass hierarchy. The reactor constraint on θ_{13} is used.

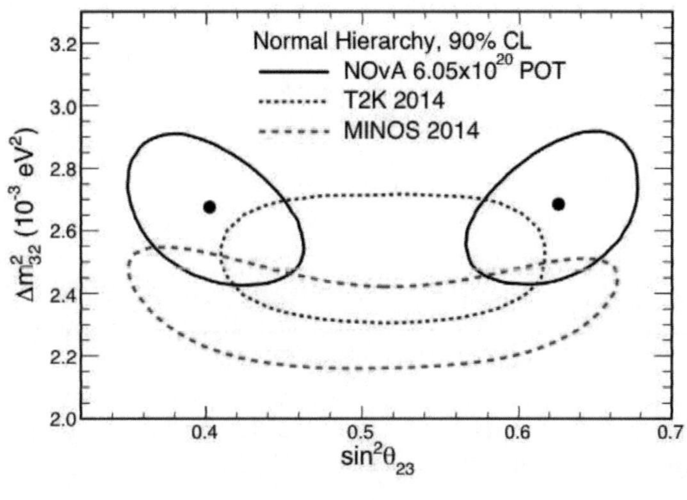

Figure 2. Confidence regions in the plane $|\Delta m^2_{32}|$ - $\sin^2\theta_{23}$ obtained in oscillation experiments NOν, T2K, and MINOS. Black dots show the best fit of the NOνA data.

that the θ_{23} values measured by Super-Kamiokande [11] and IceCube [12] are consistent with the maximal mixing.

2.3. θ_{13} from reactor experiments

The disappearance probability of reactor antineutrinos, as a function of neutrino propagation distance L and energy E can be written:

$$P(\bar{\nu}_e \to \bar{\nu}_e) \simeq 1 - \sin^2 2\theta_{13} \sin^2(\frac{\Delta m^2_{31} L}{4E}). \quad (1)$$

For the antineutrino average energy of 4 MeV, the disappearance of $\bar{\nu}_e$ reaches a maximum at $L \simeq 2$ km. Three reactor experiments, Daya Bay, RENO, and Double Chooz, have been built to measure θ_{13}. Their first results released in 2012 discovered the large value of $\theta_{13} \sim 9^{circ}$. The most precise measurement of θ_{13} was provided by Daya Bay [13]: $\sin 2\theta_{13} = 0.0841 \pm 0 : 0027(\text{stat}) \pm 0.0019(\text{syst})$. Combined analysis of the data from three reactor experiments obtained in [14] is shown in Fig. 3.

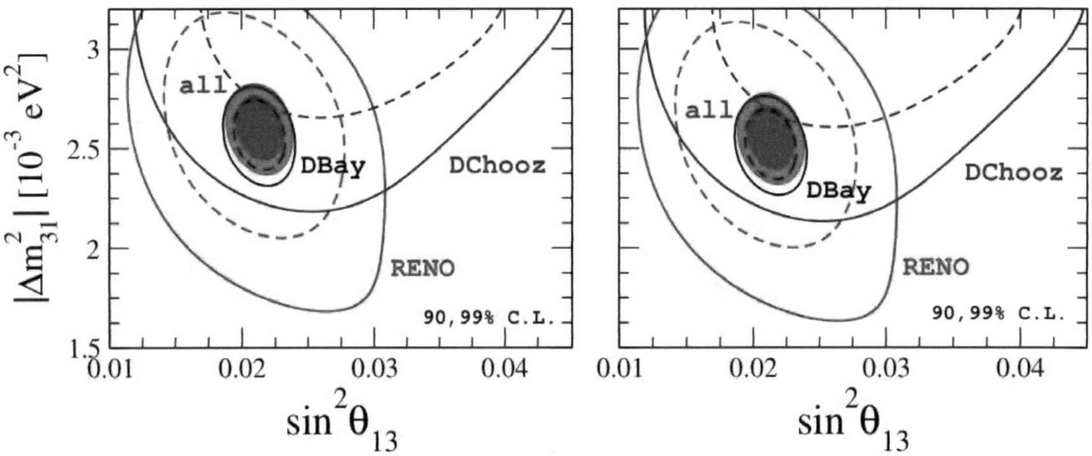

Figure 3. Allowed regions in the plane $|\Delta m^2_{31}|$ - $\sin^2\theta_{13}$ from reactor experiments. The dashed line shows 90% CL, the solid line corresponds to 99% CL. Colored regions show the combination of all experiments. The left (right) panel corresponds to the normal (inverted) mass hierarchy.

3. Future long baseline projects

A rich oscillation experimental program is under preparation to answer the fundamental questions. Is leptonic CP violating or not? What is the neutrino mass hierarchy? What is the value of θ_{23}? Some answers, or hints, can be obtained from running experiments T2K, NOνA, IceCube, Super-Kamiokande, and reactor experiments. However, only the next generation of long baseline accelerator experiments, DUNE, T2HK and reactor experiment JUNO can establish CP violation and determine the mass hierarchy.

3.1. JUNO

The JUNO (Jiangmen Underground Neutrino Observatory) experiment [15] has two main goals: the determination of mass hierarchy and precise measurement of oscillation parameters Δm^2_{21}, $\Delta m^2 31$, and $\sin^2\theta_{12}$. The JUNO detector, a 20 kt spherical unsegmented liquid scintillator detector, will be located at a distance of 53 km from two sites of Yangon and Tanisha, where 6 and 4 nuclear reactors are to be built, respectively. According to the construction plan, the total power of 35.8 GW will be available. The detector will be instrumented by more than 17000 20-inch and about 34000 3-inch photo multiplies (PMTs) ensuring a 77% photocathode coverage. To discriminate between the neutrino hierarchies at a $\geq 3\sigma$ level, the detector energy resolution is required to be $3\%/\sqrt{E(\text{MeV})}$ and the absolute energy scale should be calibrated with a precision of 1%. It is expected that after 6 years of data taking JUNO can distinguish between the true and wrong hierarchy hypothesis at a 4σ significance, assuming no systematics. If systematic uncertainties are considered (non equal baselines from reactors to the detector, backgrounds, the shape of the reactor antineutrino spectrum), the significance to determine the mass hierarchy is reduced to about 3σ. The civil construction begun in 2015 and the experimental hall is expected to be ready in 2018. JUNO plans to start data taking in 2020.

3.2. DUNE

The main scientific goals of the Deep Underground Neutrino Experiment (DUNE) [16] are the sensitive test of CP violation in the leptonic sector, determination the neutrino mass hierarchy, and precise measurements of neutrino oscillation parameters. The proposed liquid Argon far neutrino detector will be built deep underground, at a depth of about 1500 m, in the Sanford Underground Research Facility (South Dakota, USA), about 1300 km from Fermilab where a high intensity wide band on-axis neutrino beam with neutrino energies of 1-6 GeV will be formed. This neutrino beam will cover the first and the second oscillation maxima which correspond to the neutrino energy of 2.5 GeV and 0.8 GeV, respectively. The far detector will consist of four cryostats instrumented with Liquid Argon Time Projection Chambers (LAr TPCs) with a fiducial mass of 40 kt. The first 10 kt DUNE liquid argon module adopts a single-phase technology pioneered by the ICARUS T600 detector. The second DUNE far detector module can be constructed on the basis of a dual-phase TPC if its performance is confirmed as a result of activities at the CERN Neutrino Platform. The LAr TPC technology offers excellent capabilities for position and energy resolution and for high-precision reconstruction of complex interaction topologies over a broad neutrino energy range and will provide a powerful complementarity to the large, underground water Cherenkov or scintillator-based detectors. According to the present timescale, DUNE begins data taking with the first 10 kton module in 2026 and the full configuration will be ready by 2029. The expected sensitivity of DUNE is shown in Fig. 4.

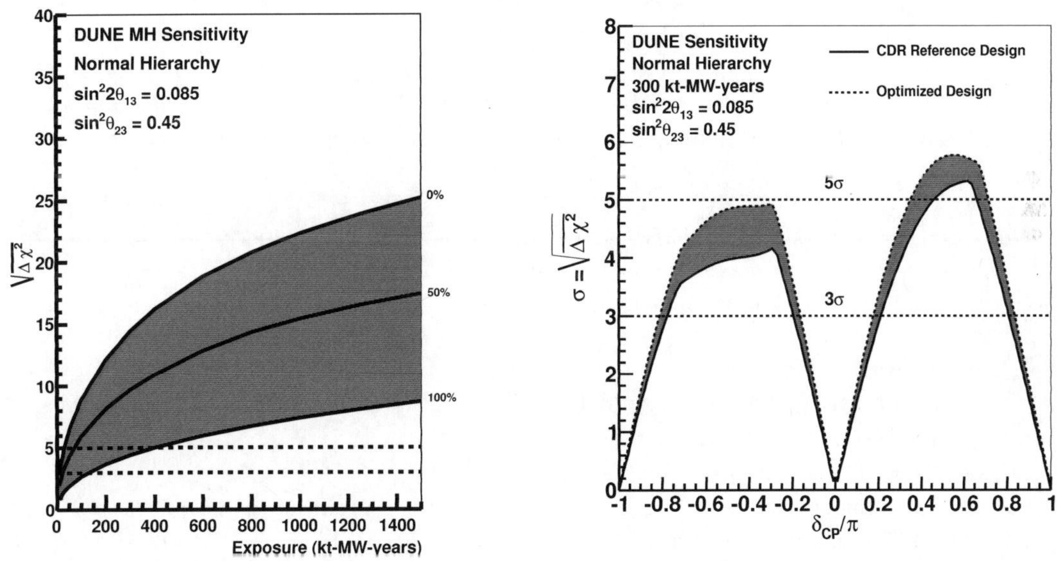

Figure 4. DUNE sensitivities to the mass hierarchy and δ_{CP} [16]. Left: minimal significance with which the mass hierarchy can be determined for all values of δ_{CP}. Right: significance with which the CP violation can be determined for an exposure of 300kt $\cdot$ MW $\cdot$ year assuming the normal mass hierarchy.

3.3. T2HK

The main goal of the T2HK (Tokai-to-Hyper-Kamiokande) experiment is the sensitive search for CP violation in neutrino oscillations [17]. A gigantic water Cherenkov Hyper-Kamiokande

IOP Conf. Series: Journal of Physics: Conf. Series **934** (2017) 012001 doi:10.1088/1742-6596/934/1/012001

detector equipped with newly developed high efficiency and high-resolution PMTs will serve as a far detector in this experiment which will use neutrino and antineutrino beams produced at J-PARC upgraded to the power of $\sim$ 1.3 MW. The baseline design includes one 260 kt Cherenkov detector at a distance of 295 km from J-PARC. The inner detector region of the tank is viewed by 40,000 PMTs that provides a 40% photo-cathode coverage. As in T2K, the 2.5° off-axis beam tuned to the first oscillation maximum will be used. Hyper-Kamiokande is expected to collect about 1000 ν_e and $\bar{\nu}_e$ events after 10 years of data taking with one tank, assuming a 2:1 ratio between running time of neutrino and antineutrino modes. Fig.5 shows the expected significance

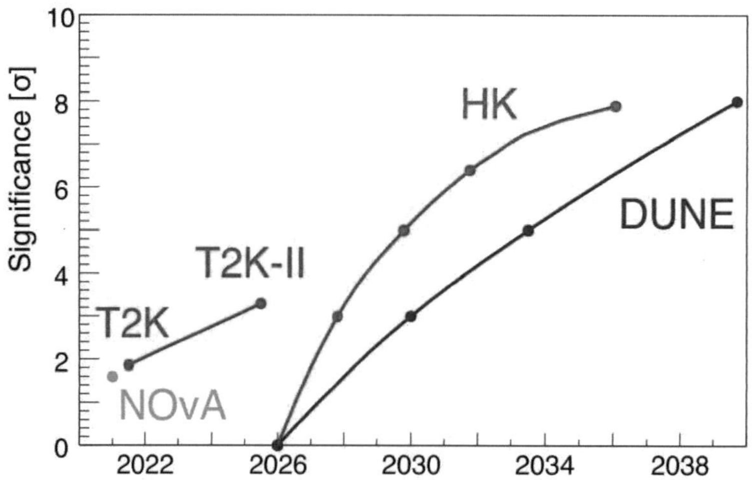

Figure 5. The expected significance of measurement of CP violation in T2HK (denoted as HK) as a function of running time with the beam power of 1.3 MW. The normal mass hierarchy, $\delta_{\mathrm{CP}} = -\pi/2$, and a 2:1 ratio of neutrino:antineutrino modes are assumed. Other experiments are also shown.

of measurement of CP violation in T2HK with one far detector as a function of running time with the beam power of 1.3 MW, for the normal mass hierarchy and $\delta_{\mathrm{CP}} = -\pi/2$. The expected sensitivities of other long baseline experiments are also shown.

4. Sterile neutrinos

A few experiments released their new results on searches for sterile neutrinos during the year.

4.1. IceCube

The IceCube experiment reported the result of the first search for atmospheric $\nu_\mu + \nu_e$ disappearance in the energy range from about 320 GeV to 20 TV [18]. At these energies, sterile neutrinos would produce distortions in the measured zenith angle distributions due to a resonant matter-enhanced oscillations during neutrino propagation through the Earth for the mass splitting of $0.01\mathrm{eV}^2 \leq \Delta m^2 \leq 10\mathrm{eV}^2$. Fig. 6 shows the IceCube result at 99% CL. The allowed region from the appearance experiments LSND and MiniBooNE is excluded at approximately 99% CL further increasing tensions between the appearance and disappearance results.

4.2. Daya Bay and MINOS

The combined analysis [19] of the MINOS and Daya Bay sterile neutrino search data is shown in Fig. 7. In this analysis muon neutrino disappearance data from MINOS are combined with

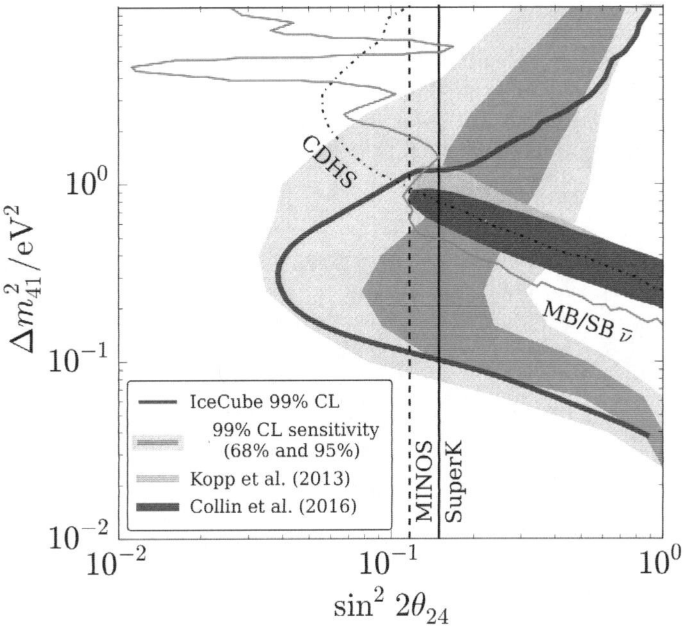

Figure 6. The IceCube 99% CL contour (red solid line) excludes the regions to the right. Also shown the allowed region from LSND/MiniBooNE.

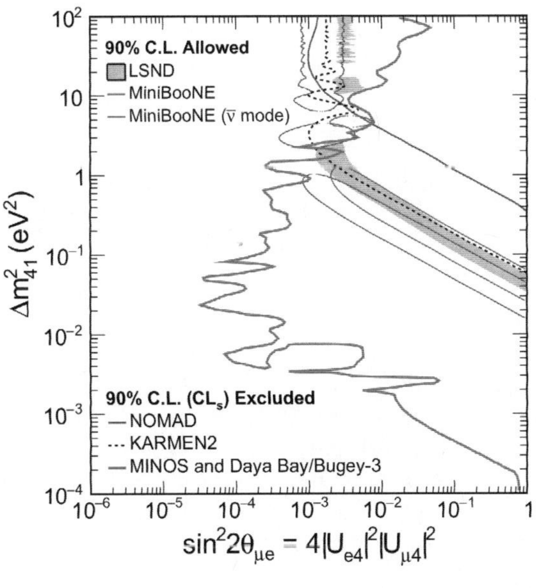

Figure 7. MINOS and Daya Bay + Bugey-3 combined 90% CL limit on $\sin^2 2\theta_{\mu e}$ compared to LSND and MiniBooNE 90% allowed regions. Regions of parameter space to the right of the red contour are excluded.

electron antineutrino disappearance data from Daya Bay and Bugey-3. As seen from Fig. 7, this result excludes at 95% CL the sterile neutrino mixing phase space allowed by LSND and MiniBooNe for $\Delta m^2 14 < 0.8$ eV^2.

4.3. NEOS
The NEOS experiment (Korea) measured the flux of $\bar{\nu}_e$ at a distance of about 22 m from a core of a 2.8 GW reactor for a period of 8 months with a signal-to-background ratio of about 22:1 [20]. The measured energy spectrum of reactor antineutrinos was compared with the simulated spectrum obtained for the 3+1 hypothesis and with the spectrum measured in the Daya Bay experiment. The result of the oscillation analysis is shown in Fig 8.

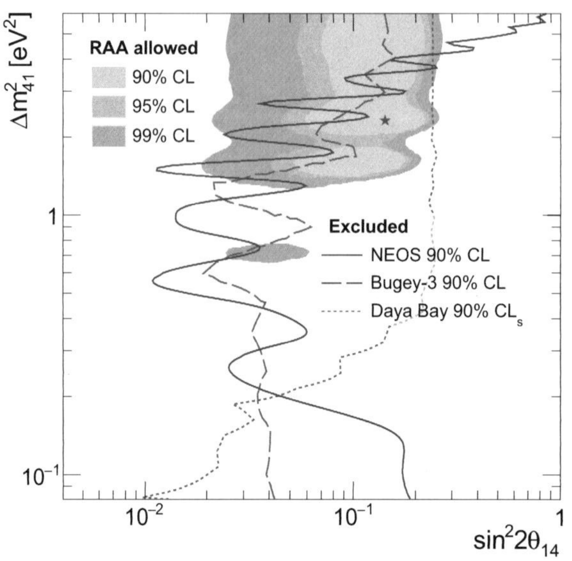

Figure 8. The NEOS 90% CL exclusion curve (solid blue line). The shaded area is the allowed region from the reactor anomaly.

No strong evidence for $3+1$ oscillations is observed and the new upper limit on θ_{14} in the region of Δm^2_{41} is obtained. The sensitivity is expected to be further increased if the reference spectrum from the RENO experiment will be used in the analysis. It should be noted the reactor experiment DANSS recently obtained the preliminary result [21] which is similar to the NEOS one shown in Fig. 8.

5. Conclusion
The known oscillation data is successfully described by the standard 3-flavor neutrino paradigm. Still missing information about unknown parameters - $\delta_{\rm CP}$, mass hierarchy - is expected to be obtained in current and future accelerator and reactor experiments. Anomalies which point to existence of sterile neutrinos are being addressed by many experiments at reactors, accelerators and with intensive neutrino sources. Very interesting physics and new surprises are ahead of us.

5.1. Acknowledgments
This work was supported in part by the RFBR/JSPS grant # 17-52-50038.

6. References
[1] Kajita T 2016 *Rev. Mod. Phys.* **88** 030501.
[2] McDonald A B 2016 *Rev. Mod. Phys.* **88** 030502.
[3] Aguilar A et al 2001 *Phys. Rev.* **D64** 112007.
[4] Abdurashitov J et al 2006 *Phys.Rev.* **C73** 045805.

[5] Hampel W et al *Phys.Lett.* 1998 **B420** 114.
[6] Mention G et al 2011 *Phys. Rev.* **D83** 073006.
[7] Abe K et al 2011 *Nucl. Instrum. Meth.* **A659** 106.
[8] Kim J 2017 Talk at NUFACT2017, 25-27 September 2017, Uppsala, Sweden.
[9] Adamson P et al 2017 *Phys. Rev. Lett.* **118** 231801.
[10] Adamson P et al 2017 *Phys. Rev. Lett.* **118** 1518002.
[11] Wendell R 2016 *PoS, ICRC2015* 1062.
[12] Aartsen M G et al 2017 *Preprint* 1707.07081 [hep-ex].
[13] An F P et al 2017 *Phys. Rev.* **D95** 072006.
[14] De Salas P F et al 2017 *Preprint* 1708.01186 [hep-ph].
[15] Adam T et al 2015 *Preprint* 1508.07166 [hep-ex].
[16] Acciarri R et al 2015 *Preprint* 1512.06148 [physics.ins-det].
[17] Abe K et al 2014 *Preprint* 1412.4673 [physics.ins-det].
[18] Aartsen M G et al 2017 *Phys.Rev.* **D95** (2017) 112002.
[19] Adamson P et al 2016 *Phys. Rev. Lett.* **117** 151801.
[20] Ko Y J et al 2017 *Phys. Rev. Lett.* **118** 121802.
[21] Danilov M 2017 Talk at the 52nd Rencontres de Moriond EW 2017, La Thuile, Italy; https://indico.in2p3.fr/event/13763/.

IOP Conf. Series: Journal of Physics: Conf. Series **934** (2017) 012002 doi:10.1088/1742-6596/934/1/012002

New limits on Heavy Neutrino from NA62

Sergey Kholodenko on behalf of the NA62 Collaboration [1]
NRC "Kurchatov Institute" - IHEP, 142280, 1 Nauki sq., Protvino, Moscow region, Russia
E-mail: sergey.kholodenko@cern.ch

Abstract. The NA62 experiment at CERN collected a large sample of charged kaon decays in flight with a minimum bias trigger in 2007. Upper limits on the rate of the charged kaon decay into a muon and a heavy neutral lepton (HNL) obtained from this data are reported for a range of HNL masses. The NA62 experiment has collected further data in 2015 with a completely new and improved detector. New limits on heavy neutrinos from kaon decays into electron and HNL will be presented.

1. Introduction

One of the possible Standard Model (SM) extensions that allow to accommodating nonzero neutrino masses to describe neutrino oscillations is the Neutrino Minimal Standard Model (νMSM) [1]. The νMSM introduce three additional neutrino mass states at 10 keV/c^2 and GeV/c^2 scales that mix with SM neutrinos. Heavy neutrinos with masses up to 493 GeV/c^2 and 388 GeV/c^2 could be produced in $K \to e\nu_e$ and $K \to \mu\nu_\mu$ decays respectively. A narrow peak in the missing mass spectra of the K_{l2} candidates would be a signature of heavy neutrino produced in $K^+ \to l^+\nu_l$.

[1] R Aliberti, F Ambrosino, R Ammendola, B Angelucci, A Antonelli, G Anzivino, R Arcidiacono, M Barbanera, A Biagioni, L Bician, C Biino, A Bizzeti, T Blazek, B Bloch-Devaux, V Bonaiuto, M Boretto, M Bragadireanu, D Britton, F Brizioli, M B Brunetti, D Bryman, F Bucci, T Capussela, A Ceccucci, P Cenci, V Cerny, C Cerri, B Checcucci, A Conovaloff, P Cooper, E Cortina Gil, M Corvino, F Costantini, A CottaRamusino, D Coward, G DAgostini, J Dainton, P Dalpiaz, H Danielsson, N De Simone, D Di Filippo, L Di Lella, N Doble, B Dobrich, F Duval, V Duk, J Engelfried, T Enik, N Estrada-Tristan, V Falaleev, R Fantechi, V Fas-cianelli, L Federici, S Fedotov, A Filippi, M Fiorini, J Fry, J Fu, A Fucci, L Fulton, E Gamberini, L Gatignon, G Georgiev, S Ghinescu, A Gianoli, M Giorgi, S Giudici, F Gonnella, E Goudzovski, C Graham, R Guida, E Gushchin, F Hahn, H Heath, T Husek, O Hutanu, D Hutchcroft, L Iacobuzio, E Iacopini, E Imbergamo, B Jen-ninger, K Kampf, V Kekelidze, S Kholodenko, G Khoriauli, A Khotyantsev, A Kleimenova, A Korotkova, M Koval,V Kozhuharov, Z Kucerova, Y Kudenko, J Kunze, V Kurochka, V Kurshetsov, G Lanfranchi, G Lamanna, G Latino,P Laycock, C Lazzeroni, M Lenti, G Lehmann Miotto, E Leonardi, P Lichard, L Litov, R Lollini, D Lomidze, A Lonardo, P Lubrano, M Lupi, N Lurkin, D Madigozhin, I Mannelli, G Mannocchi, A Mapelli, F Marchetto, R Marchevski, S Martellotti, P Massarotti, K Massri, E Maurice, M Medvedeva, A Mefodev, E Menichetti, E Migliore,E Minucci, M Mirra, M Misheva, N Molokanova, M Moulson, S Movchan, M Napolitano, I Neri, F Newson, A Norton, M. Noy, T Numao, V Obraztsov, A Ostankov, S Padolski, R Page, V Palladino, C Parkinson, E Pe-dreschi, M Pepe, M Perrin-Terrin, L Peruzzo, P Petrov, F Petrucci, R Piandani, M Piccini, J Pinzino, I Polenkevich, L Pontisso, Yu Potrebenikov, D Protopopescu, M Raggi, A Romano, P Rubin, G Ruggiero, V Ryjov, A Salamon, C Santoni, G Saracino, F Sargeni, V Semenov, A Sergi, A Shaikhiev, S Shkarovskiy, D Soldi, V Sougonyaev, M Sozzi, T Spadaro, F Spinella, A Sturgess, J Swallow, S Trilov, P Valente, B Velghe, S Venditti, P Vicini, R Volpe, M Vormstein, H Wahl, R Wanke, B Wrona, O Yushchenko, M Zamkovsky, A Zinchenko

IOP Conf. Series: Journal of Physics: Conf. Series **934** (2017) 012002 doi:10.1088/1742-6596/934/1/012002

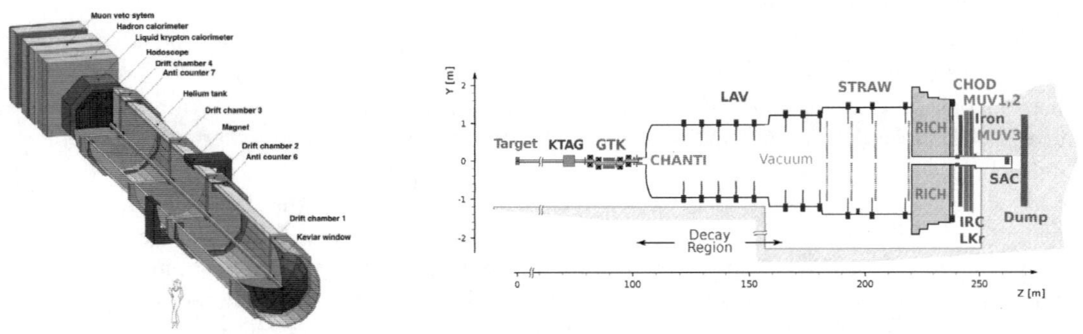

Figure 1. Schematic view of the NA62 experimental layout in 2007 (left) and 2015 (right).

Possible heavy neutrino production in $K^+ \to l^+N$ decay is determined by decay to SM neutrino $K^+ \to l^+\nu_l$, mixing matrix element $|U_{l4}|^2$ and $\rho_l(m_N)$ —kinematic enhancement factor that depends on heavy neutrino mass [2]:

$$\mathcal{B}(K^+ \to l^+N) = \mathcal{B}(K^+ \to l^+\nu) \cdot \rho_l(m_N) \cdot |U_{l4}|^2 \quad (1)$$

Previous searches of possible heavy neutrino production were performed with stopped kaon decays [3], [4]. The upper limits (UL) on mixing matrix element was set on level of 10^{-6} for heavy neutrino masses from 70 to 330 MeV/c^2 and on the level of 10^{-8} for heavy neutrino masses from 175 to 300 MeV/c^2.

This proceedings summarizes results of two searches: analysis of the $K_{\mu 2}$ decay using the data collected in 2007 [5] and a preliminary result of the K_{e2} decay analysis using the data collected during NA62 2015 pilot run.

2. The NA62 experiment at CERN

The NA62 experiment is the successor to the NA48 series of kaon decay in-flight experiments. The experiment is located at CERN North Area High Intensity Facility and operates with 75 GeV/c hadron beam (with 6% kaons) produced on the beryllium target by primary 400 GeV/c protons extracted from CERN SPS.

2.1. The NA62 detector in 2007

During the 2007-08 run (NA62-R_K phase) the main purpose of the experiment was to measure the ratio $R_K = \Gamma(K^\pm \to e^\pm\nu)/\Gamma(K^\pm \to \mu^\pm\nu)$ [6]. The detector was based on the beam line and apparatus of the previous NA48/2 experiment (figure 1 left). The detailed description of the detector and beam line could be found in Ref. [7]. The hadron beam with 6% kaons enters a 114 m long decay volume (fiducial volume) with the vacuum inside. The secondary charged particles produced by kaon decays were registered by the spectrometer with momentum resolution $\sigma/p = 0.48\% \oplus 0.009\% \cdot p$ that consists of a dipole magnet and four drift chambers and was housed in a helium filled tank. The fast trigger signal was produced by the hodoscope CHOD that consists of two planes of scintillation bars. The electromagnetic liquid krypton calorimeter LKr with energy resolution $\sigma_E/E = 3.2\%/\sqrt{E} \oplus 9\%/E \oplus 0.42\%$ and $\sigma_{x,y} = 0.42cm/\sqrt{E} \oplus 0.06$ cm spatial resolution provides high granularity in the transverse plane. Muon veto system MUV located at the end of the detector was used for the muon identification. The beam intensity was reduced compared to NA48/2 by a factor of 10 to provide high efficiency operation with minimum bias trigger conditions.

IOP Conf. Series: Journal of Physics: Conf. Series **934** (2017) 012002 doi:10.1088/1742-6596/934/1/012002

2.2. The NA62 detector in 2015

The main goal of the NA62 experiment is to measure branching ratio of ultra rare decay $K^+ \to \pi^+\nu\bar{\nu}$ with 10% accuracy. The experiment operates with high intensity hadron beam with total particle rate of 750 MHz, of which about 6% is from K$^+$. The detailed description of the NA62 detector and beam line can be found in Ref. [8]. The schematic view of the experimental setup is shown in figure 1 (right). There are significant improvements with respect to 2007 setup that include the precise measurement of the beam kaon time and momenta by the KTAG (differential Cherenkov detector) and GTK. The drift chambers from the NA48/2 experiment were replaced with four stations of straw tubes that are operating in the vacuum. Hermetic photon veto system (LAV, LKr, IRC and SAC). New detectors for particle identification: a RICH detector and muon veto system consist of fast muon hodoscope MUV3 and two hadron calorimeters MUV1 and MUV2.

3. Search for heavy neutrinos in K_{l2} decays

A signature for the possible heavy neutrino production in K_{l2} decay is a narrow spike in the missing mass $m_{miss} = \sqrt{(p_K - p_\mu)^2}$ spectrum, where p_K and p_l are the kaon and secondary lepton 4-momentum respectively. The average beam kaon momentum was measured during the data taking using fully reconstructed $K^+ \to \pi^+\pi^+\pi^-$ decay.

The analyses are based on single-track events. With an assumption that mixing matrix element $|U_{l4}|^2 < 10^{-4}$ and that heavy neutrino decays only to SM particles, the mean free path of heavy neutrino produced in NA62 is greater than 10 km and their decays in-flight could be neglected.

A search for heavy neutrinos production in kaon decays performed using two data samples:

- $K_{\mu 2}$ data sample with 5.977×10^7 kaon decays in fiducial volume collected in 2007 using a minimum bias single track topology with positively charged muon in the final state.
- K_{e2} data sample with 3.01×10^8 kaon decays in fiducial volume recorded during 5 days of operation in NA62 2015 pilot run. The beam intensity varies from 0.4% to 1.3% of the nominal 45 MHz K^+ rate. Trigger conditions require a beam particle to be identified as a K^+, single positively charged secondary track and anti-coincidence with a signal from muon hodoscope (MUV3).

3.1. Event selection

Both analyses require single positively charged tracks. Vertex position reconstructed as a point of a closest distance of approach of the track to the K^+ beam axis reconstructed from $K^+ \to \pi^+\pi^+\pi^-$ decay.

In $K_{\mu 2}$ data sample analysis muons are positively identified by MUV signals associated with the track in time and space. Events with clusters in the LKr that are not associated with the track are rejected. The number of background events from the muon halo evaluated with a control data sample, and from kaon decays, evaluated with simulation.

In K_{e2} data sample analysis, electrons are identified using $0.9 < E/p < 1.15$ criteria and identification algorithm based on the RICH hit pattern. Events with hits in any other photon veto detectors or with clusters not associated with the track are rejected. The background in each mass hypothesis is evaluated from side bands of the data m_{miss} distribution.

The missing mass spectrum of selected K_{l2} candidates from both data and simulation are presented in figure 2.

3.2. New limits on Heavy Neutrino production

The search for the possible heavy neutrino production with $K^+ \to l^+N$ decay performed for different heavy neutrino masses. A signal region for the $K_{\mu 2}$ data sample selected in

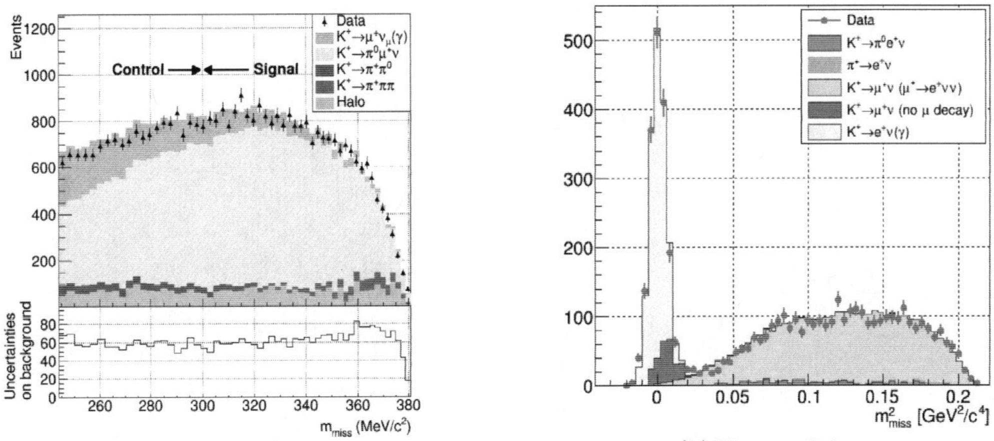

(a) $K_{\mu 2}$ candidates (b) K_{e2} candidates

Figure 2. Missing mass spectrum distribution of $K_{\mu 2}$ (a) and K_{e2} (b) candidates and background estimation from MC.

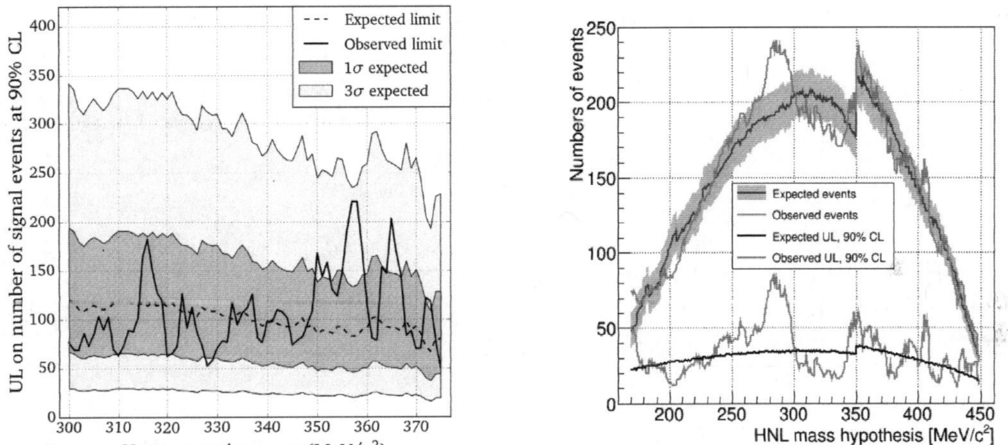

$K_{\mu 2}$ K_{e2}

Figure 3. Observed and expected number of events passing selection criteria of $K_{\mu 2}$ (left) and K_{e2} (right) events as a function of heavy neutrino mass.

$300 < m_{miss} < 375$ MeV/c^2 while the region $m_{miss} < 300$ MeV/c^2 used for the control purpose.

For the K_{e2} data sample a signal region selected in $170 < m_{miss} < 448$ MeV/c^2. The analyses performed with 1 MeV/c^2 step in the signal regions with an additional condition on the reconstructed mass to be within $\pm\sigma_h^l$ window of the assumed heavy neutrino mass, where σ_h^l depends on heavy neutrino mass resolution obtained from simulation. The statistical analysis is performed using Rolke-Lopez method [9] to define 90% confidence intervals for the number reconstructed events with $K^+ \to l^+ N$ decays. The upper limits on the signal events, the expected and observed limits at 90% CL are presented in figure 3 as a function of heavy neutrino (HNL) mass.

The upper limits (ULs) on mixing parameter $|U_{l4}|^2$ for each heavy neutrino mass hypotheses

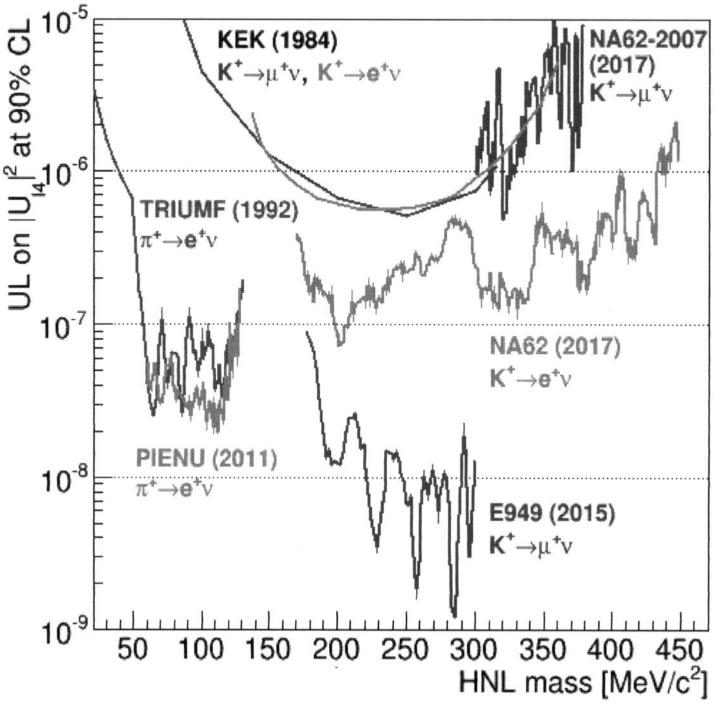

Figure 4. Upper Limit on $|U_{l4}|^2$ at 90% CL from the present NA62 analyses and comparison with the limits using π^+ and K^+ decays from other experiments.

calculated and presented in figure 4 with a comparison to other analyses of heavy neutrinos searched performed in π and K decays.

The new NA62 results improves the existing limits for both $|U_{\mu 4}|^2$ and $|U_{e4}|^2$ in the analyzed signal regions.

References

[1] Asaka T and Shaposhnikov M 2005 The nuMSM, dark matter and baryon asymmetry of the universe *Phys. Lett.* B **620** 17 [hep-ph/0505013]
[2] Shrock R E 1980 New Tests For, and Bounds On, Neutrino Masses and Lepton Mixing *Phys. Lett.* **96B** 159
[3] Hayano R S *et al.* 1982 Heavy Neutrino Search Using K(mu2) Decay *Phys. Lett.* **49**1305
[4] Artamonov A V *et al.* (E949 Collaboration) 2015 Search for heavy neutrinos in $K^+ \to \mu^+\nu_H$ decays *Phys. Rev.* D **91** no.5, 052001 *Preprint* arXiv:1411.3963 [hep-ex]
[5] Lazzeroni C *et al.* (NA62 Collaboration) 2017 Search for heavy neutrinos in $K^+ \to \mu^+\nu_\mu$ decays *Phys. Lett.* B **772** 712 *Preprint* arXiv:1705.07510 [hep-ex]
[6] Lazzeroni C *et al.* (NA62 Collaboration) 2013 Precision Measurement of the Ratio of the Charged Kaon Leptonic Decay Rates *Phys. Lett.* B **719** 326 *Preprint* arXiv:1212.4012 [hep-ex]
[7] Fanti V *et al.* (NA48 Collaboration) 2007 The Beam and detector for the NA48 neutral kaon CP violations experiment at CERN *Nucl. Instrum. Meth.* A **574** 433
[8] Cortina Gil E *et al.* (NA62 Collaboration) 2017 The Beam and detector of the NA62 experiment at CERN *JINST* **12** no.05, P05025, *Preprint* arXiv:1703.08501 [physics.ins-det]
[9] Rolke W A and Lopez A M 2001 Confidence intervals and upper bounds for small signals in the presence of background noise *Nucl. Instrum. Meth.* A **458** 745 [hep-ph/0005187]

IOP Conf. Series: Journal of Physics: Conf. Series **934** (2017) 012003 doi:10.1088/1742-6596/934/1/012003

CeSOX: An experimental test of the sterile neutrino hypothesis with Borexino

M Gromov[1,†], M Agostini[2,9], K Altenmüller[3], S Appel[3], V Atroshchenko[4], Z Bagdasarian[5], D Basilico[6], G Bellini[6], J Benziger[7], D Bick[8], G Bonfini[9], D Bravo[6,10], B Caccianiga[6], F Calaprice[11], A Caminata[12], S Caprioli[6], M Carlini[9], P Cavalcante[9,10], A Chepurnov[1], K Choi[13], O Cloué[29], L Collica[6], M Cribier[29], D D'Angelo[6], S Davini[12], A Derbin[14], X F Ding[2,9], A Di Ludovico[11], L Di Noto[12], I Drachnev[2,9,14], M Durero[29], S Farinon[12], V Fischer[29], K Fomenko[15], A Formozov[15,6], D Franco[16], F Gabriele[9], J Gaffiot[29], C Galbiati[11], M Gschwender[30], C Ghiano[12], M Giammarchi[6], A Goretti[11], D Guffanti[2,9], C Hagner[8], T Houdy[16], E Hungerford[17], Aldo Ianni[9,18], Andrea Ianni[11], N Jonquères[31], A Jany[19], D Jeschke[3], V Kobychev[20], D Korablev[15], G Korga[17], V Kornoukhov[21], D Kryn[16], T Lachenmaier[30], T Lasserre[29], M Laubenstein[9], E Litvinovich[4,21], F Lombardi[9,22], P Lombardi[6], L Ludhova[5,23], G Lukyanchenko[4], L Lukyanchenko[4], I Machulin[4,21], G Manuzio[12], S Marcocci[2,9,12], J Maricic[13], G Mention[29], J Martyn[24], E Meroni[6], M Meyer[25], L Miramonti[6], M Misiaszek[19], V Muratova[14], R Musenich[12], B Neumair[3], L Oberauer[3], B Opitz[8], V Orekhov[4], F Ortica[26], M Pallavicini[12], L Papp[3], Ö Penek[5,23], N Pilipenko[14], A Pocar[27], A Porcelli[24], G Ranucci[6], A Razeto[9], A Re[6], M Redchuk[5,23], A Romani[26], R Roncin[9,16], N Rossi[9], S Rottenanger[30], S Schönert[3], L Scola[29], D Semenov[14], M Skorokhvatov[4,21], O Smirnov[15], A Sotnikov[15], L F F Stokes[9], Y Suvorov[28,4], R Tartaglia[9], G Testera[12], J Thurn[25], M Toropova[4], A Trantel[30], E Unzhakov[14], C Veyssiére[29], A Vishneva[15], M Vivier[29], R B Vogelaar[10], F von Feilitzsch[3], H Wang[28], S Weinz[24], M Wojcik[19], M Wurm[24], Z Yokley[10], O Zaimidoroga[15], S Zavatarelli[12], K Zuber[25] and G Zuzel[19]

[1]Lomonosov Moscow State University Skobeltsyn Institute of Nuclear Physics, 119234 Moscow, Russia
[2]Gran Sasso Science Institute (INFN), 67100 L'Aquila, Italy
[3]Physik-Department and Excellence Cluster Universe, Technische Universität München, 85748 Garching, Germany
[4]National Research Centre Kurchatov Institute, 123182 Moscow, Russia
[5]Institut für Kernphysik, Forschungszentrum Jülich, 52425 Jülich, Germany
[6]Dipartimento di Fisica, Universitá degli Studi e INFN, 20133 Milano, Italy
[7]Chemical Engineering Department, Princeton University, Princeton, NJ 08544, USA
[8]Institut fär Experimentalphysik, Universität Hamburg, 22761 Hamburg, Germany
[9]INFN, Laboratori Nazionali del Gran Sasso, 67100 Assergi (AQ), Italy
[10]Physics Department, Virginia Polytechnic Institute and State University, Blacksburg, VA 24061, USA
[11]Physics Department, Princeton University, Princeton, NJ 08544, USA
[12]Dipartimento di Fisica, Universitá degli Studi e INFN, 16146 Genova, Italy

[13]Department of Physics and Astronomy, University of Hawaii, Honolulu, HI 96822, USA
[14]St. Petersburg Nuclear Physics Institute NRC Kurchatov Institute, 188350 Gatchina, Russia
[15]Joint Institute for Nuclear Research, 141980 Dubna, Russia
[16]AstroParticule et Cosmologie, Université Paris Diderot, CNRS/IN2P3, CEA/IRFU, Observatoire de Paris, Sorbonne Paris Cité, 75205 Paris Cedex 13, France
[17]Department of Physics, University of Houston, Houston, TX 77204, USA
[18]Laboratorio Subterráneo de Canfranc, Paseo de los Ayerbe S/N, 22880 Canfranc Estacion Huesca, Spain
[19]M. Smoluchowski Institute of Physics, Jagiellonian University, 30059 Krakow, Poland
[20]Kiev Institute for Nuclear Research, 03028 Kiev, Ukraine
[21]National Research Nuclear University MEPhI (Moscow Engineering Physics Institute), 115409 Moscow, Russia
[22]Physics Department, University of California, San Diego, CA 92093, USA
[23]RWTH Aachen University, 52062 Aachen, Germany
[24]Institute of Physics and Excellence Cluster PRISMA, Johannes Gutenberg-Universität Mainz, 55099 Mainz, Germany
[25]Department of Physics, Technische Universität Dresden, 01062 Dresden, Germany
[26]Dipartimento di Chimica, Biologia e Biotecnologie, Universitá degli Studi e INFN, 06123 Perugia, Italy
[27]Amherst Center for Fundamental Interactions and Physics Department, University of Massachusetts, Amherst, MA 01003, USA
[28]Physics and Astronomy Department, University of California Los Angeles (UCLA), Los Angeles, California 90095, USA
[29]Commisariat à l'Énergie Atomique et aux Énergies Alternatives, Centre de Saclay, IRFU, 91191 Gif-sur-Yvette, France
[30]Kepler Center for Astro and Particle Physics, Universität Tübingen, 72076 Tübingen, Germany
[31]Commisariat à l'Énergie Atomique et aux Énergies Alternatives, Centre de Saclay, DEN/DM2S/SEMT/BCCR, 91191 Gif-sur-Yvette, France

E-mail: †gromov@physics.msu.ru

Abstract. The third phase of the Borexino experiment that's referred to as SOX is devoted to test the hypothesis of the existence of one (or more) sterile neutrinos at a short baseline ($\sim 5-10\,\mathrm{m}$). The experimental measurement will be made with artificial sources namely with a ^{144}Ce$-^{144}$Pr antineutrino source at the first stage (CeSOX) and possibly with a ^{51}Cr neutrino source at the second one. The fixed ^{144}Ce$-^{144}$Pr sample will be placed beneath the detector in a special pit and the initial activity will be about $100-150\,\mathrm{kCi}$. The start of data taking is scheduled for April 2018. The article gives a short description of the preparation for the first stage and shows the expected sensitivity.

1. Introduction

A number of extensions of the Standard model in elementary particle physics assume an existence of one or more singlet neutrino flavours called sterile. Depending on their masses these particles might play an important role in cosmological processes and astrophysical phenomena. However the influence is minimal when the masses lie in the interval from 1 eV to 1 keV [1]. Additional attention to such kind of light sterile neutrinos is due to the opportunity to explain a few unexpected experimental results at short baseline ($L/E_\nu \sim 1\,\mathrm{m/MeV}$) using extra neutrino types. The unexpected results are well known as the accelerator, gallium and reactor antineutrino anomalies.

For the first time the accelerator anomaly had been observed in the LSND experiment. An excess of $\overline{\nu}_e$ events was registered in a $\overline{\nu}_\mu$ at $\approx 3.8\sigma$ beam [2, 3, 4, 5]. Two similar experiment, KARMEN and MiniBooNE, reported contradictory results. The former didn't find the excess in the transition $\overline{\nu}_\mu \longrightarrow \overline{\nu}_e$ [6] whereas the latter demostrated the excess signals at 2.8σ and 3.4σ

for oscillations $\overline{\nu}_\mu \longrightarrow \overline{\nu}_e$ and $\nu_\mu \longrightarrow \nu_e$ respectively [7, 8].

The gallium anomaly or ν_e disappearance had been revealed in two radiochemical solar neutrino experiments SAGE and GALLEX during processing of calibration data that were acquired with ^{51}Cr and ^{36}Ar artificial radioactive sources [9, 10, 11, 12, 13, 14]. It is reported in [15] that the significance of the anomaly is $\approx 2.9\sigma$.

After revaluation of reactor antineutrino spectra [16, 17] a deficit of counting rate was noticed in almost all reactor neutrino experiments [18] and a recent calculation [15] indicates the effect at $\approx 2.8\sigma$. However the reactor anomaly is strongly weakened by the results of Daya Bay [19].

2. CeSOX overview

The third phase of the Borexino experiment or CeSOX (**S**hort-distance **O**scillations with Bore**X**ino and a **Ce**rium source) [20] is dedicated to the search for sterile neutrinos by means of a compact artificial antineutrino source ^{144}Ce$-^{144}$Pr [21] with characteristic dimensions of 15 cm. The Borexino detector is a large ultra low background liquid scintillator detector. The Borexino target mass for CeSOX equals ~ 240 t and the fiducial volume has a radius of 4 m. Along with that the spatial resolution is about $10\,\text{cm}/\sqrt{E(\text{MeV})}$. It's planned to place the source right beneath the detector center in a special pit. Such configuration of the experiment will allow to make a measurement at distances 4.5 – 12.5 m from the source. Thus one will be observed not only a deficit of counting rate as it occurs in a standard neutrino disappearance experiment but a spectrum distortion with distance (so-called oscillation waves) as well. This complex approach will provide a clear experimental 2D[1] or 3D[2] pattern [21] in case of the existence of sterile neutrinos with $\Delta m^2_{14} \in (0.5, 5.0)\,\text{eV}^2$ (3+1 model, NH). A few characteristic 2D patterns obtained from the Monte Carlo simulation are shown in Figure 1.

The inverse β-decay reaction, $\overline{\nu}_e + p \longrightarrow e^+ + n$, is applied for antineutrino detection. The process has a clear signature consisting of two consecutive events with a delay of $258.7 \pm 0.8(\text{stat}) \pm 2.0(\text{sys})\,\mu$s [23] between them. The former is a prompt event and it's a electron-pósitron annihilation. The released energy is directly related to the $\overline{\nu}_e$ energy as $E_{\text{prompt}} = E_{\overline{\nu}_e} - 0.789\,\text{MeV}$. The latter is a delayed event and this is a 2.22 MeV de-excitation γ-ray that appears as a result of a neutron capture on a proton (on a ^{1}H nucleus).

The inverse β-decay has one significant disadvantage consisting in the presence of an energy threshold $E_{thr} = 1.806\,\text{MeV}$. For selection a radioactive source it means one need to find a source with the β-decay energy of a few MeV and the life-time of more than a year simultaneously. To satisfy the requirements a two-component antineutrino source was chosen. It consists of a long-lived nuclide with a low β-decay energy (^{144}Ce, $Q = 318\,\text{keV}$, $T_{1/2} = 285\,\text{days}$) and a daughter nuclide with a high β-decay energy (^{144}Pr, $Q_1 = 2996\,\text{keV}$, $Q_2 = 2301\,\text{keV}$, $T_{1/2} = 17.3\,\text{min}$). Thus only ^{144}Pr component will be observed in the CeSOX experiment. Herewith two ^{144}Pr decay branches will be measured namely a non-unique first-forbidden transition $0^- \longrightarrow 0^+$ (97.9%) with endpoint energy 2996 keV and a unique first-forbidden transition $0^- \longrightarrow 2^+$ (1.0%) with endpoint energy 2301 keV.

The ^{144}Ce$-^{144}$Pr source has an activity of $(3.7 - 5.5) \cdot 10^{15}$ Bq (100 – 150 kCi). Taking into account the half-life of the source the CeSOX will take data about 1.5 yr. During this period 10^4 events will be acquired. All measurements will be mostly background free. Based on the data of the geo-neutrino study with Borexino [24] it can be argued the background equils $\sim 15\,\text{ev/yr}$ and it is negligible for CeSOX.

[1] The dependence of the counting rate on the L/E_ν, where L is a distance from the source to an antineutrino event and E_ν is an energy of the corresponding event.

[2] The dependence of the counting rate on the energy of the antineutrino event and the distance to the source.

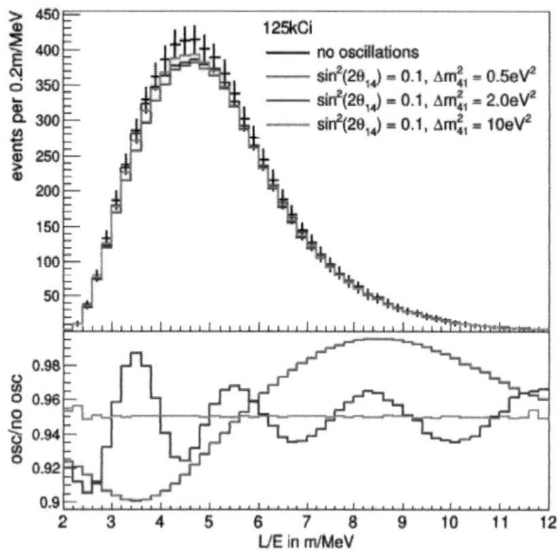

Figure 1. Top plot: the expected counting rate of antineutrino events in the Borexino detector as a function of the ratio L/E_ν for three characteristic sets of the oscillation parameters and for the non-oscillation case. Bottom plot: The ratio of the counting rates that may be observed with and without neutrino oscillations to the sterile component as a function of L/E_ν. There are two clear 2D oscillometric patterns in case of the specific values of the parameters on the plot.

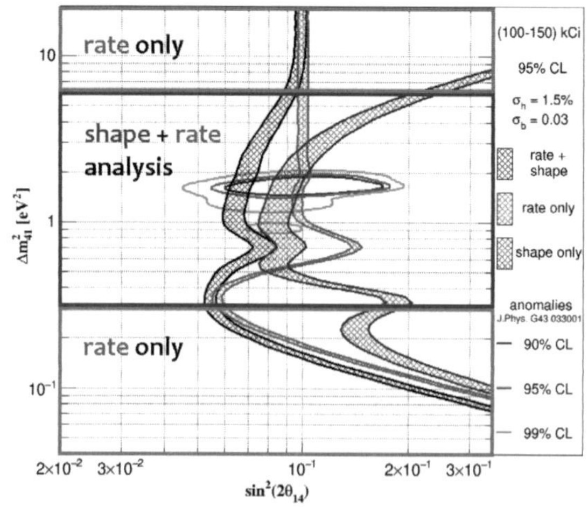

Figure 2. The expected sensitivity of the SOX experiment to the sterile neutrino oscillation parameters for the 3+1 scenario. The possible results are shown taking into account the following systematic uncertainties: a total uncertainty of the normalization rate of 1.5% and an absolute error of 0.03 on the ^{144}Pr electron spectrum shape factor b. The anomalies are taken from [22].

3. Source production and transportation

The ^{144}Ce$-^{144}$Pr antineutrino source is made by extracting cerium from exhausted nuclear fuel. ^{144}Ce is produced in the form of a chemical compound CeO_2 (powder). Due to the cerium β-decay the compound transforms into Pr_2O_3 with the release of oxygen O_2. A special stainless steel capsule was made to withstand high temperature (500 °C) and pressure (up to 6 bar). All additional radioactive backgrounds of the source are almost completely suppressed with a thick tungsten container (minimum thickness 19 cm).

The source will be manufactured by the PA «Mayak» company in Russia and delivered to Gran Sasso in April 2018. During transportation the source capsule and shielding will be inside an extra container (TN MTR). The root to Gran Sasso includes a way to St. Petersbourg (Russia) by train, then to Le Harve (France) by ship, to Saclay (France) by truck and finally to Gran Sasso (Italy) by truck as well. It's expected that the transportation will take three weeks and the source will lose $\sim 5\%$ of its initial activity.

4. Sensitivity and features of the CeSOX experiment

As it is mentioned above the main idea of the experiment is an observation of the 2D or 3D oscillation pattern. The sensitivity of CeSOX is shown in Figure 2. It's clearly seen that the experiment may exclude a large part of the region of the allowed parameters and actually demonstrate the failure of the hypothesis of the sterile neutrino existence at $\sim 3\sigma$. But the results of the investigation largely depend on the precise characterization of the ^{144}Ce$-^{144}$Pr source. This dependence is expressed in the counting rate $N(E_\nu, L, t)$

$$N(E_\nu, L, t) \sim A(t) \times S_\nu(E_\nu, b) \sim \frac{P(t)}{< E(b) >} \times S_\nu(E_\nu, b), \quad (1)$$

where $A(t)$ is an activity of the ^{144}Ce-^{144}Pr source, $S_\nu(E_\nu, b)$ is a shape of the ^{144}Pr spectrum, $P(t)$ is a thermal power of the source, $< E(b) >$ is a mean energy per decay. The b factor is a parameter of the weak finite-size correction $C(Z, W)$ in case of the widespread parameterization: $C(Z, W) \equiv 1+a \cdot W + b/W + c \cdot W^2$, where a, b and c are parameters, Z is a charge of the nucleus, $W = E_e/m_e + 1$ is a total energy of the β^- particle in units of the electron rest mass. There are two physical quantities, b and $P(t)$, whose influence on the sensitivity of the measurements is crucial. Figures 3 and 4 illustrate that fact.

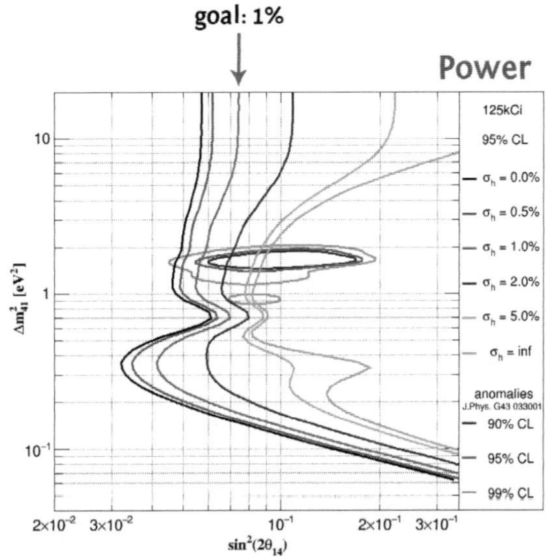

Figure 3. The impact of the calorimetric measurement uncertainty on the sensitivity of the SOX experiment.

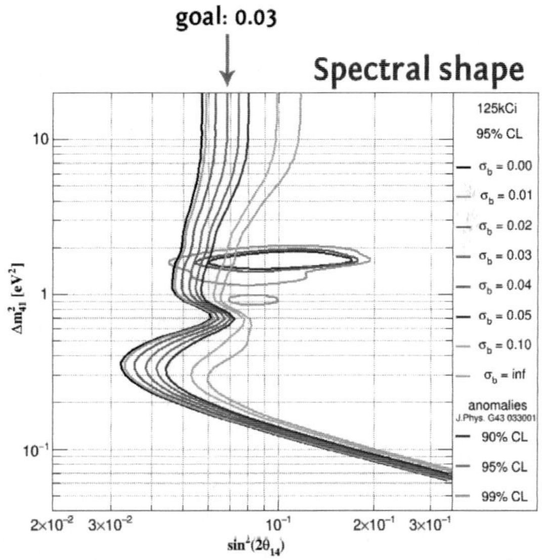

Figure 4. The impact of the uncertainty related to the spectral shape of the main ^{144}Pr β^--decay branch on the sensitivity of the SOX experiment. All curves correspond to 95% CL, unless otherwise explicitly stated.

The thermal power will be measured with two independent calorimeters immediately after delivery of the source to Gran Sasso and also instantly after the end of data-taking. For the precision knowledge of the ^{144}Pr shape factor five spectroscopic experiments are ongoing within the Borexino collaboration. The previous shape factor measurements differ by 10%.

Among other factors affecting the sensitivity it can be mentioned the following: spatial and energy uncertainties, the precise knowledge of the Inner Vessel shape, the Monte Carlo simulation quality and the efficiency of the selection cuts. To minimize the uncertainties introduced by these factors the new comprehensive calibration campaign is scheduled for January and February 2018. A lot of different radioactive sources will be applied: ^{241}Am-^{9}Be+Ni (neutrons), ^{68}Ga-^{68}Ge (positrons), ^{40}K (γ), ^{54}Mn (γ), ^{65}Zn (γ), ^{85}Sr (γ) and ^{222}Rn+^{14}C (α, β and γ) as well.

5. Conclusion

In two years CeSOX may exclude the most part of the region of the allowed parameters. If the oscillation pattern is observed additional measurements with a neutrino source ^{51}Cr might be performed.

Acknowledgments

The author acknowledges the continual support to the SOX project coming from INFN and the Laboratori Nazionali del Gran Sasso. A special thank goes also to CEA-Saclay for their crucial role in the development of the ^{144}Ce$-^{144}$Pr antineutrino source, and to all members of the Borexino collaboration for their active role in the development of the project. The SOX project is funded by the European Research Council project ERC-Adv 320873, P.I. Marco Pallavicini. The research was partly supported by the grants of the Russian Foundation for Basic Research (project № 16-29-13014) and the Russian Science Foundation (project № 17-12-01009).

References

[1] Gorbunov D 2014 *Phys. Usp.* **57** 503
[2] Athanassopoulos C *et al.* (LSND Collaboration) 1995 *Phys. Rev. Lett.* **75** 2650
[3] Athanassopoulos C *et al.* (LSND Collaboration) 1996 *Phys. Rev. C* **54** 2685
[4] Athanassopoulos C *et al.* (LSND Collaboration) 1996 *Phys. Rev. Lett.* **77** 3082
[5] Aguilar A *et al.* (LSND Collaboration) 2001 *Phys. Rev.* D **64** 112007 (*Preprint* hep-ex/0104049v3)
[6] Armbruster B *et al.* (KARMEN Collaboration) 2002 *Phys. Rev.* D **65** 112001 (*Preprint* hep-ex/0203021v1)
[7] Aguilar-Arevalo A *et al.* (MiniBooNE Collaboration) 2009 *Phys. Rev. Lett.* **102** 101802 (*Preprint* 0812.2243v2)
[8] Aguilar-Arevalo A *et al.* (MiniBooNE Collaboration) 2013 *Phys. Rev. Lett.* **110** 161801 (*Preprint* 1303.2588v2)
[9] Abdurashitov J *et al.* (SAGE Collaboration) 2006 *Phys. Rev.* C **73** 045805 (*Preprint* nucl-ex/0512041v1)
[10] Abdurashitov J *et al.* (SAGE Collaboration) 2009 *Phys. Rev.* C **80** 015807 (*Preprint* 0901.2200v3)
[11] Laveder M 2007 *Nucl. Phys. Proc. Suppl.* **168** 344
[12] Giunti C and Laveder M 2009 *Mod. Phys. Lett.* A **22** 2499
[13] Acero M, Giunti C and Laveder M 2008 *Phys. Rev.* D **78** 073009 (*Preprint* 0711.4222v3)
[14] Giunti C and Laveder M 2011 *Phys. Rev.* C **83** 065504 (*Preprint* 1006.3244v3)
[15] Giunti C 2017 Status of the sterile neutrinos *Recent Developments in Neutrino Physics and Astrophysics* (Assergi and L'Aquila, Italy)
[16] Mueller T *et al.* 2011 *Phys. Rev.* C **83** 054615 (*Preprint* 1101.2663v3)
[17] Huber P 2011 *Phys. Rev.* C **84** 024617 erratum: 2012 *Phys. Rev.* C **85** 029901
[18] Mention G, Fechner M, Lasserre T, Mueller T, Lhuillier D, Cribier M and Letourneau A 2011 *Phys. Rev.* D **83** 073006 (*Preprint* 1101.2755v4)
[19] An F *et al.* (Daya Bay Collaboration) 2017 *Phys. Rev. Lett.* **118** 251801 (*Preprint* 1704.01082v2)
[20] Bellini G *et al.* (Borexino Collaboration) 2013 *J. High Energy Phys.* **2013** 38 (*Preprint* 1304.7721v2)
[21] Cribier M *et al.* 2011 *Phys. Rev. Lett.* **107** 201801 (*Preprint* 1107.2335v2)
[22] Gariazzo S, Giunti C, Laveder M, Li Y and Zavanin E 2016 *J. Phys.* G **43** 033001 (*Preprint* 1507.08204v2)
[23] Bellini G *et al.* (Borexino Collaboration) 2013 *JCAP* **2013** 49 (*Preprint* 1304.7381v2)
[24] Agostini M *et al.* (Borexino Collaboration) 2015 *Phys. Rev.* D **92** 031101 (*Preprint* 1506.04610v2)

Neutrino scattering and the reactor antineutrino anomaly

Estela Garcés[1], Blanca Cañas[2], Omar Miranda[1] and Alexander Parada[2]

[1]Departamento de Física, Centro de Investigación y de Estudios Avanzados del IPN, Apdo. Postal 14-740, 07000 Ciudad de México, México

[2] Universidad Santiago de Cali, Campus Pampalinda, Calle 5 No. 6200, 760001, Santiago de Cali, Colombia

E-mail: egarces@fis.cinvestav.mx

Abstract. Low energy threshold reactor experiments have the potential to give insight into the light sterile neutrino signal provided by the reactor antineutrino anomaly and the gallium anomaly. In this work we analyze short baseline reactor experiments that detect by elastic neutrino electron scattering in the context of a light sterile neutrino signal. We also analyze the sensitivity of experimental proposals of coherent elastic neutrino nucleus scattering (CENNS) detectors in order to exclude or confirm the sterile neutrino signal with reactor antineutrinos.

1. Introduction

The reactor antineutrino anomaly [1] and the so called gallium anomaly [2, 3] are around 3σ hints that could not be explained in the context of the standard 3 neutrino oscillation picture. On the other side experimental results of solar, atmospheric and accelerator neutrino experiments can be very well explained by the standard 3 neutrino oscillation scheme [4], however the tension produced by these anomalies could be solved by the existence of a sterile neutrino.

In this contribution we discuss our results on the analysis of reactor antineutrino electron scattering data, including a sterile neutrino. The Texono [5], MUNU [6], Rovno [7] and Krasnoyarsk [8] experiments provide the current measurement of antineutrino electron scattering cross sections at very short baselines.

With the recent discovery of CENNS by the COHERENT Collaboration [9], there is new interest in using CENNS experiments as a tool to probe the Standard Model, making possible the study of neutrino magnetic moment, neutrino non standard interactions, among other interesting neutrino properties, and as we propose in this work of sterile neutrinos.

We analyze the potential of the antineutrino neutrino electron scattering and CENNS at reactor experiments, as alternatives to the detection by inverse beta decay (IBD) where the rector anomaly has been reported. This work is mainly based on [10], the authors refer the reader to it for any further details and for updated versions.

2. The 3+1 neutrino oscillation scheme

The survival probability in the 3+1 neutrino oscillation scheme for short baseline $\bar{\nu}_e$ can be approximated as

$$P^{\mathrm{SBL}}_{\bar{\nu}_e \to \bar{\nu}_e} = \sin^2 2\theta_{ee} \sin^2\left(\frac{\Delta m^2_{41} L}{4E}\right), \tag{1}$$

where

$$\sin^2 2\theta_{ee} = 4|U_{e4}|^2(1 - |U_{e4}|^2), \tag{2}$$

the two oscillation parameters that can be studied in short baseline reactor neutrinos are $\sin^2 2\theta_{ee}$ and Δm^2_{41}.

3. Antineutrino electron scattering measurement

The differential cross section for the antineutrino scattering off electrons, at tree level can be written as [11]

$$\frac{d\sigma}{dT} = \frac{2G_F^2 m_e}{\pi}\left[g_R^2 + g_L^2(1 - \frac{T}{E_\nu})^2 - g_L g_R m_e \frac{T}{E_\nu^2}\right], \tag{3}$$

m_e is the electron mass, G_F is the Fermi constant, and $g_L = 1/2 + \sin^2\theta_{\mathrm{W}}$, $g_R = \sin^2\theta_{\mathrm{W}}$ are the Standard Model couplings. When we perform a chi squared analysis of the current neutrino-electron data [5, 6, 7, 8], and using the most recent predictions for the reactor neutrino flux [12] we find that due to huge uncertainties the exclusion region in the 3+1 oscillation parameter space is not competitive with current global fits. However this is enough to constrain part of the region allowed by the gallium anomaly [2, 3] .

Future measurements from the GEMMA experiment[13] could be very useful to constrain the sterile signal using the antineutrino electron scattering as detection reaction.

4. Sensitivities of CENNS reactor experiments to sterile neutrinos

The coherent elastic neutrino nucleus scattering is a standard model prediction and was proposed in the early seventies by Freedman [14, 15]. In the last few decades several experimental collaborations have been in the quest towards its detection [16, 17, 18, 19], recently it was finally achieved by the first time by the COHERENT experiment using neutrinos from a spallation source.

Table 1. List of CENNS experiments analyzed, reactor composition, energy threshold and baseline.

	^{235}U	^{239}Pu	^{238}U	^{241}Pu	T_{thres}	Baseline
TEXONO(1kg) [16]	0.55	0.32	0.07	0.06	100 eV	28 m
RED100 [20]	0.54	0.33	0.07	0.06	500 eV	19 m
MINER [18]	1.0	-	-	-	10 eV	1-3 m
CONNIE [17]	$\simeq 1.0$	–	–	–	50 eV	30 m

The cross section for CENNS is given by

$$\left(\frac{d\sigma}{dT}\right)^{\mathrm{coh}}_{\mathrm{SM}} = \frac{G_F^2 M}{2\pi}\left[1 - \frac{MT}{E_\nu^2} + \left(1 - \frac{T}{E_\nu}\right)^2\right]\left\{[Z g_V^p + N g_V^n]^2\right\}, \tag{4}$$

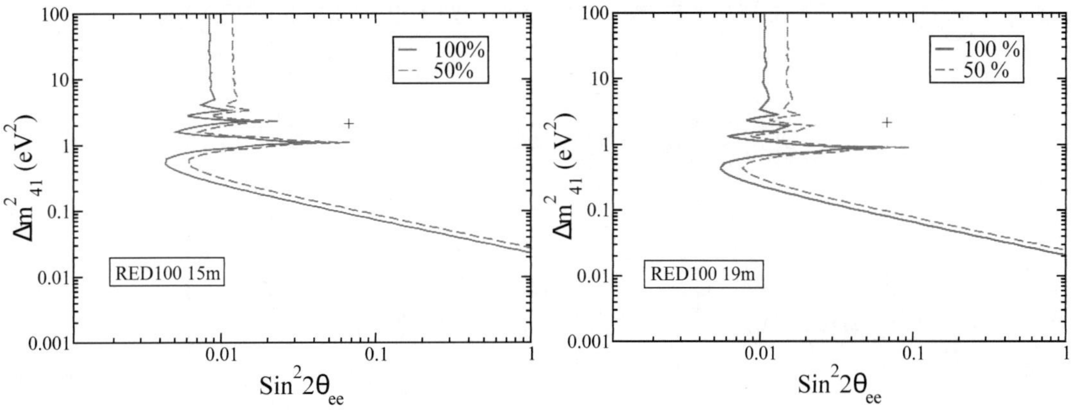

Figure 1. Exclusion regions for the RED100 proposal with a baseline of 15(19) m. The solid (green) line correspond to a detector with 100% efficiency and the dashed (red) lines a 50% efficiency. The current best fit point for the sterile analysis is shown as a reference [10].

G_F ins the Fermi constant, M is the nucleus mass, E_ν the antineutrino energy and T the recoil energy in the nucleus. The neutral current vector couplings are [21]

$$g_V^p = \rho_{\nu N}^{NC}\left(\frac{1}{2} - 2\hat{\kappa}_{\nu N}\hat{s}_Z^2\right) + 2\lambda^{uL} + 2\lambda^{uR} + \lambda^{dL} + \lambda^{dR}$$
$$g_V^n = -\frac{1}{2}\rho_{\nu N}^{NC} + \lambda^{uL} + \lambda^{uR} + 2\lambda^{dL} + 2\lambda^{dR}, \quad (5)$$

where $\rho_{\nu N}^{NC} = 1.0082$, $\hat{s}_Z^2 = \sin^2\theta_W = 0.23126$, $\hat{\kappa}_{\nu N} = 0.9972$, $\lambda^{uL} = -0.0031$, $\lambda^{dL} = -0.0025$, and $\lambda^{dR} = 2\lambda^{uR} = 7.5 \times 10^{-5}$ [22].

We calculate the number of expected events in each detector, by

$$N_{\text{events}}^{\text{SM}} = t\phi_0 \frac{M_{\text{detector}}}{M} \int_{E_{\nu\text{min}}}^{E_{\nu\text{max}}} \lambda(E_\nu) P_{\nu_\alpha \to \nu_\alpha}^{\text{SBL}} dE_\nu \int_{T_{\text{min}}}^{T_{\text{max}}(E_\nu)} \left(\frac{d\sigma}{dT}\right)_{\text{SM}}^{\text{coh}} dT, \quad (6)$$

in the above equation, M_{detector} is the detector mass, ϕ_0 the incoming neutrino flux and t period of time the experiment is working data recording, $\lambda(E_\nu)$ accounts for he neutrino spectrum, E_ν is the neutrino energy, and T recoil energy of the nucleus. $P_{\nu_\alpha \to \nu_\alpha}^{\text{SBL}} = 1$ is the SM case and the $3+1$ oscillation is given by Eq. (1).

The nuclear recoil energy can have a maximum value depending on the neutrino energy and its mass

$$T_{\text{max}}(E_\nu) = 2E_\nu^2/(M + 2E_\nu). \quad (7)$$

We will assume one year of data taking and compare statistically the case of no oscillations, $P_{\nu_\alpha \to \nu_\alpha}^{\text{SBL}} = 1$ the one in Eq. (1). We will use the prescriptions for the experiments summarized in Table (1) with one year of data taking. We show in Fig. (1) the case of the RED100 experiment with two different baselines and two possible efficiencies. The expectative to improve current constraints sterile neutrino mixing is very promising.

In Fig(2) we compare the antineutrino rate measured if a 5% decrease in the ^{235}U is considered [23] (without sterile neutrino) versus the expected ratio assuming a sterile neutrino $\Delta m^2 = 1.7$ eV ^{2} and $\sin^2 2\theta_{ee} = 0.062$ [24]. It is illustrated how experiments with different baselines, thresholds, and fuel proportions can discriminate possible explanations of the reactor anomaly.

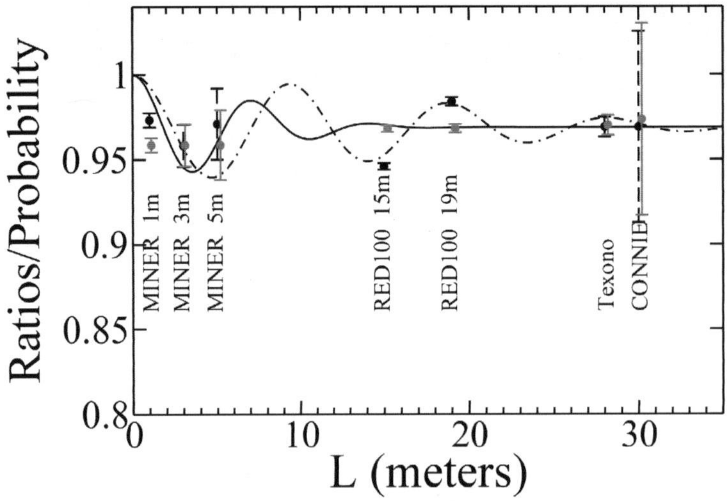

Figure 2. Ratios R of predicted to expected rates for different proposed CENNS experiments. Black dots show the expected ratio for the case of a sterile neutrino with $\sin^2\theta_{ee} = 0.062$ and $\Delta m^2 = 1.7$ eV. The blue dots give the ratio for the case of a decrease in the ^{235}U of 5% [23]. The black and dotted lines represent the average probabilities for a mean energies of 4 MeV, and 6.5 MeV and $\sigma = 15\%$ of resolution. The error bars take into account the statistical errors.

5. Conclusions

In this contribution we obtain an exclusion region to the mixing of a sterile neutrino from reactor antineutrino electron scattering data. We analyze the TEXONO, Rovno, MUNU and Krasnoyarsk experiments taking into account the most recent theoretical prediction of the reactor neutrino flux. We find that given the poor statistics the limit is not competitive with the current best fit point of sterile neutrinos but it is possible to exclude part of the region allowed by the gallium anomaly. We also study the future sensitivity of some CENNS experiments, using reactor neutrinos and we find that CENNS experiments could provide a way to finally discover or rule out the existence of a sterile neutrino.

5.1. Acknowledgments

The authors thank Dmitri Akimov and Alexis Aguilar-Arevalo for useful discussions; E.A.G. would like to thank Gabriel Lopez-Castro and Red temática de Altas Energía (RED FAE) for support to attend the ICPPA 2017 conference. This work has been supported by CONACyT. E. A. G. thanks CONACYT Project No. FOINS-296-2016 (Fronteras de la Ciencia).

References

[1] Mention G, Fechner M, Lasserre T, Mueller T A, Lhuillier D, Cribier M and Letourneau A 2011 *Phys. Rev.* D **83** 073006
[2] Laveder M 2007 *Nucl. Phys. Proc. Suppl.* **168** 344 doi:10.1016/j.nuclphysbps.2007.02.037
[3] Giunti C and Laveder M 2011 *Phys. Rev.* C **83** 065504 doi:10.1103/PhysRevC.83.065504 (*Preprint* 1006.3244)
[4] de Salas P F, Forero D V, Ternes C A, Tortola M and Valle J W F 2017 Status of neutrino oscillations 2017 *Preprint* 1708.01186
[5] Deniz M *et al.* [TEXONO Collaboration] 2010 *Phys. Rev.* D **81** 072001
[6] Amsler C *et al.* [MUNU Collaboration] 1997 *Nucl. Instrum. Meth.* A **396** 115
[7] Derbin A I, Chernyi A V, Popeko L A, Muratova V N,Shishkina G A and Bakhlanov S I 1993 *J. Exp. Theor. Phys Lett.* **57** 768, *Pisma Zh. Eksp. Teor. Fiz.* **57** 755

[8] Vidyakin G S, Vyrodov V N, Gurevich I I, Kozlov Y V, Martemyanov V P, Sukhotin S V, Tarasenkov V G and Turbin E V *et al.* 1992 *J. Exp. Theor. Phys Lett.* **55** 206, *Pisma Zh. Eksp. Teor. Fiz.* **55** 212
[9] Akimov D *et al.* 2017 *Science* 357 (6356) 1123-1126
[10] Cañas B C, Garcés E A, Miranda O G and Parada A 2017 The reactor antineutrino anomaly and low energy threshold neutrino experiments *Preprint* 1708.09518
[11] Vogel P and Engel J 1989 *Phys. Rev.* D **39** 3378
[12] Mueller T A, Lhuillier D, Fallot M, Letourneau A, Cormon S, Fechner M, Giot L and Lasserre T *et al.* 2011 *Phys. Rev.* C **83** 054615
[13] Beda A G, Brudanin V B, Egorov V G, Medvedev D V, Pogosov V S, Shirchenko M V and Starostin A S 2012 *Adv. High Energy Phys.* **2012** 350150
[14] Freedman D Z 1974 *Phys. Rev.* D **9** 1389
[15] Drukier A and Stodolsky L 1984 *Phys. Rev.* D **30** 2295
[16] Wong H T 2008 *Mod. Phys. Lett.* A **23** 1431
[17] Aguilar-Arevalo A *et al.* [CONNIE Collaboration] 2016 *J. Instrum.* **11** 07024
[18] Dutta B, Mahapatra R, Strigari L E and Walker J W 2016 *Phys. Rev.* D **93** 013015
[19] Collar J I, Fields N E, Hai M, Hossbach T W, Orrell J L, Overman C T, Perumpilly G and Scholz B 2015 *Nucl. Instrum. Meth.* A **773** 56
[20] Akimov D Y *et al.* [RED Collaboration] 2013 *J. Instrum.* **8** 10023
[21] Barranco J, Miranda O G and Rashba T I 2005 *J. High Energy Phys.* **0512** 021
[22] Beringer J *et al.* [Particle Data Group Collaboration] 2012 *Phys. Rev.* D **86** 010001
[23] Giunti C 2017 *Phys. Lett.* B **764** 145
[24] Gariazzo S, Giunti C, Laveder M and Li Y F 2017 *J. High Energy Phys.* **1706** 135

IOP Conf. Series: Journal of Physics: Conf. Series **934** (2017) 012005 doi:10.1088/1742-6596/934/1/012005

Development of the reactor antineutrino detection technology within the iDream project.

M Gromov[2] , D Kuznetsov[1], A Murchenko[1], G Novikova[4], B Obinyakov[1], A Oralbaev[1], K Plakitina[1], M Skorokhvatov[1,3], S Sukhotin[1], A Chepurnov[2], A Etenko[1].

[1]National Research Center "Kurchatov institute",123182, Moscow, Russia.
[2]Skobeltsyn Institute of Nuclear Physics, Moscow State University, 119991, Moscow, Russia
[3]National Research Nuclear University MEPhI (Moscow Engineering Physics Institute), 115409 , Moscow, Russia
[4] Institute for Nuclear Research, Russian Academy of Sciences, 117312, Moscow, Russia

Abstract. The iDREAM (industrial Detector for reactor antineutrino monitoring) project is aimed at remote monitoring of the operating modes of the atomic reactor on nuclear power plant to ensure a technical support of IAEA non-proliferation safeguards. The detector is a scintillator spectrometer. The sensitive volume (target) is filled with a liquid organic scintillator based on linear alkylbenzene where reactor antineutrinos will be detected via inverse beta-decay reaction. We present first results of laboratory tests after physical launch. The detector was deployed at sea level without background shielding. The number of calibrations with radioactive sources was conducted. All data were obtained by means of a slow control system which was put into operation.

1. Introduction

The detector iDREAM (industrial Detector for REactor Antineutrino Monitoring) was developed to demonstrate and to prove the ability of the remote control of the atomic reactors by means of the antineutrino flux irradiated from the core of the reactor. Additional information about the design and the other properties have been described in the following paper [1].

2. Physical Start-up of the detector.

Main investigations of the methods of development of a Liquid Organic Scintillator (LOS) on the base of Linear Alkyl Benzene (LAB) have been finished by the middle of 2017 and the LOS without gadolinium for the target of the detector has been synthesized. As a first scintillation admixture was used PPO ($C_{15}H_{11}ON$) in the following concentration 3 g/l, as a second admixture was bis-MSB ($C_{24}H_{22}$) with concentration 20 mg//l.

2.1. Measurements of the main physical parameters of the detector.A subsection

The target of the detector was filled with the scintillator, the circle volume was filled with pure LAB. The target volume PMTs were installed and connected to the DAQ system. On the first stage a setup of working voltage of PMTs was conducted. The high voltage (HV) was adjusting individually for each PMT in accordance with Monte-Carlo simulation. For the PMTs voltage set up ^{137}Cs (E_γ=662 keV) was used. The source was placed in the center of the target.

After the completion of the setup the HV was applied for all PMTs. The spectrum of the total signal is shown on the figure 1. A blue line is representing the cumulative spectrum of the ^{137}Cs with background (BG), a red line is BG, a green line is the "pure" source spectrum (the BG subtracted). To achieve the peak position, the standard deviation and the energy resolution of the detector, the "pure" spectrum was fitted with the Gaussian distribution, so the values for the energy distribution with

IOP Conf. Series: Journal of Physics: Conf. Series **934** (2017) 012005 doi:10.1088/1742-6596/934/1/012005

energy 662 keV are 612.9 and 47.55 for the peak position and the standard deviation respectively, thus the energy resolution is 7.8%.

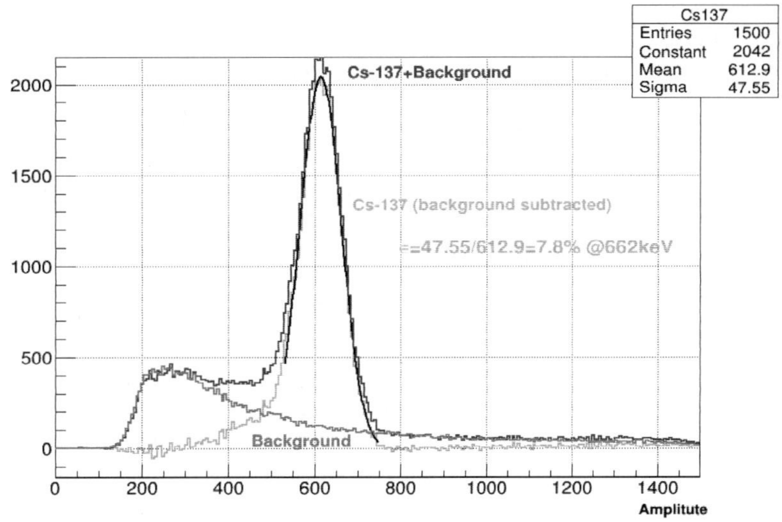

Figure 1. Energy spectrum of the ^{137}Cs in the center of the iDREAM target.

For the detector calibration the ^{207}Bi source was used and it let to determine the detector response in the energy region from 500 keV to 2500 keV. The obtained energy spectrum of ^{207}Bi (BG subtracted) are represented in figure 2, where is also represented ^{137}Cs spectrum. Both of the sources were placed in the center of the detector.

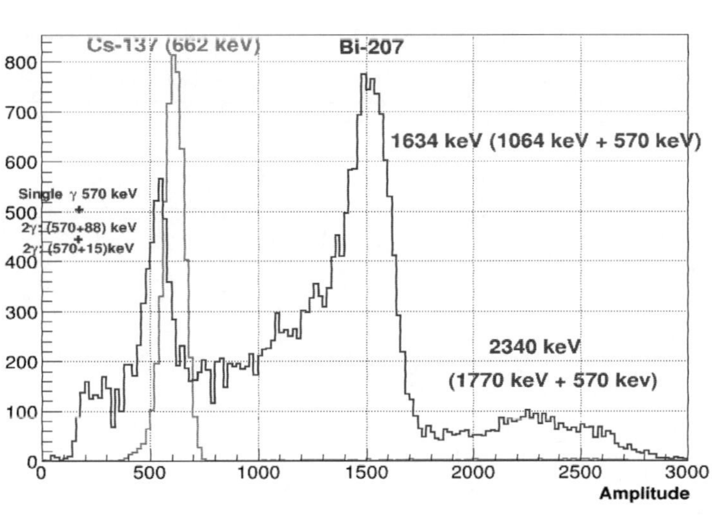

Figure 2. Energy spectrum of ^{137}Cs and ^{207}Bi in the center of the detector.

The peak 662 keV corresponds to a single gamma. Spectrum region around 1634 keV and 2340 keV forms with two cascade gammas with energies (1064 keV + 570 keV) and (1770 keV + 570 keV) respectively.

Around the energy ~600 keV , the spectrum of ^{207} Bi consists of three types of radiation:

- single gamma with energy 570 keV (~6.7%);
- two gammas: 570 keV + 88 keV = 685 keV (~7.4 %);
- two gammas 570 keV + 15 keV = 585keV (~2.3%).

The average irradiated energy in this region is ~612keV.

For setups, which use LOS as a target is typical a little nonlinearity of the response function of the detector especially below ~1MeV. The response function of the iDREAM was based on the measurements of the calibration of ^{207}Bi and ^{137}Cs within the 0.5-2.5 MeV energies. The results are represented in figure 3, where a true energy of gammas is on the X-axis and the ratio of visible to true energy is on the Y-axis.

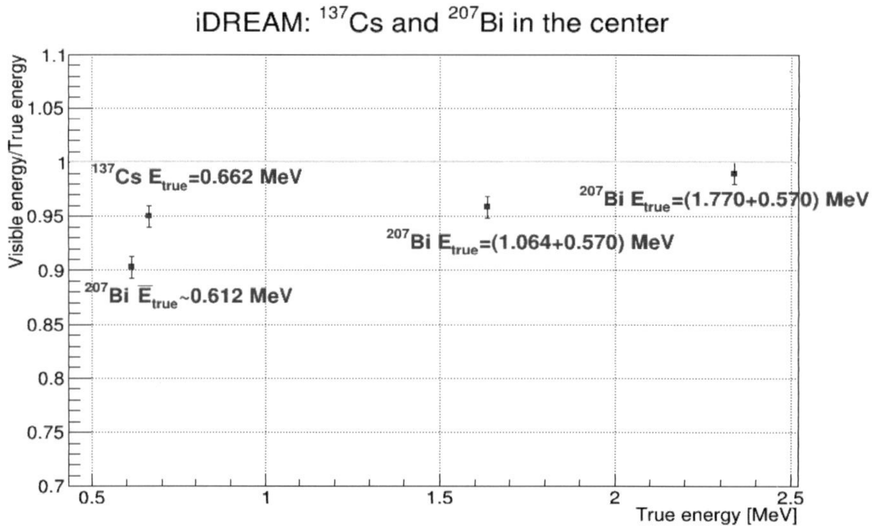

Figure 3. The ratio visible energy to true energy for ^{207}Bi and ^{137}Cs in the iDREAM.

We can see on the figure 3, that besides "usual" nonlinearity an additional decrease of the ratio for the gammas of the ^{207}Bi is observed. It can be explained, that ^{207}Bi irradiates two gammas, each of them gives an impact in the nonlinearity effect during the registration in the liquid scintillator.

The important characteristic is a nonuniformity of a light gathering in the detector. To measure this parameter we used ^{137}Cs, which was moving along the center vertical axis with a pitch of 10 cm and also along the vertical axis in the circle volume close to the target volume wall.

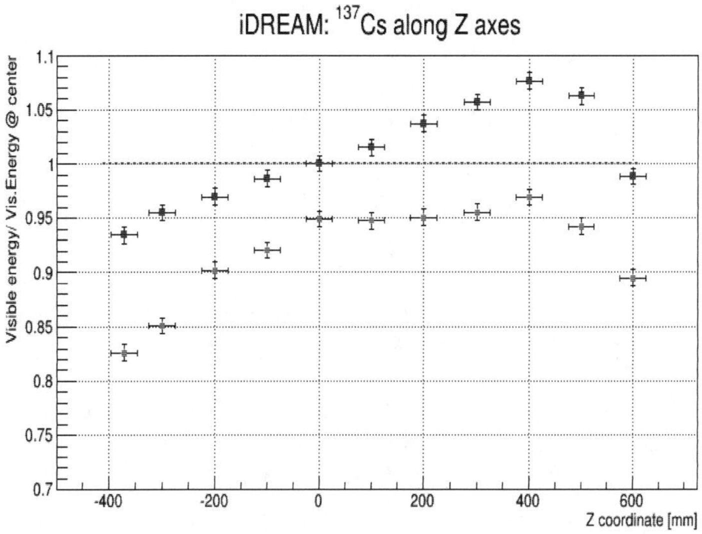

Figure 4. The nonuniformity of the light gathering along the central axis and the axis in the circle volume.

Final results are represented in figure 4. Blue points represent the calibration data along the central axis, red points represent the data along the axis in the circle volume. The maximum difference of the measured energy in the target is 13%.

3. Conclusion.

The physical startup of the detector iDREAM has been conducted successfully. We have measured the general parameters of the detector, such as the energy resolution (7.8%), the response function (figure 2) and the nonuniformity of the energy scale (figure 3) The results correspond with the parameters which were factored in the design.

Acknowledgments.
This work was supported by NRC " Kurchatov institute" within the R&D "The development of the Slow-Control system of the detector iDREAM for the remote monitoring of the atomic reactors".

References

[1] Gromov M B *et al.* 2015 *Moscow Univ. Phys.* **70** 190

IOP Conf. Series: Journal of Physics: Conf. Series **934** (2017) 012006 doi:10.1088/1742-6596/934/1/012006

Neutrino Physics at Kalinin Nuclear Power Plant: 2002 – 2017

I Alekseev[1,2,3], V Belov[4], V Brudanin[4], M Danilov[2,3,5], V Egorov[4,6], D Filosofov[4], M Fomina[4], Z Hons[4,7], S Kazartsev[4,6], A Kobyakin[1,3], A Kuznetsov[4], I Machikhiliyan[1,2], D Medvedev[4], V Nesterov[1], A Olshevsky[4], N Pogorelov[1], D Ponomarev[4], I Rozova[4], N Rumyantseva[4], V Rusinov[1], A Salamatin[4], Ye Shevchik[4], M Shirchenko[4], Yu Shitov[4,8], N Skrobova[1,3,5], A Starostin[1,2], D Svirida[1,2], E Tarkovsky[1], I Tikhomirov[1], J Vlášek[4,9], I Zhitnikov[4], D Zinatulina[4]

[1] Alikhanov Institute for Theoretical and Experimental Physics, B. Cheremushkinskaya 25, Moscow, 117218, Russia
[2] National Research Nuclear University MEPhI (Moscow Engineering Physics Institute), Kashirskoe highway 31, Moscow, 115409, Russia
[3] Moscow Institute of Physics and Technology, 9 Institutskiy per., Dolgoprudny, Moscow Region, 141701, Russia
[4] Joint Institute for Nuclear Research, Joliot-Curie 6, Dubna, Moscow region, 141980, Russia
[5] Lebedev Physical Institute of the Russian Academy of Sciences, 53 Leninskiy Prospekt, Moscow, 119991, Russia
[6] Dubna State University, Universitetskaya 19, Dubna, Moscow Region, 141982, Russia
[7] Nuclear Physics Institute, Řež 130, 250 68 Řež, Cz, Czech Republic
[8] Imperial College London, South Kensington Campus, SW7 2AZ, London, United Kingdom
[9] Czech Technical University in Prague, Zikova 1903/4, 166 36 Prague 6, Czech Republic

E-mail: starostin@itep.ru; igor.alekseev@itep.ru; dmitry.svirida@itep.ru

Abstract. The results of the research in the field of neutrino physics obtained at Kalinin nuclear power plant during 15 years are presented. The investigations were performed in two directions. The first one includes GEMMA I and GEMMA II experiments for the search of the neutrino magnetic moment, where the best result in the world on the value of the upper limit of this quantity was obtained. The second direction is tied with the measurements by a solid scintillator detector DANSS designed for remote on-line diagnostics of nuclear reactor parameters and search for short range neutrino oscillations. DANSS is now installed at the Kalinin Nuclear Power Plant under the 4-th unit on a movable platform. Measurements of the antineutrino flux demonstrated that the detector is capable to reflect the reactor thermal power with an accuracy of about 1.5% in one day. Investigations of the neutrino flux and their energy spectrum at different distances allowed to study a large fraction of a sterile neutrino parameter space indicated by recent experiments and perform the reanalysis of the reactor neutrino fluxes. Status of the short range oscillation experiment is presented together with some preliminary results based on about 170 days of active data taking during the first year of operation.

1. Introduction

The experimental neutrino physics is inseparably linked with nuclear reactors where powerful flows of antineutrino are generated in β-decays of nuclear fragments. A light water power reactor (LWR) with thermal power of 3 GW produces an antineutrino flux ~$2 \cdot 10^{13}$ ν/cm^2·s at a distance of 15 m from the core center. The nuclear reactors were used for studying antineutrino fundamental properties at Savannah River Laboratory (the United States) in 1953–1955 [1] where in the experiment of the Reines and Cowen group interactions of neutrino with matter were observed for the first time. Starting from late 1970s, reactor experiments were initiated in Europe (Gösgen, Switzerland; Bugey and CHOOZ, France) and in the Soviet Union (Rovno and Krasnoyarsk Nuclear Power Plants). ITEP and LNP JINR started experiments on neutrino physics at Kalininskaya nuclear power plant (KNPP) fifteen years ago. KNPP is located in Tver region, about 350 km from Moscow. There are 4 energy units with thermal power of 3100 MW each. The scope of our projects includes both fundamental and applied directions. Experimental setups are GEMMA I and GEMMA II for searching of the neutrino magnetic moment and the DANSS detector (Detector of Anti Neutrino based on Solid Scintillator) for the remote measurements of the reactor parameters in real scale of time and for probing of short-range reactor antineutrino oscillations to the sterile state.

2. Neutrino Magnetic Moment

In recent years a series of remarkable results were obtained in looking for neutrino oscillations in the experiments with atmospheric, solar and reactor neutrinos. Joint analysis of experimental data provided possibility to conclude that neutrino has a finite mass and to define the process of neutrino states mixing. However, some fundamental neutrino properties are not determined up to now. One of them is a neutrino magnetic moment (NMM). Minimal Standard Model predicts very small value of NMM for a massive neutrino, which cannot be observed in present experiments:

$$\mu_\nu = \frac{3eG_F}{8\pi^2\sqrt{2}} \cdot m_\nu \approx 3 \cdot 10^{-19} \mu_B \cdot \frac{m_\nu}{1eV} \tag{1}$$

Here μ_B is Bohr magneton ($\mu_B = eh/2m_e$) and m_ν is a neutrino mass. On the other hand there is a number of extensions of the theory beyond the SM where the Majorana neutrino magnetic moment could be at the level of $10^{-(10-12)}\mu_B$ irrespective of the neutrino mass [2, 3, 4], whereas the Dirac neutrino NMM could not exceed 10^{-14} μ_B [5, 6, 7]. Therefore, observation of the NMM value higher then 10^{-14} μ_B would be an evidence for New Physics and, in addition, would indicate undoubtedly that the neutrino is a Majorana particle. That is why it is rather important to make laboratory NMM measurements sensitive enough to reach the ~$10^{-11}\mu_B$ region.

A laboratory measurement of NMM is based on its contribution to the neutrino-electron scattering. For nonzero NMM neutrino-electron cross section is given by the sum of standard weak interaction cross section $(d\sigma/dT)_{weak}$ and the electromagnetic one $(d\sigma/dT)_{EM}$ [8]:

$$\frac{d\sigma_W}{dT} = \frac{G_F^2 m_e}{2\pi}\left[\left(1-\frac{T}{E_\nu}\right)^2\left(1+2\sin^2\theta_W\right)^2 + 4\sin^2\theta_W - 2\left(1+\sin^2\theta_W\right)\sin^2\theta_W\frac{m_e T}{E_\nu^2}\right] \tag{2}$$

$$\frac{d\sigma_{EM}}{dT} = \left(\frac{\mu_\nu}{\mu_B}\right)\frac{\pi\alpha^2}{m_e^2}\left(\frac{1}{T}-\frac{1}{E_\nu}\right)^2, \tag{3}$$

where E_ν is the incident neutrino energy, T is the electron recoil energy. At low recoil energy ($T<<E_\nu$) the value of $(d\sigma/dT)_{weak}$ becomes almost constant while $(d\sigma/dT)_{EL}$ increases as T^{-1}, so that lowering of the detector threshold leads to considerable increase of μ_ν effect with respect to the weak irremovable contribution. Dependence of differential cross sections of antineutrino-electron scattering on electron recoil energy is shown in figure 1.

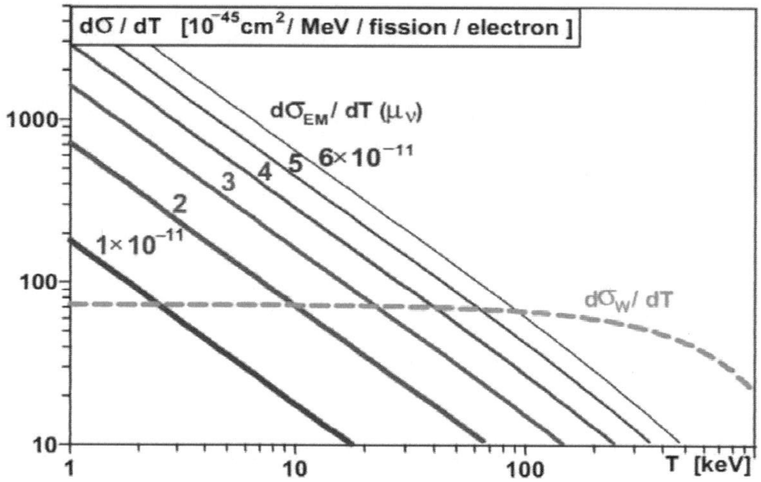

Figure 1. Weak (W) and electromagnetic (EM) cross-sections calculated for several NMM values.

3. Experiments GEMMA I and GEMMA II

Our group was the first to propose a germanium detector for the search for NMM in order to decrease the detection threshold and thereby increase sensitivity of the measurements. In our GEMMA spectrometer [9], we used a 1.5 kg HPGe detector with the energy threshold as low as 2.8 keV. Various methods are implemented for the background suppression. The detector is placed inside a cup-shaped NaI crystal with 14 cm thick walls and surrounded by 5 cm of electrolytic copper and 15 cm of lead. This active and passive shielding reduces the external γ-background in the region from 11.2 to 55 keV to the level of ~2 counts/keV/kg/day. Being located just under the reactor #2 of KNPP (at the distance of 13.9 m from the reactor core center) the detector is well shielded against the hadronic component of cosmic rays by the reactor body and technologic equipment (overburden ~70 m. w. e.). The muon component is reduced by a factor of 10 at ±20° with respect to vertical line and 3 at 70°÷80°. Nevertheless a part of residual muons is captured in the massive shielding and produce neutrons that scatter elastically in the Ge detector and raise the low energy background. To suppress this effect the spectrometer is covered with additional plastic scintillator plates which produce relatively long μ-veto signals (figure 2). The data set was collected in operating ("on") and shut down ("off") states of the reactor during four years. Analysis of the data resulted in the upper limit on the value of NMM $\mu_\nu < 2.9\cdot10^{-11}\mu_B$. Up to now it's the best result in the world [10].

At present time we prepare experiment GEMMA II. The measurements will be performed at third unit of KNPP. The detector is installed directly under the core of an industrial 3.1 GW reactor. The distance to the core center as small as 10.7 m can be reached by means of a movable platform (figure 3). At that distance the density of the antineutrino flow is $5.2\cdot10^{13}$ cm^{-2}c^{-1}. Four point-contact HPGe detectors (~400 g each) produced at LNP JINR will be used for antineutrino registration. The energy threshold is 300 – 350 eV. Sensitivity down to $\mu_\nu \leq 1\cdot10^{-11}$ μ_B is expected to be achieved within 3 years of operation.

4. The DANSS project

The DANSS project is aimed to create a relatively compact neutrino spectrometer for carrying out fundamental and applied experiments. The detector DANSS is based on a cubic meter of plastic scintillator and designed for remote on-line diagnostics of nuclear reactor parameters and search for short range neutrino oscillations. It is located at KNPP under 3.1 GW LWR unit #4, which provides ~50 m w. e. shield against cosmic muons (6-times muon reduction and no cosmic neutrons). The detector is placed on a movable platform which allows to change the distance to the reactor core center in the range 10.7–12.7 m. 2500 scintillator strips (10×40×1000 mm^3) with thin Gd-containing surface

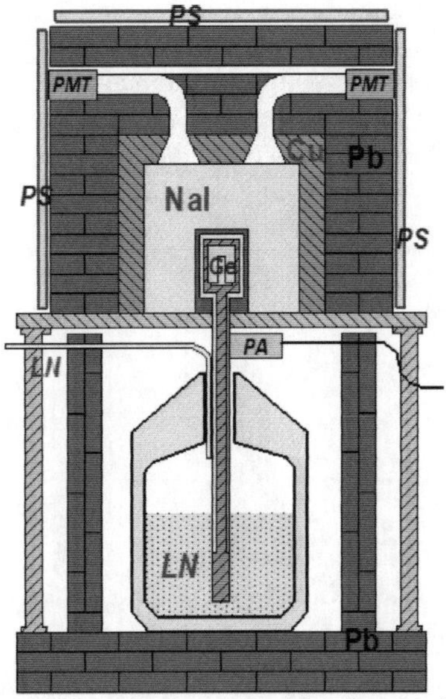

Figure 2. Ge detector inside the active and passive (Cu, Pb) shielding

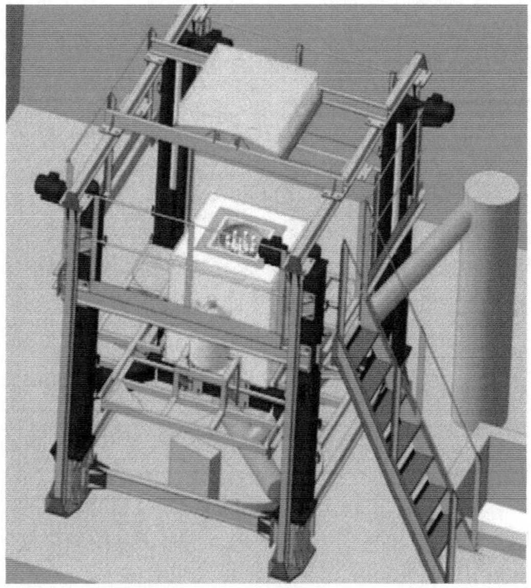

Figure 3. The set up GEMMA II at movable platform under 4-th unit of KNPP

coating are read out individually by SiPMs and in groups of 50 by PMTs. In addition to the overburden by the reactor the detector has multilayer passive shielding and active muon veto. Passive shielding consists of: electrolytic copper frame ~5 cm, borated polyethylene 8 cm, lead 5 cm, borated polyethylene 8 cm. Active muon veto is formed by double layer 3 cm PS plates from 5 sides of the detector (figure 4). Details of the DANSS detector could be found elsewhere [11].

To detect antineutrinos, the reaction of inverse beta decay (IBD) on a free proton of hydrogen in plastic is used:

$$\bar{\nu}_e + \mathrm{p} \rightarrow e^+ + \mathrm{n}. \quad (4)$$

The reaction is well studied theoretically and has the largest cross section in comparison with other neutrino processes. In this reaction practically all antineutrino energy above the energy threshold 1.806 MeV is transferred to the positron. And the positron kinetic energy T is equal to $E_\nu - 1.806$ [MeV]. The problem of separating the reaction (4) from the background is significantly simplified by employing the method of delayed coincidences of the positron ("prompt") and neutron ("delayed") signals. The prompt signal is produced immediately and consists of the positron track ionization and Compton scattering of the two γ-quanta, coming from the annihilation of the stopped positron. The neutron undergoes moderation and then it is captured by gadolinium, which is included in the strip coating. This capture produces a flash of γ-rays with the total energy of about 8 MeV. The time difference between the prompt and the delayed signal is in the tens of microsecond's range, which produces a very good reaction signature. In our experiment separate recording of positron and neutron candidates is carried out, while the building of the delayed coincidence pairs is performed during "off line" data treatment.

5. Measurement of the reactor thermal power

The possibility of the reactor parameters monitoring with an antineutrino detector is based on the unique dependence of antineutrino spectra and flux densities on the condition and characteristics of nuclear reactors. The counting rate of antineutrino detector N_ν can be explicitly expressed in terms of the reactor power:

$$N_{\nu} = \gamma(1+k)W_{th} \quad (5)$$

where N_{ν} is the counting rate of the neutrino detector, W_{th} is the thermal power of a nuclear reactor, γ is a proportionality factor, and $(1+k)$ is the coefficient taking into account the nuclear fuel composition.

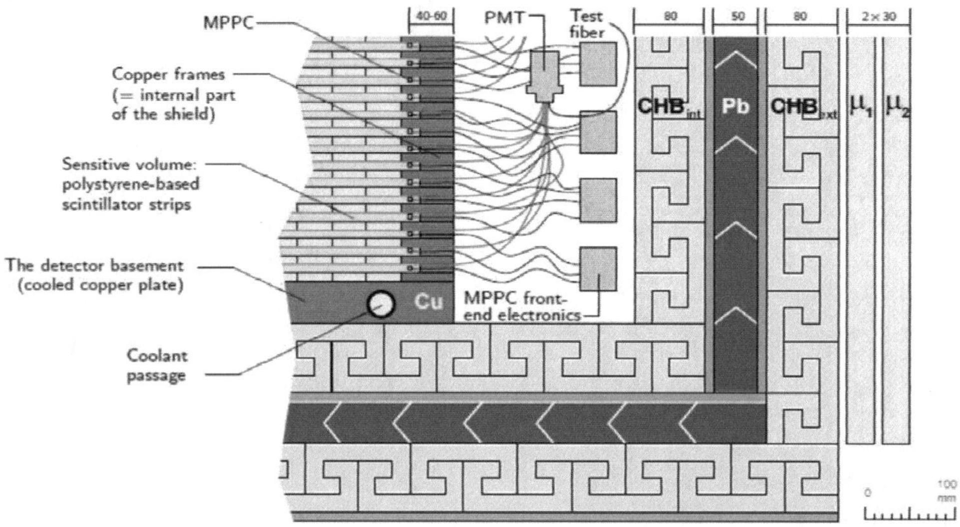

Figure 4. Composition of the detector shielding.

It follows from (5) that the reactor power determination accuracy depends on the statistical accuracy of N_{ν} measurements, the systematic error in determination of the proportion factor γ, while the coefficient $(1+k)$ varies in the course of the operation period. To measure the reactor power, one should know the factor γ. It can be found by calibration of the detector. For detector near reactor core the neutrino counting rate of ~10^4 events/day can be achieved. In this case proportionality factor γ can be determined with about 0.3% accuracy during 10 days of the calibration. The coefficient $(1 + k\text{ (T)})$ is known from calculations with similar accuracy $\sigma_{(1+k)}$ ~ 0.3% [12]. W_{th} can be determined using calorimetric measurements with the accuracy of about 0.5% [13]. This leads to total systematic error $\sigma_{\Sigma} = (\sigma_{\gamma}^2 + \sigma_k^2 + \sigma_{th}^2)^{0.5} \approx 0.6$ %. DANSS detector currently takes statistics at full speed of about 5000 antineutrino events per day with background < 1%. Combining this with the statistical error of 1.4% one gets the total error in the W_{th} determination of about ~1.5% in one day of measurement (figure 5).

6. Searching for "sterile" neutrino

There are several indications in favor of existence of the 4th neutrino flavor – "sterile", which also can be seen in oscillations [14]. The oscillation parameters ($\Delta m^2_{new} > 1.5$ eV^2 and $\sin^2(2\vartheta_{new}) = 0.14\pm0.08$) proposed by G. Mention et al. [15] assume relatively short oscillation range. The ideal distance for direct search for such oscillations would be around five to twenty meters from the reactor core. Since the detector DANSS is located just below the core of an industrial reactor it possesses a number of significant advantages in searching for "sterile" neutrino. Firstly, the detector position below the reactor provides high neutrino flux and an overburden of about 50 m w. e., protecting from cosmic muons and fast neutrons. Secondly, the lifting gear allows the detector movement by nearly 2 m and gives opportunity to measure neutrino spectrum at various distances with the same detector. Both spectra are measured in the same experiment with the same detector. No dependence on the theory, absolute detector efficiency or other experiments.

The experiment was started in April 2016. Data are taken at 3 distances 10.7 m (Up), 11.7 m (Middle), and 12.7 m (Down) from the reactor (center to center). Details of the data set and its treatment could be found in [16]. In 'Up' detector position about 5000 neutrino events/day in fiducial volume of 78% is recorded with the background not exceeding 1%. The 'Up', 'Middle' and 'Down' positron spectra at total statistics 674489 events collected till May 2017 are shown in figure 6. The spectra shows only pure kinetic energy of the positron (annihilation photons not included). In figure 7 Down/Up positron spectrum ratio is shown. As it seen from the plot the ratio does not contradict to a straight line with current statistics.

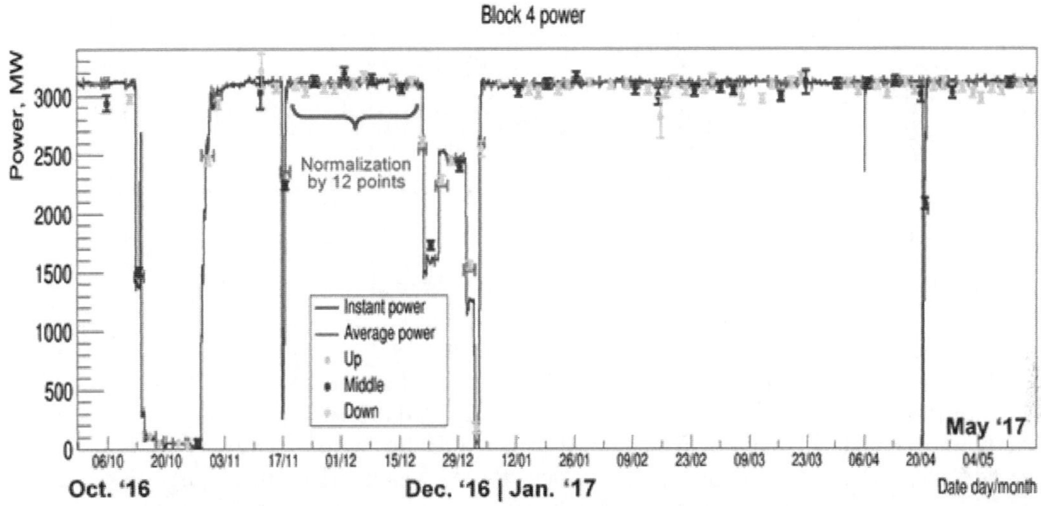

Figure 5. IBD flux measured by the DANSS perfectly matches the actual reactor power monitor data, reported by the KNPP.

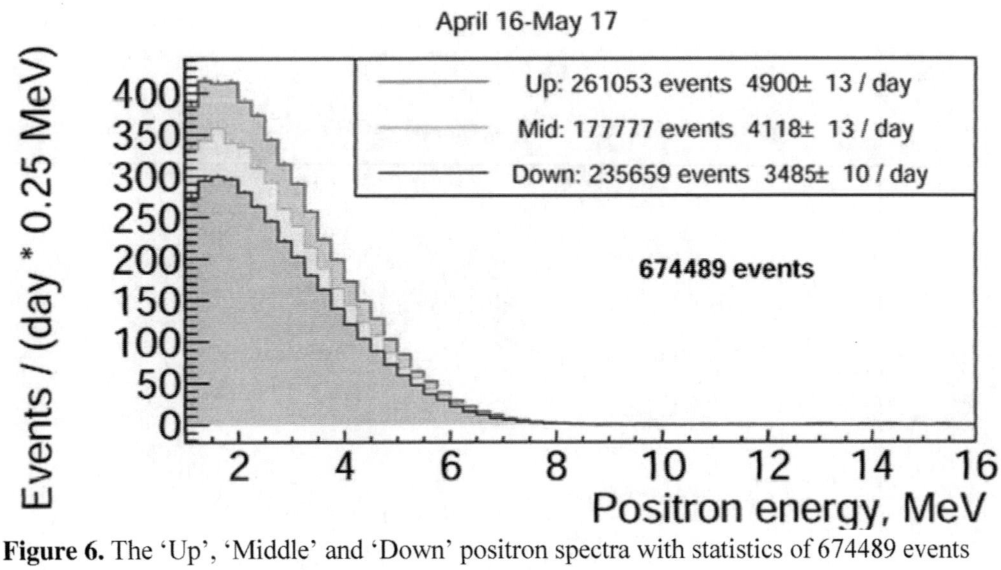

Figure 6. The 'Up', 'Middle' and 'Down' positron spectra with statistics of 674489 events

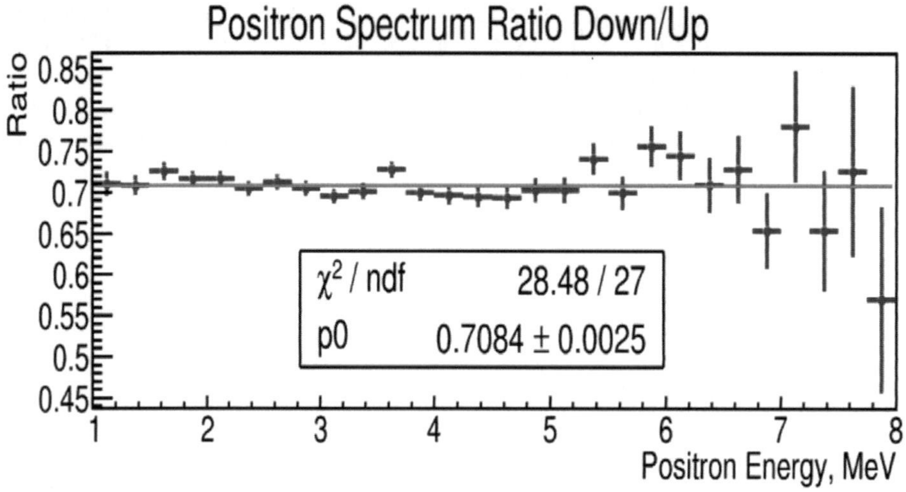

Figure 7. The 'Up'/ 'Down' positron spectrum ratio with statistics by May 2017.

The exclusion region for the parameters of "sterile" neutrinos was calculated using Gaussian CLs method [17] which is more conservative than usual Confidence Interval method. A large fraction of allowed parameter space is already excluded by preliminary DANSS results based on the data sample collected before May 2017 (figure 8). The analysis is performed using only ratio of positron spectrum at different distance (independent on ν spectrum, detector efficiency). Theoretical curves for each Δm^2 and $\sin^2(2\vartheta)$ calculated based on: model neutrino spectrum from Huber and Mueller [18], fuel burning profile from NPP, detector size and detector energy resolution. Systematics studies include variations

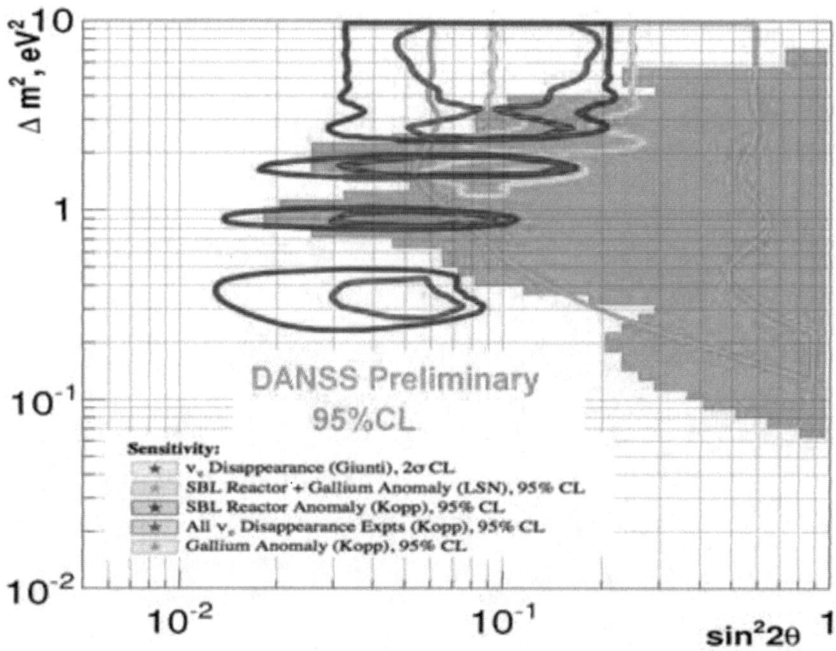

Figure 8. DANSS preliminary (95% CL) results for exclusion region for parameters of "sterile" neutrinos. Compilation of allowed region was taken from [19].

This summer DANSS took data during 40 days of the scheduled reactor shutdown, which will allow even better background estimates. Further plans include improvements on MC for perfect reflection of detector energy response and elaboration of analysis methods for better sensitivity.

Acknowledgments
The authors are grateful to the directorates of ITEP and JINR for constant support of this work. The authors appreciate the administration of the KNPP and the staff of the KNPP Radiation Safety Department for permanent assistance in the experiment. This work is supported in part by the Russian State Corporation ROSATOM (state contracts H.4x.44.90.13.1119 and H.4x.44.9B.16.1006) and by the Russian Foundation for Basic Research, grant 09-02-00449. Young JINR physicists were supported by JINR grants 14-202-(07,08), 15-203-(02,03,07,10), 16-202-(03,04), 16-203-(02,03) and Czech Ministry of Education, Youth and Sports INGO II-LG14004. DANSS detector operation and data analysis becomes possible due to the valuable support from Russian Science Foundation grant 17-12-01145.

References

[1] Reines F and Cowen C L 1953 *Phys. Rev.* **92** 830
[2] Voloshin M B, Vysotsky M I and Okun L B 1986 *Sov. Phys. JETP* **64** 446
[3] Fukugita M and Yanagida T 1987 *Phys. Rev. Lett.* **58** 1807
[4] Pakvasa S and Valle J 2003 Neutrino properties before and after KamLAND (*Preprint* hep-ph/0301061)
[5] Bell N F, Cirigliano V, Ramsey-Musolf M J, Vogel P and Wise M B 2005 How magnetic is the Dirac neutrino? *Phys. Rev. Lett.* **95** 151802 (*Preprint* hep-ph/0504134)
[6] Bell N F, Cirigliano V, Ramsey-Musolf M J, Vogel P and Wise M B 2006 Magnetic moments of Dirac neutrinos *AIP Conf. Proc.* **842** 874-76 (*Preprint* hep-ph/0601005)
[7] Giunti C and Studenikin A 2010 *J. Phys.: Conf. Series* **203** 012100 (*Preprint* hep-ph/2010.1502)
[8] Vogel P and Engel J 1989 *Phys. Rev.* D **39** 3378
[9] Beda A *et al.* 2007 *Phys. At. Nucl.* **70** 1873 (*Preprint* hep-ex/0705.4576)
[10] Beda A *et al.* (GEMMA Collaboration) 2012 The results of search for the neutrino magnetic moment in GEMMA experiment *Adv. High Energy Phys.* **2012** 350150
[11] Alekseev I *et al.* 2016 *JINST* **11** 11011
[12] Klimov Yu *et al.* 1994 Neutrino method remote measurement of reactor power and power output *Atomic Energy* **76** 123
[13] Djurcic Z *et al.* 2008 Uncertainties in the antineutrino production of nuclear reactor (*Preprint* hep-ex/0808.0747)
[14] Abazajian K N *et al.* 2012 Light Sterile Neutrinos: A White Paper FERMILAB-PUB-12-881-PPD (*Preprint* hep-ph/1204.5379)
[15] Mention G *et al.* 2011 *Phys. Rev.* D **83** 073006 (*Preprint* hep-ex/1101.2755)
[16] Alekseev I *et al.* 2016 *Preprint* ins-det/1606.02896
[17] Qian X *et al.* 2014 *Preprint* hep-ex/1407.5052
[18] Huber P 2011 *Phys. Rev.* C **84** 024617
Mueller T A *et al.* 2011 *Phys. Rev.* C **83** 054615
[19] Achenfelter J *et al.* 2015 *Preprint* ins-det/1512.02202

IOP Conf. Series: Journal of Physics: Conf. Series **934** (2017) 012007 doi:10.1088/1742-6596/934/1/012007

Heavy neutrinos effects for oscillation of light neutrinos at short distances

V V Khruschov[1,2] **and S V Fomichev**[1,3]

[1]NRC Kurchatov Institute, Academician Kurchatov Place 1, Moscow, 123182 Russia
[2]Center for Gravitation and Fundamental Metrology, VNIIMS, Ozernaya Str. 46, Moscow, 119361 Russia
[3]Moscow Institute of Physics and Technology (State University), Institutskii Lane 9, Dolgoprudnyi, Moscow Region, 141700 Russia

E-mail: khruschov_vv@nrcki.ru; fomichev_sv@nrcki.ru

Abstract. The transition and survival probabilities for the different light neutrinos are calculated and displayed with account of the contributions of heavy neutrinos. It is shown that in the case of the mixing matrix of a definite type the explanation of the neutrino anomalies at short distances is possible. A new parametrization and a certain form of the mixing matrix for light and heavy neutrinos are used with allowance for the possible violation of CP invariance.

The problem of new particles, which are additional to the Standard Model (SM) particles is amongst the most important problems of the particle physics. New particles are necessary for numerous models beyond the SM, and they include supersymmetric particles, new particles for Grand Unification Theories, new particles of phenomenological models, etc. As an example, let us point at sterile neutrinos (SNs) and dark matter particles (DMPs). There are many indications of their existence in cosmology, astrophysics and neutrino physics. For instance, a number of neutrino ground-based experiments at short distances point to the presence of neutrino- and antineutrino-flux anomalies (so-called LNSD, Reactor and Gallium anomalies) [1–3]. If it is confirmed at a high confidence level, it will be beyond the Minimally Extended Standard Model (MESM). MESM is the SM with three active neutrinos having different masses. A possible explanation of short distance anomalies is SNs existence with masses about 1 eV. DMPs can be among new particles too [4–6]. Characteristics of DMPs are still unknown, although their existence is most probably. Below we will call new fermionic and electrically neutral particles with masses more than active neutrinos masses as heavy neutrinos (HNs). The active neutrinos may be called as light neutrinos (LNs). HNs with masses less than half of the Z-boson mass are surely SNs. The most suitable for explanation of some cosmological and astrophysical data are SNs and DMPs with eV-keV-MeV-GeV masses [5, 7–9]. Intensive experimental and theoretical studies are being currently performed in order to solve the problem of SNs. It is expected that experimental data will make it possible to confirm or disprove the existence of the aforementioned anomalies in the near future [1, 10–13].

Below we consider a phenomenological model to describe neutrino anomalies at short distances and other effects of HNs. This model includes three active neutrinos ν_a ($a = e, \mu, \tau$) and four HNs, which consist of a sterile neutrino ν_s and three extra HNs (ξ, N and S). We consider this model in two options, which can be called as a low mass option and a high mass option. Let us present three generations of known quarks and leptons and a new generation of four HNs in the form of a 4×4-matrix C. Here we regard only left chiral states of the C particles. The multiplet C_L represents the sixteen-dimensional representation of the new $SO(10)_L$ group [14]:

$$C_L = \begin{pmatrix} (\nu_e)_L & (\nu_\mu)_L & (\nu_\tau)_L & (\nu_s)_L \\ (e^-)_L & (\mu^-)_L & (\tau^-)_L & (\xi)_L \\ (u)_L & (c)_L & (t)_L & (N)_L \\ (d')_L & (s')_L & (b')_L & (S)_L \end{pmatrix}, \quad (1)$$

where $(d')_L$, $(s')_L$ and $(b')_L$ are left chiral states of d, s and b quarks mixed by the Cabibbo-Kobayashi-Maskawa matrix. The new neutral particles are distributed among two doublets of χ and X states, which differ considerably by their mass values: $\chi = \begin{pmatrix} \nu_s \\ \xi \end{pmatrix}$ and $X = \begin{pmatrix} N \\ S \end{pmatrix}$.

It is known that the mixing of neutrino states can be described in terms of the Pontecorvo–Maki–Nakagawa–Sakata matrix $U_{\text{PMNS}} \equiv U \equiv V \cdot P$ as $\psi_L^a = U_i^a \psi_L^i$, where $\psi_L^{a,i}$ are left-handed chiral fields with flavor index $a = e, \mu, \tau$ or mass index $i = 1, 2, 3$, respectively [15]. For three LNs, the matrix V can be expressed in terms of standard parametrization [16] through mixing angles θ_{ij} and CP phase $\delta \equiv \delta_{CP}$ associated with violation of the CP invariance in the lepton sector for Dirac or Majorana neutrinos, while $P = \text{diag}\{1, e^{i\alpha}, e^{i\beta}\}$, with $\alpha \equiv \alpha_{CP}$ and $\beta \equiv \beta_{CP}$ the phases associated with CP violation only for Majorana neutrinos. At present, there are experimental data for the neutrino mixing angles and the mass-squared differences Δm_{12}^2 and Δm_{13}^2 (where $\Delta m_{ij}^2 = m_i^2 - m_j^2$), which specify three-flavor neutrino oscillations in a vacuum. These data were obtained from a global analysis of the most recent high-precision measurements of neutrino oscillation parameters [17]. The CP phases α_{CP} and β_{CP} are still not known now. The absolute values of the neutrino masses can be ordered by two ways, namely, *a*) $m_1 < m_2 < m_3$ and *b*) $m_3 < m_1 < m_2$, that is, it may be either the normal ordering/hierarchy case [NH, case (a)] or inverted ordering/hierarchy case [IH, case (b)]. In the following, we restrict ourselves to the NH case only, by setting $\delta_{CP} = -\pi/2$.

At the same time indications exist that the neutrino fluxes for some processes have anomalies, which manifest themselves at short distances (more precisely, at distances L such that the values of parameter $\Delta m^2 L/E$, where E is the neutrino energy, are about unity). The experiments at such distances are referred to as short-baseline (SBL) experiments. Data on the anomalies concern both the appearance of electron neutrinos or antineutrinos in the fluxes of muon neutrinos or antineutrinos, respectively, and the disappearance of electron or muon neutrinos or antineutrinos. The anomalies observed to date in the neutrino fluxes data can be explained by existence of light SNs. The Universe also contains heavy DMPs, which can include HNs.

We will consider the mixing matrix for LNs and HNs in the form of 7×7-matrix, which can be named as the generalized mixing matrix U_{mix}. We consider here only particular form of U_{mix} and present U_{mix} as $V \cdot P$, where diagonal matrix P is $\text{diag}\{1, e^{i\phi_1}, \ldots, e^{i\phi_6}\}$ with Majorana CP-violating phases ϕ_i, $i = 1, \ldots, 6$. We keep a succession for notations and denote fifteen Dirac CP phases as δ_i and κ_j and twenty one mixing angles as θ_i and η_j, with $\delta_1 \equiv \delta_{CP}$, $\theta_1 \equiv \theta_{12}$, $\theta_2 \equiv \theta_{23}$ and $\theta_3 \equiv \theta_{13}$. We choose an approximation when the mixing between two X neutrinos, as well as their mixing with other neutral particles are suppressed. In this case we can reduce the 7×7 matrix U_{mix} to 6×6 matrix $\widetilde{U}$, which was considered in detail in Ref. [18] for the case of three active and three sterile neutrinos. We will use the indices 4, 5, 6 and 7 to distinguish among the extra mass states and then denote by f the set of indices distinguishing the extra flavor states ν_s, ξ and N among h_f and by i' the set of indices 4, 5 and 6 for mass states $h_{i'}$. The matrix U_{mix} can then be expressed in terms of 3×3 matrices R, T, V and W as

$$\begin{pmatrix} \nu_a \\ h_f \\ S \end{pmatrix} = \begin{pmatrix} R & T & 0 \\ V & W & 0 \\ 0 & 0 & 1 \end{pmatrix} \begin{pmatrix} \nu_i \\ h_{i'} \\ h_7 \end{pmatrix}. \quad (2)$$

For matrix $\widetilde{U}$ we will consider, taking into account additional physical assumptions, only some particular cases rather than the most general form. Firstly, the matrix R will be represented in the form of $R = U + \Delta U$, where the matrix ΔU, as well as the matrices T and V in Eq. (2) should be small as compared with U. In order to estimate quantitatively both the mixing between LNs and HNs and the HNs-induced corrections to LNs mixing, we assume that $\Delta U = -\epsilon U$ and $\epsilon = 1 - \varkappa$. Then $R = \varkappa U$, where U is well known unitary 3×3 matrix of LNs mixing. Taking into account on the basis of data available from astrophysical and laboratory observations that the mixing between LNs and HNs is small and can be determined by the small parameter ϵ, we choose the matrices T and V as $T = \sqrt{1-\varkappa^2}\, a$ and $V = \sqrt{1-\varkappa^2}\, bU$, where a and b are arbitrary unitary 3×3 matrices. Then the unitary 6×6 matrix $\widetilde{U}$ can be written in the form of

$$\widetilde{U} = \begin{pmatrix} R & T \\ V & W \end{pmatrix} \equiv \begin{pmatrix} \varkappa U & \sqrt{1-\varkappa^2}\, a \\ \sqrt{1-\varkappa^2}\, bU & \varkappa c \end{pmatrix}, \tag{3}$$

with $c = -b \cdot a$. Taking into account the results of Ref. [18], the following forms for the matrices a and b can be used, namely

$$a = \begin{pmatrix} \cos\eta_2 & \sin\eta_2 & 0 \\ -\sin\eta_2 & \cos\eta_2 & 0 \\ 0 & 0 & e^{-i\kappa_2} \end{pmatrix}, \qquad b = -\begin{pmatrix} \cos\eta_1 & \sin\eta_1 & 0 \\ -\sin\eta_1 & \cos\eta_1 & 0 \\ 0 & 0 & e^{-i\kappa_1} \end{pmatrix}, \tag{4}$$

where κ_1 and κ_2 are new mixing phases and η_1 and η_2 are new mixing angles. In our calculations, we will use the following test values: $\kappa_1 = \kappa_2 = -\pi/2$, $\eta_1 = 5^\circ$, $\eta_2 = \pm 30^\circ$, and also assume that the small parameter ϵ satisfies the condition $\epsilon \lesssim 0.03$. The neutrino masses will be specified by means of the set $\{m\} = \{m_i, m_{i'}, m_7\}$ ordered normally. We will use estimations of LNs mass values from Refs. [9, 18] (in eV units) as $m_1 \approx 0.0016$, $m_2 \approx 0.0088$ and $m_3 \approx 0.0497$.

Let us consider two options for our model, which differ by mass values of HNs h_4, h_5, h_6 and h_7. In the framework of the first option we adopt the mass value of h_4 of the order of 1 eV and the mass value of h_5 of the order of 10 keV. Then it becomes possible to explain both the occurrence of SBL anomalies in neutrino data [3] and the detection of 3.55 keV line in γ-spectra of some astrophysical sources [19–21]. Next, we take the mass values of h_6 and h_7 of the order of 1 MeV and 1 GeV, respectively. In this case one can explain the cosmological origin of the baryonic matter in the Universe, as well as a number of the observations concerning the dark matter [5, 22]. In the framework of the second option we eliminate existence of most light HNs and choose the HNs masses to be higher [8, 23, 24]. So, it is convenient to introduce four test sets of mass values, two sets for χ_1 and χ_2 and two sets for X_1 and X_2, namely, $\{m\}_{LM\chi} = \{1\,\text{eV}, 10\,\text{keV}\}$, $\{m\}_{HM\chi} = \{50\,\text{keV}, 10\,\text{MeV}\}$, $\{m\}_{LMX} = \{1\,\text{MeV}, 1\,\text{GeV}\}$ and $\{m\}_{HMX} = \{500\,\text{GeV}, 2\,\text{TeV}\}$. The low mass case with two sterile neutrinos, which are nearly degenerate with respect to their mass values has been studied in Ref. [18]. This case may be denoted as the low mass twins option (LMTO). Here we consider other two options for the absolute mass values of four HNs, namely, the low mass option (LMO) and the heavy mass option (HMO). Below they are presented in eV units for complete sets of seven LNs and HNs as $\{m\}_{LMO} = \{0.0016, 0.0088, 0.0497, 1, 1\times10^4, 1\times10^6, 1\times10^9\}$ and $\{m\}_{HMO} = \{0.0016, 0.0088, 0.0497, 5\times10^4, 1\times10^7, 5\times10^{11}, 2\times10^{12}\}$. The equations for the flavor amplitudes of neutrino propagation in a vacuum are well known (see, e.g., Ref. [18]). Moreover, one can obtain analytical expressions for transitions probabilities between neutrino flavors with the help of these equations [25], which can be used to control the accuracy of numerical results obtained on the basis of the equations for the probability amplitudes.

In the present study, we focus primarily on the possibility of describing the anomalies that were found in data from the LSND experiment on the oscillations of accelerator muon neutrinos and antineutrinos, which were tested afterwards and will be still tested in a number of future

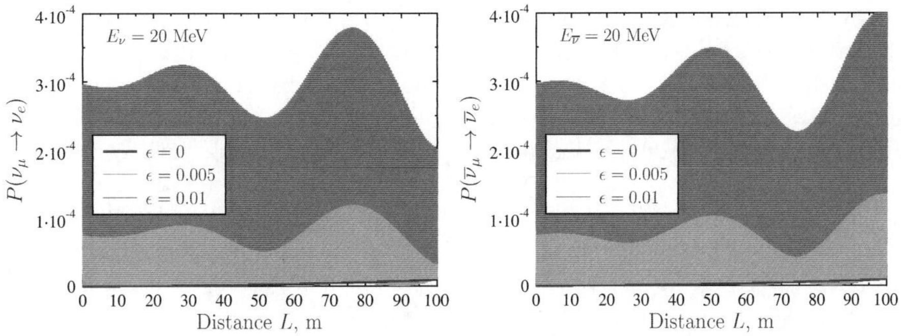

Figure 1. Appearance probabilities for electron neutrinos/antineutrinos from the beams of muon neutrinos/antineutrinos at various values of ϵ at $\eta_2 = \pi/6$ and for the LMO mass values.

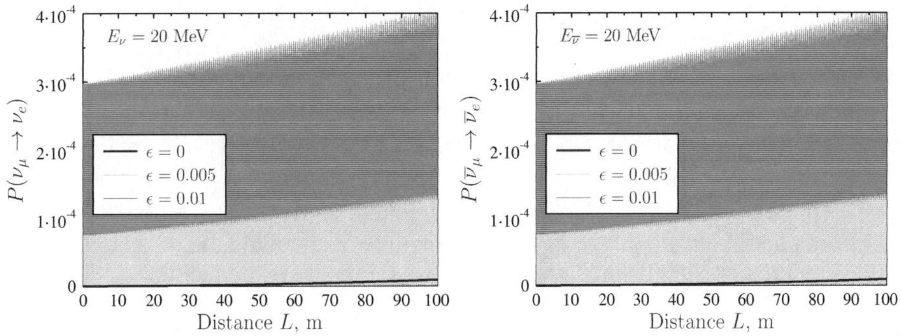

Figure 2. The same as in figure 1 for the HMO mass values.

accelerator experiments [26]. A typical ratio of the distance travelled by a neutrino before detection to the neutrino energy is either several meters per MeV or one meter per several MeV. Attempts of a simultaneous description of all data in these processes lead to difficulties. In particular, problems associated with different values of the excess of both the ν_e and $\overline{\nu}_e$ yields in the MiniBooNE experiment can be solved upon admitting *CP* violation [27–29]. It should be noted that the Reactor and Gallium anomalies manifesting themselves as the disappearance of electron neutrinos and antineutrinos can be described within our model by appropriate choosing a value of ϵ. The appearance probabilities for electron neutrinos and antineutrinos in accelerator beams of muon neutrinos and antineutrinos are shown in figures 1 – 2 for the LMO and HMO neutrino mass values, respectively, at the neutrino energy 20 MeV. Shown in the same figures for the sake of comparison are the dependences obtained for the probabilities for only three LNs, that is, in the absence of HNs contributions (at $\epsilon = 0$).

In figure 1, the results for $P(\nu_\mu \to \nu_e)$ and $P(\overline{\nu}_\mu \to \overline{\nu}_e)$ are shown for the mixing matrix in the form of the present paper and for the LMO mass values. In this case, there is distinctive difference between the left-hand and right-hand panels, so there is the difference between the appearance probabilities of electron neutrinos and antineutrinos in beams of muon neutrinos and antineutrinos, respectively. The main difference between the mixing matrix in the form of Ref. [18] and in the form of the present paper is in the introducing the new *CP* violation with two phases κ_1 and κ_2. However, it is more important that there is a difference in the mass values of the HNs. For the LMO mass values used in the results of figure 1, only one mass ($m_4 = 1$ eV) is of the order of 1 eV, while other masses are much higher (e.g., $m_5 = 10$ keV). So, one light sterile neutrino with the mass of the order of 1 eV is sufficient for explanation of the possible

LNSD anomaly in neutrino data. In figure 2, the results for $P(\nu_\mu \to \nu_e)$ and $P(\overline{\nu}_\mu \to \overline{\nu}_e)$ are shown in the same form for the HMO mass values. In this case there are no distinctive features between the left-hand and right-hand panels, so there is no difference between the appearance probabilities of electron neutrinos and antineutrinos. The appearance probability of electron neutrino/antineutrino also strongly oscillates as a function of distance in both figures 1 and 2, with smooth envelopes which practically are the same for neutrino and antineutrino in figure 2. But the scale of the electron neutrino/antineutrino yields is the same as for the case of figure 1.

In conclusion, the properties of LNs and HNs have been examined in this paper on the basis of the phenomenological neutrino model. Specifically, this has been done by using two options of our model, namely, the LMO and the HMO. The properties of oscillations of LNs and HNs in a vacuum at test values of the model parameters have been calculated. All calculations have been performed for the case of normal hierarchy for the LNs masses, taking into account possible *CP* violation in the lepton sector and setting the Dirac *CP* phase in the U_{PMNS} matrix for active neutrinos to $-\pi/2$. The graphical dependences of the appearance probabilities for electron neutrinos/antineutrinos on the distance from the source of muon neutrinos/antineutrinos have been determined for neutrino energy 20 MeV. The results obtained allow to interpret experimental neutrino oscillation data permitting the existence of the LNSD anomaly at a neutrino energy of about 40 MeV and a distance of about 30 m from the source. The great advantage of our calculations lies in revealing of the asymmetry between the appearance probabilities for electron neutrinos and antineutrinos (see figure 1) due to another source of *CP* violation with respect to δ_{CP}. The model in question may also be used to describe some astrophysical data for the cases where heavy neutrinos are involved, as well as in predicting the results of new experiments relating to the sterile neutrino existence problem [1, 10–13].

References

[1] Abazajian K N *et al.* *Preprint* hep-ph:1204.5379
[2] Kopp J, Machado P A N, Maltoni M and Schwetz T 2013 *J. High Energy Phys.* JHEP05(2013)050
[3] Gariazzo S, Giunti C, Laveder M and Lie Y F *Preprint* hep-ph:1703.00860v3
[4] Vogelsberger M *et al.* 2014 *Nature* **509** 177
[5] Demiański M and Doroshkevich A G *Preprint* astro-ph.CO:1511.07989v6
[6] Abazajian K N *Preprint* hep-ph:1705.01837v2
[7] Undagoitia T M and Rauch L 2016 *J. Phys.* G **43** 013001
[8] Argüelles C R, Krut A, Rueda J A and Ruffini R *Preprint* astro-ph.GA:1606.07040v2
[9] Yudin A V, Nadyozhin D K, Khruschov V V and Fomichev S V 2016 *Astron. Lett.* **42** 800
[10] Gavrin V N, Gorbachev V V, Veretenkin E P and Cleveland B T *Preprint* nucl-ex:1006.2103v2
[11] Bellini G et al. *J. High Energy Phys.* JHEP08(2013)038
[12] Serebrov A P et al. 2015 *Tech. Phys.* **60** 1863 (*Preprint* physics.ins-det:1501.04740)
[13] Dentler M, Hernández-Cabezudo A, Kopp J, Maltoni M and Schwetz T *Preprint* hep-ph:1709.04294
[14] Khruschov V V *Preprint* hep-ph:1609.01858
[15] Bilenky S M and Pontekorvo B M 1977 *Sov. Phys. Usp.* **20** 776
[16] Patrignani C *et al.* [Particle Data Group] 2016 *Chin. Phys.* C **40** 100001
[17] Esteban I, Gonzalez-Garcia M C, Maltoni M, Martinez-Soler I and Schwetz T *Preprint* hep-ph:1611.01514v2
[18] Khruschov V V, Fomichev S V and Titov O A 2016 *Phys. At. Nucl.* **79** 708 (*Preprint* hep-ph:1612.06544v1)
[19] Bulbul E *et al.* 2014 *Astrophys. J.* **789** 13
[20] Boyarsky A, Ruchayskiy O, Iakubovskyi D and Franse J 2014 *Phys. Rev. Lett* **113** 251301
[21] Horiuchi S *et al.* *Preprint* astro-ph.CO:1512.04548
[22] Drewes M and Garbrecht B *Preprint* hep-ph:1502.00477
[23] Vogelsberger M *et al.* *Preprint* astro-ph.CO:1512.05349v2
[24] Bertuzzo E, Barros C J C and di Cortona G G *Preprint* hep-ph:1707.00725
[25] Bilenky S M 2015 *Phys. Part. Nucl. Lett.* **12** 453 (*Preprint* hep-ph:1502.06158)
[26] Gariazzo S, Giunti C, Laveder M, Li Y F and Zavanin E M *Preprint* hep-ph:1507.08204
[27] Maltoni M and Schwetz T 2007 *Phys. Rev.* D **76** 093005
[28] Palomares-Ruiz S, Pascoli S and Schwetz T 2005 J. High Energy Phys. JHEP09(2005)048
[29] Karagiorgi G *et al.* 2007 *Phys. Rev.* D **75** 013011 [Erratum: 2009 *Phys. Rev.* D **80** 099902(E)].

IOP Conf. Series: Journal of Physics: Conf. Series **934** (2017) 012008 doi:10.1088/1742-6596/934/1/012008

Influence of hadronic interaction models on characteristics of the high-energy atmospheric neutrino flux

A D Morozova[1,*], A A Kochanov[2,3], T S Sinegovskaya[4] and S I Sinegovsky[3,5]

[1] Lomonosov Moscow State University, Leninskie Gory, 1, Moscow, 119991, Russia
[2] Institute of Solar-Terrestrial Physics SB RAS, Lermontov str. 126a, Irkutsk, Russia
[3] Irkutsk State University, Gagarin blv. 20, Irkutsk, 664003, Russia
[4] Irkutsk State Transport University, Chernyshevskiy str. 15, Irkutsk, 664074, Russia
[5] Joint Institute for Nuclear Research, Joliot-Curie, 6, Dubna, Moscow region, 141980, Russia

E-mail: *refriz@yandex.ru, kochanov@iszf.irk.ru, tanya@api.isu.ru, sinegovsky@api.isu.ru

Abstract. The high-energy conventional atmospheric neutrino fluxes are calculated with the hadronic interaction models: Kimel & Mokhov, QGSJET II-03(04), SIBYL 2.1(2.3), EPOS LHC. The influence of hadron-nuclear interactions on the neutrino flux ratios, $\nu/\bar{\nu}$, $(\nu_\mu + \bar{\nu}_\mu)/(\nu_e + \bar{\nu}_e)$, is studied. A comparison of calculations obtained with use of two different approaches, $\mathcal{Z}(E,h)$-functions method and the Matrix Cascade Equations (MCEQ), demonstrates close agreement in whole but some of partial contrubutions. The comparison of calculated muon neutrino spectra with the latest experimental data justifies reliability of performed computation which describes correctly the atmospheric neutrino production. The calculation made with the model EPOS LHC, combined with Hillas & Gaisser parametrization of the cosmic ray spectrum, is in close agreement with the best fit of IceCube for energy spetrum of atmospheric muon neutrinos in the energy range 1 – 500 TeV.

1. Introduction

High-energy neutrinos from decays of mesons and baryons are produced in collisions of cosmic rays with the the Earth's atmosphere forming unavoidable background for detecting of neutrinos from astrophysical sources. The calculation of the energy spectrum and zenith-angle distribution of the atmospheric neutrinos became really urgent problem since detecting in the IceCube experiment of 82 events, with energy deposition in the range 30 TeV–2 PeV, from neutrinos of cosmic origin [1–5].

We calculate the atmospheric neutrino spectra and neutrino flux ratios at energies $10 - 10^7$ GeV with the hadronic model by Kimel & Mokhov [6], and at $10^2 - 10^7$ GeV using high-energy set of hadronic models QGSJET II [7,8], SIBYLL 2.1 [9], SIBYLL 2.3 [10], EPOS LHC [11,12]. Two known paramerizations of the cosmic ray spectrum, by Zatsepin & Sokolskaya [13] and by Hillas & Gaisser [14], are used in the computation.

The neutrino-to-antineutrino flux ratios $\Phi_\nu/\Phi_{\bar{\nu}}$ and flavor ratios $\Phi(\nu_\mu + \bar{\nu}_\mu)/\Phi(\nu_e + \bar{\nu}_e)$ depending on cross sections of the kaon production in hadron-nucleus collisions also bear the imprint of the cosmic ray elemental composition. The flavor ratio is important because of high sensitivity to the neutrino source addition – a rare decay mode, charmed particle decays, etc.

2. Methods of the calculation

Decay modes $\pi_{\mu 2}$, $K_{\mu 2}$ and $K^0_{\mu 3}$ dominate the flux of muon neutrinos. Also we consider minor contributions: $K^{\pm}_{\mu 3}$ (the fraction of 0.0335), μ_{e3} and those arise from decay chains $K \to \pi \to \nu_\mu$. Dominant sources of the electron neutrinos are three-particle decays of kaons: K^0_{e3} (0.405), $K^{\pm}_{e3}$ (0.0507). The semileptonic decays of K^0_S are a significant source of the ν_e flux though fraction of the decay mode $K^0_S \to \pi^{\pm} + e^{\mp} + \bar{\nu}_e(\nu_e)$ is very small ($7.04 \cdot 10^{-4}$). This mode gives a considerable contribution to the atmospheric ν_e flux at high energies, reaching 30% at $E_\nu = 500$ TeV for zenith angle $\theta = 0^\circ$. Close to the vertical $(\nu_e + \bar{\nu}_e)$ flux from the K^0_S decay becomes nearly equal to that from K^0_L one at $E_\nu \approx 1$ PeV (see figure 1). The decay mode $K^0_S \to \pi^{\pm} + \mu^{\mp} + \bar{\nu}_\mu(\nu_\mu)$ (the fraction of $4.69 \cdot 10^{-4}$) contributes up to 10% of the $(\nu_\mu + \bar{\nu}_\mu)$ flux.

Main calculations were made with $\mathcal{Z}(E, h)$-method [15, 16], allowing the computation of atmospheric fluxes of hadrons, muons [17, 18] and neutrinos [19] in case of nonpower cosmic-rays spectrum, nonscaling behavior of inclusive cross-sections and growing with energy inelastic cross-sections of hadron-nucleus collision. Characteristics of the atmospheric neutrino flux, calculated with $\mathcal{Z}(E, h)$-method, we compare to those derived within the framework of different approach, Matrix Cascade Equations [20, 21]. All calculations necessary for this comparison we performed independently using the free access package MCEq [22]. Results of this comparison are shown in figures 1–3 (see also [23]).

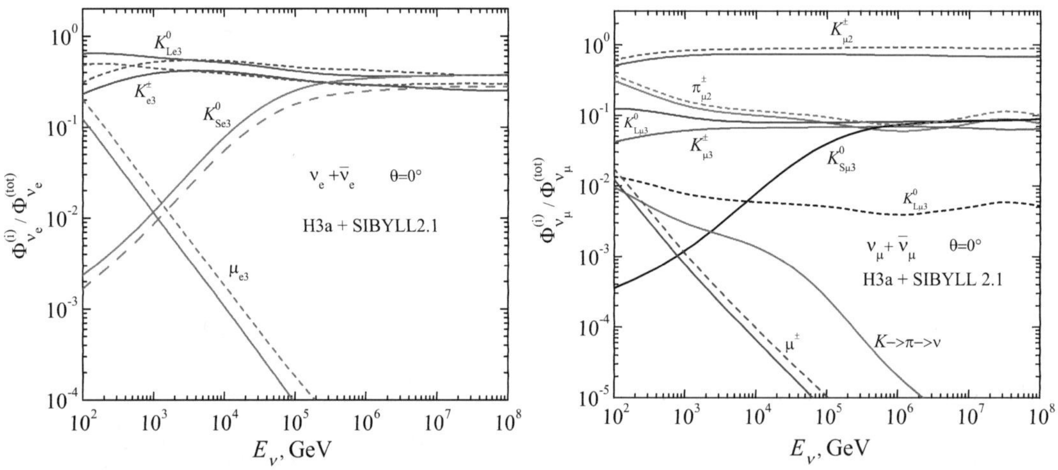

Figure 1. Partial contributions to the atmospheric $(\nu_e + \bar{\nu}_e)$ flux (left) and $(\nu_\mu + \bar{\nu}_\mu)$ (right) with SYBILL 2.1. Solid lines – $\mathcal{Z}(E, h)$-method, dashed – MCEq one.

3. Results and discussion

The neutrino flux ratios $\Phi_\nu/\Phi_{\bar{\nu}}$, $R_{\nu_\mu/\nu_e} = \Phi(\nu_\mu + \bar{\nu}_\mu)/\Phi(\nu_e + \bar{\nu}_e)$ at high energies depend on cross sections of kaons production in the atmospheric hadron cascade. Elemental composition of cosmic rays also impacts on these ratios through the primary proton-to-neutron ratio as well to π^+/π^-, K^+/K^- и π/K ratios of conventional neutrino sources.

Calculated neutrino flux ratios for the set of hadronic models are shown in figures 2, 3. The calculation of flavor ratio R_{ν_μ/ν_e} of atmospheric neutrino fluxes near vertical direction is performed for two models of the cosmic ray spectrum (figure 3, left panel): curves 1–5 are

IOP Conf. Series: Journal of Physics: Conf. Series **934** (2017) 012008 doi:10.1088/1742-6596/934/1/012008

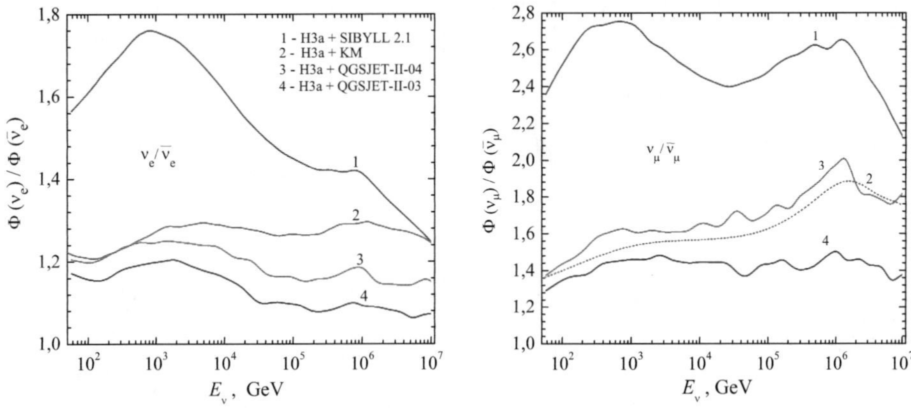

Figure 2. The neutrino-to-antineutrino ratio $\nu_e/\bar{\nu}_e$ (left) и $\nu_\mu/\bar{\nu}_\mu$ (right), calculated with Hillas & Gaisser cosmic ray spectrum (H3a) for zenith angle $\theta = 0°$.

obtained with the MCEq technique [21] for Hillas & Gaisser cosmic ray spectrum (H3a ≡ HGm); curves 6, 7 represent the calculations for Zatsepin & Sokolskaya spectrum (ZS) with $\mathcal{Z}(E, h)$-method. Right panel of figure 3 displays R_{ν_μ/ν_e} averaged over zenith angles, and also only point derived in IceCube experiment [24] is plotted (•) on the figure ($R_{\nu_\mu/\nu_e} = 16.9$ at $E_\nu = 1.7$ TeV). The square at upper boundary of the experimental error marks value obtained by Honda [25], the symbol ⋆ close to lower error boundary marks the Bartol group calculation (borrowed from [24]).

Impact of semileptonic decay of K_S^0 on the flavor ratio of atmospheric neutrinos at 50−100 TeV is unexpectedly strong: this contribution in SIBYLL 2.1 leads to lowering of R_{ν_μ/ν_e} by factor ~ 1.28 at ~ 100 TeV (the curve 4 in figure 3, right) as compared to that if no K_S^0 decay is taken into consideration (upper dashed curve). Probably, neglect this contribution led to high value of R_{ν_μ/ν_e} in the calculation [26], performed with usage of CORSIKA 6.990 (no K_S^0 decay option).

The $(\nu_\mu + \bar{\nu}_\mu)$ flux calculation for Hillas & Gaisser cosmic ray spectrum H3a (HGm) with

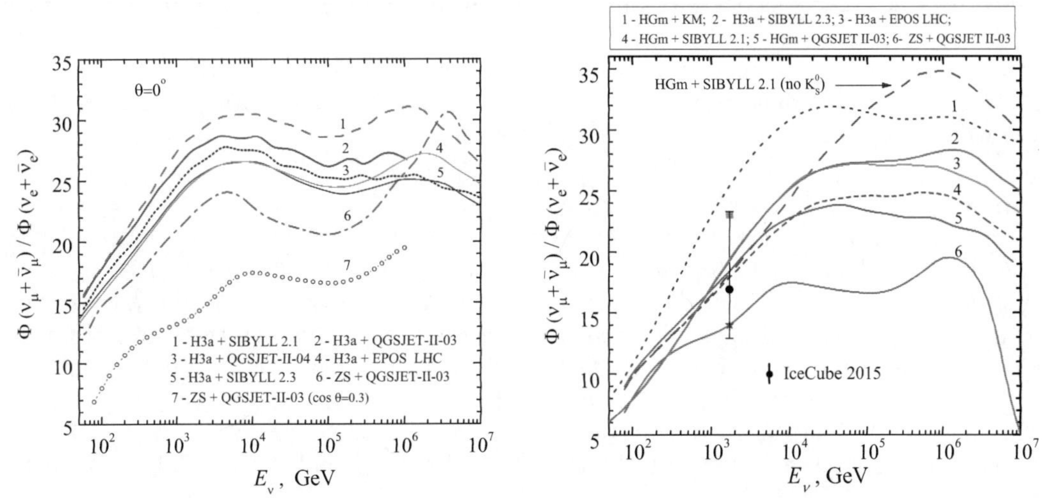

Figure 3. Atmosperic neutrino flavor ratio at $\theta = 0°$ (left) and zenith angle averaged one (right) calculated for the hadronic models KM, SIBYLL 2.1 (2.3), QGSJET II-03 (04) and EPOS LHC.

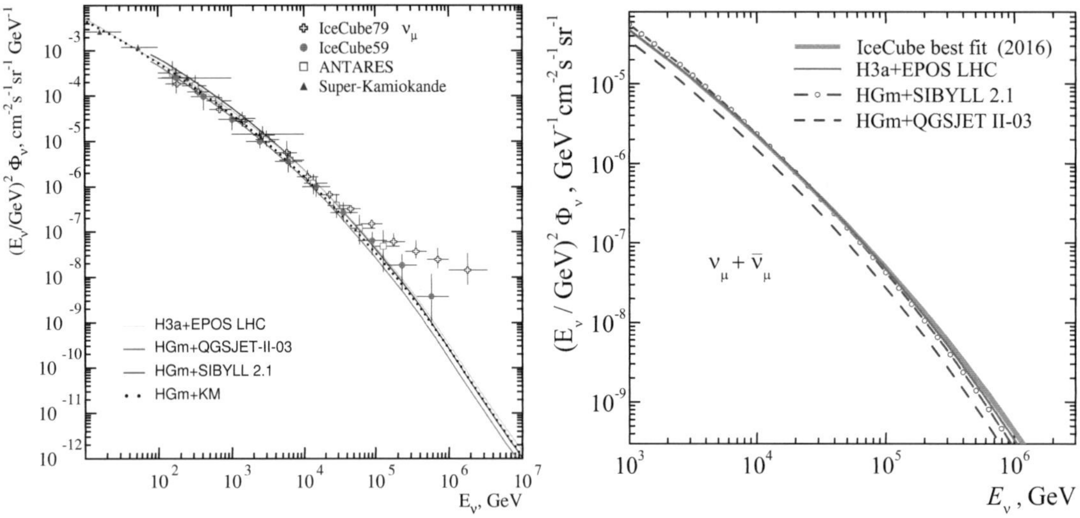

Figure 4. Energy spectrum of atmospheric muon neutrinos $(\nu_\mu + \bar{\nu}_\mu)$ (zenith-angle averaged). Left: symbol – experimental data of IceCube [27, 28], ANTARES [29], Super-Kamiokande [30]; curves – the calculations. Right: the calculations as compared to IceCube best fit for the $(\nu_\mu+\bar{\nu}_\mu)$ spectrum [4] (green band). Closest curve to the best fit is derived for EPOS LHC (red line).

use of hadronic models Kimel & Mokhov (KM), QGSJET II-03, SIBYLL 2.1 and EPOS LHC is shown in figure 4. The calculation for EPOS LHC [11,12] is performed using the program MCEq by Fedynitch et al. [22]) (green line). At energies $E_\nu \leq 100$ GeV the prediction made with KM (dotted) is rather close to that with EPOS LHC, and above 10 TeV the former one is close to that of SIBYLL 2.1 (dashed line). The data obtained with the 79-strings configuration of IceCube detector [28] display the change of spectral index of the muon neutrino flux, caused by addition of the astrophysical neutrinos to the atmospheric ones (figure 4, left); the former dominate at energies above 500 TeV. The best fit of the IceCube data [4] on the atmospheric muon neutrino spectrum, averaged over zenith angles, is shown in right panel of figure 4 by thick green line (width corresponds to one-sigma error of the measured spectrum). The rest three curves present calculations, performed with EPOS LHC (red line), SIBYLL 2.1 и QGSJET II-03 for the Hillas & Gaisser cosmic-rays spectrum. The hadronic interaction model EPOS LHC in combination with Hillas & Gaisser parametrisation of the cosmic ray spectrum is in closest agreement with the the IceCube best fit in the energy range $1 - 500$ TeV.

4. Conclusions

The atmospheric neutrino spectra and the flavor ratio in the energy range $10^2 - 10^8$ GeV are calculated with high-energy hadron interaction models (QGSJET-II, SIBYLL 2.1, EPOS LHC, etc) using two parametrizations of the primary cosmic ray spectrum, Zatsepin & Sokolskaya and Hillas & Gaisser. Calculation results show rather weak dependence on the model of primary cosmic rays in the energy range of $10 - 10^5$ GeV, while high-energy interaction models lead to the discrepancy in the calculated neutrino fluxes, sequent from differences in K-meson production. On the contrary, the neutrino flavor ratio is more sensitive to the cosmic-rays spectrum.

Both calculation schemes used in the study, $\mathcal{Z}(E, h)$ [19] and MCEq [20], demonstrate consistent results at least in the range of 100 GeV – 1 PeV. Compatibility of two methods allows one to examine calculation results made for the same primary spectrum and hadronic model, identifying possible sources of differences – a model of the atmosphere, secondary particles spectra in kaon decays, minor neutrino sources.

The comparison of calculated neutrino spectra with available experimental data justifies reliability of the computation that describes correctly the neutrino production in the Earth's atmosphere. The calculation of the atmospheric muon neutrino flux, performed with the hadronic interaction model EPOS LHC in combination with Hillas & Gaisser parametrization of the cosmic ray spectrum, agrees well with the the IceCube best fit in the energy range 1 – 500 TeV.

Acknowledgements

This work was supported partly by the Russian Federation Ministry of Education and Science, projects 3.904.2017/4.6, 14.B25.31.0010.

References

[1] Aartsen M G *et al.* 2013 *Science* **342** 1242856
[2] Aartsen M G *et al.* 2014 *Phys. Rev. Lett.* **113** 101101
[3] Aartsen M G *et al.* 2015 *Proc. Sci.* PoS(ICRC2015)1081 (*Preprint* 1510.05223)
[4] Aartsen M G *et al.* 2016 *Astrophys. J.* **833** 3
[5] Aartsen M G *et al.* 2017 *Proc. Sci.* PoS(ICRC2017)981
[6] Kimel L R and Mokhov N V 1974 *Izv. Vyssh. Uchebn. Zaved.* **10** 17
[7] Kalmykov N N, Ostapchenko S S and Pavlov A I 1997 *Nucl. Phys. Proc. Suppl.* B **52** 17
[8] Ostapchenko S 2006 *Phys. Rev.* **74** 014026
[9] Ahn E J *et al.* 2009 *Phys. Rev.* D **80** 094003
[10] Riehn F *et al.* 2015 *Proc. Sci.* PoS(ICRC2015)558
[11] Pierog T *et al.* 2015 *Phys. Rev.* C **92** 034906 (*Preprint* 1306.0121)
[12] Pierog T 2017 *EPJ Web Conf.* **145** 18002
[13] Zatsepin V I and Sokolskaya N V 2006 *Astron. Astrophys.* **458** 1
[14] Gaisser T K 2012 *Astropart. Phys.* **35** 801
[15] Naumov V A and Sinegovskaya T S 2000 *Phys. Atom. Nucl.* **63** 1927
[16] Kochanov A A, Sinegovskaya T S and Sinegovsky S I 2008 *Astropart. Phys.* **30** 219
[17] Sinegovsky S I *et al.* 2010 *Int. J. Mod. Phys.* A **25** 3733
[18] Kochanov A A, Sinegovskaya T S, Sinegovsky S I 2013 *J. Exp. Theor. Phys.* **116** 395
[19] Sinegovskaya T S, Morozova A D, Sinegovsky S I 2015 *Phys. Rev.* D **91** 063011
[20] Fedynitch A *et al.* 2015 *EPJ Web Conf.* **99** 08001 (*Preprint* 1503.00544)
[21] Fedynitch A *et al.* 2015 *Proc. Sci.* PoS(ICRC2015)1129
[22] Fedynitch A *https://github.com/afedynitch/MCEq*
[23] Morozova A D *et al.* 2017 *J. Phys. Conf. Ser.* **798** 012101
[24] Aarsten M G *et al.* 2015 *Phys. Rev.* D **91** 122004
[25] Honda M *et al.* 2007 *Phys. Rev.* D **75** 043006
[26] Fedynitch A, Becker Tjus J, Desiati P 2012 *Phys. Rev.* D **86** 114024
[27] Aartsen M G *et al.* 2015 *Eur. Phys. J.* C **75** 116
[28] Aarsten M G *et al.* *Preprint* 1705.07780.
[29] Adrian-Martinez S *et al.* 2013 *Eur. Phys. J.* C **73** 2606
[30] Richard E *et al.* 2016 *Phys. Rev.* D **94** 052001

Search for heavy neutrino in leptonic decays of K^+

Alexander Sadovsky for the OKA collaboration
NRC "Kurchatov Institute"-IHEP, Protvino, pl. Nauki 1, Russia
E-mail: Alexander.Sadovskiy@ihep.ru

Abstract. A high statistics data sample of the $K^+ \to \mu^+\nu_\mu$ decay was recorded in 2012 by the OKA collaboration. The missing mass analysis was performed to search for a decay channel $K^+ \to \mu^+\nu_H$ with a stable heavy neutrino in the final state. The obtained missing mass spectrum did not reveal statistically significant peaks corresponding to stable heavy neutrinos in the mass range $(220 < m_{\nu_H} < 375)$ MeV/c^2. Instead, we update upper limits on the branching ratio and on the value of the mixing element $|U_{\mu H}|^2$.

1. Introduction

It was proposed several decades ago [1], that experimental investigation of two body weak leptonic decays of charged kaons (and pions) could provide sensitive test for neutrino mass and mixing with help of peak search method on a missing mass plot, especially in hypothetical cases in which the spectrum of neutrino masses can be thought to be extended up to several MeV/c^2. Several approaches were exploited to search for heavy neutrinos (HN): via modification of beta decay spectrum, attempts to measure an excess in yield due to possible decay products of HN (or appearance in beam dump experiments), with the peak search at missing mass distributions, see reviews [2,3]; they gave negative results on admixture of heavy neutrinos to either electron or to muon neutrinos. A revised view on cosmological considerations [4] consistent with observed non zero masses and oscillations of active neutrinos put new arguments to search for heavy (sterile) neutrinos with emphasis to the mass range from $\sim$ 150 MeV/c^2 to 100 GeV/c^2 [5, 6]. Two nowadays experiments [7, 8] analyzed their data on $K \to \mu\nu$ decay to search for stable[1] heavy neutrino in mass range of few hundred MeV/c^2 and improved upper limit on the $|U_{\mu H}|^2$ known from the early experiment [9]. To contribute to this series of investigations we analyzed large data sample obtained in 2012 by the OKA experiment.

2. The OKA setup

The OKA is a fixed target experiment dedicated to investigation of kaon decays it is located at NRC "Kurchatov Institute"-IHEP in Protvino (Russia). A secondary hadron beam with enhanced fraction of kaons is obtained from the U-70 Proton Synchrotron through the beam line equipped by two (obtained from CERN SPS [11]) RF separator cavities according to Panofsky scheme [10].

The OKA setup makes use of two magnetic spectrometers placed along the beam line with an 11 m long Decay Volume (DV) enclosed between them, see figure 1. The DV is filled with helium

[1] Or assuming "disappearance" decay modes like $\nu_H \to \nu_\xi\nu_\varepsilon\bar{\nu}_\varepsilon$ into active neutrinos (ν_e, ν_μ, ν_τ instead of ν_ξ, ν_ε).

and contains 11 rings of guard system (GS) at its outer surface, made of 670 Lead-Scintillator sandwiches. It is complemented by an electromagnetic calorimeter BGD (with a central opening).

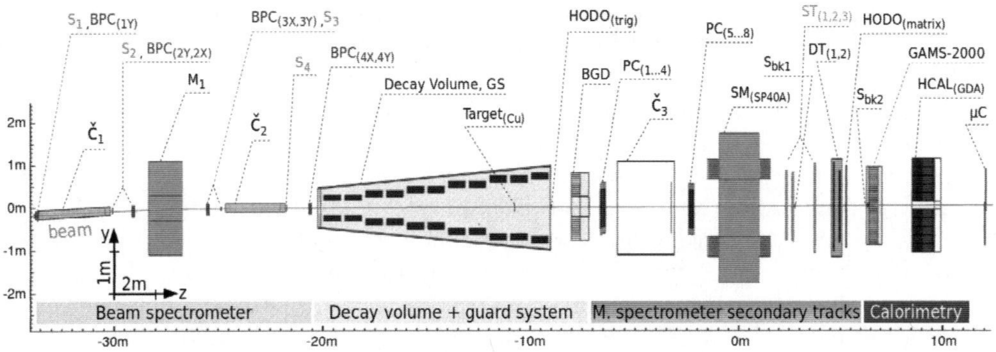

Figure 1. Schematic elevation view of the OKA setup. See text for details.

The first magnetic spectrometer measures incoming beam particles. It consists of magnet M_1 surrounded by a set of (beam) proportional chambers $BPC_{(1Y,2Y,2X,3X,3Y,4X,4Y)}$. The second spectrometer is dedicated to charged decay products. It consists from a wide aperture 200×140 cm^2 spectrometric magnet and surrounded by tracking stations: proportional chambers $PC_{(1,...,8)}$, straw tubes $ST_{(1,2,3)}$ and by drift tubes $DT_{(1,2)}$. In addition a matrix hodoscope $HODO_{(matrix)}$ is used to improve time resolution and to link x–y projections of a track.

At the end of the setup there are two calorimeters: the electromagnetic one (GAMS-2000 – made of lead glass blocks), and the hadron one ($HCAL_{(GDA)}$ – made of 100 iron-scintillator sandwiches) and, finally, four partially overlapping (1×1 m^2 scintillator plate each) muon counters μC behind the HCAL.

Trigger logic used in the analysis is organized upon a following set of detectors:
– beam trigger part is based on four scintillation counters $S_{(1)}$, $S_{(2)}$, $S_{(4)}$ (200×200×1 mm^3 each), and a thicker one, 60×85×6 mm^3, delivering timing, $S_{(3)}$;
– the identification of beam kaon is done with a set of two threshold Cherenkov counters $\check{C}_1$, $\check{C}_2$, while $\check{C}_3$ was not used in this run;
– suppression of events with undecayed beam particles is done in anti-coincidence with one of two scintillator counters S_{bk1}, S_{bk2} (80 mm and 90 mm in diameter).

More details on the OKA setup can be found elsewhere [12–14].

3. Event selection

As already mentioned, to search for heavy neutrino we follow well established strategy [1, 7–9]. We perform peak search (at few hundreds MeV/c^2 range) on the squared missing mass spectrum from the $K^+ \to \mu^+\nu_\mu$ exclusive decay channel. In this report we discuss measurements performed in November 2012 run with the momentum of secondary hadron beam of 17.7 GeV/c. Beam intensity ($S_1 \cdot S_2 \cdot S_3 \cdot S_4$) was $\sim 2 \cdot 10^6$ per spill. The fraction of kaons in the beam after kaon separators was $\sim$ 12.5%. An average absolute number of kaons is ~250k/spill.

One of the tasks persuaded during data taking was connected to the Primakoff effect; for that a thin copper target was installed inside the DV near its exit for the half time of the run. This part of the run was included in analysis for the sake of statistics, hence dedicated Monte-Carlo simulation (MC) for that period was performed.

Two prescaled triggers are used for further analysis. The first one selects beam kaons which decay in the vicinity of the DV, $\mathtt{Tr_{Kdecay}} = \mathtt{S_1} \cdot \mathtt{S_2} \cdot \mathtt{S_3} \cdot \mathtt{S_4} \cdot \check{\mathtt{C}}_1 \cdot \bar{\check{\mathtt{C}}}_2 \cdot \bar{\mathtt{S}}_{\mathtt{bk}}$. The second one, $\mathtt{Tr_{K\to\mu X}} = \mathtt{Tr_{Kdecay}} \cdot \mu\mathtt{C}$, additionally requires secondary muon by inclusion of signal from muon

IOP Conf. Series: Journal of Physics: Conf. Series **934** (2017) 012009 doi:10.1088/1742-6596/934/1/012009

counters μC. The prescale factors for $\mathtt{Tr}_{\mathtt{Kdecay}}$ and $\mathtt{Tr}_{\mathrm{K}\to\mu\mathrm{X}}$ are 1/10 and 1/4, while the proportion between numbers written with these two triggers turned out to be equal. The total number of $\sim 1.6 \times 10^{10}$ events with kaon decays are logged.

The event selection for $K^+ \to \mu^+\nu$ decay channel chosen in off-line analysis such as to select events with single beam track and single secondary reconstructed track (both with measured momentum). The sole secondary segment after the second magnet SM is well matched to showers of the muon type (i.e. one or two adjacent cells with the MIP energy deposition) in both GAMS-2000 and HCAL calorimeters. The missing mass resolution is obtained with the sufficiently high number of points on the four track segments in two magnetic spectrometers. The measured momentum of incoming into the OKA setup kaon is consistent with $\approx$ 17.7 GeV/c (beam setting). The required momentum of a secondary muon required to be below 16.4 GeV/c. To ensure good decay vertex reconstruction, there are requirements for *a*) minimal angle between vertex segments of 3 mrad, and *b*) minimal (matching) distance between them < 1 cm. The vertex position is required to be inside the DV situated by 2.5σ margins (in terms of z-vertex resolution) between the entrance foil of the DV and the position of the copper target. Other decay channels are suppressed by requiring the total energy deposition in GS to be below 50 MeV/c^2 and to be below 100 MeV/c^2 in BGD, respectively, while the total energy deposition in GAMS-2000 and HCAL should be consistent with that of a single muon. After applying these cuts, 25.7×10^6 $K^+ \to \mu^+\nu$ decays were selected for subsequent analysis.

4. Search for heavy neutrono

Search for peaks in the missing mass distribution, $m_\nu^2 = m_{miss}^2 = (p_K - p_\mu)_i \cdot (p_K - p_\mu)^i$, $i = 1, 2, 3, 4$, which may be responsible for heavy neutrino in the final state of the $K^+ \to \mu^+\nu_H$ decay requires description of the expected signal shape. For that a detailed GEANT-3.21 Monte-Carlo simulation (MC) is done with the subsequent off-line reconstruction and analysis. It provides realistic background description, from which the expected signal must be distinguished. Noticeable background processes were simulated with the MC and weighted according to their matrix elements and branching ratios: $K^+ \to \mu^+\nu_\mu$, $K^+ \to \mu^+\nu_\mu\gamma$, $K^+ \to \mu^+\nu_\mu\pi^0$, $K^+ \to \pi^+\pi^0$, $K^+ \to e^+\nu_e\pi^0$, also processes when kaon either scatters or interacts while passing the setup. A series of heavy neutrino masses were simulated to parametrize the detector response.

The experimental data and main backgrounds from MC, passing the reconstruction and selections cuts, are shown in the $(m_{miss}^2; p_\mu)$ plots of figure 2; additionally MC signals from five different cases of assumed heavy neutrino mass are shown superimposed. The $K^+ \to \mu^+\nu_\mu\gamma$ and $K^+ \to \mu^+\nu_\mu\pi^0$ can not be suppressed without significant loss in statistics for $K^+ \to \mu^+\nu_H$. However, it is evident that any features of background processes like peaks or abrupt changes in population of distribution over m_{miss}^2 would introduce bias to the result of peak search procedure, contaminating possible signal from heavy neutrino by its shape. The abrupt drop of the m_{miss}^2 distribution from the $K^+ \to \pi^+\pi^0$ can be eliminated, to a large extent, by introducing kinematic acceptance limits with a smooth curves in $p_\mu : m_{miss}^2$, as indicated at figure 2 (a). The selection of these limits partially improves the resolution at $m_{miss}^2 = 0$. The corresponding m_{miss}^2 distributions are shown in figure 3 (a). The normalization is done to the experimental data at $m_{miss}^2 = 0$, relative normalization of different MC background processes are done according to known branching ratios with respect to the $K^+ \to \mu^+\nu_\mu$. A set of reconstructed heavy neutrino signals was produced by the MC simulation with a neutrino mass step of 20 MeV/c^2. It was found that the signal, as a function of m_{miss}^2, is well approximated by the Gaussian shape, with an integral error of $\sim$2%. An interpolation between the generated set of masses is done with polynomial curves for the obtained signal width and for the total efficiency ε_H (including acceptance), which are shown at the right chart of figure 3.

Three approaches were used to extract possible signal from heavy neutrino.

The first approach does not use information from MC background processes, but the

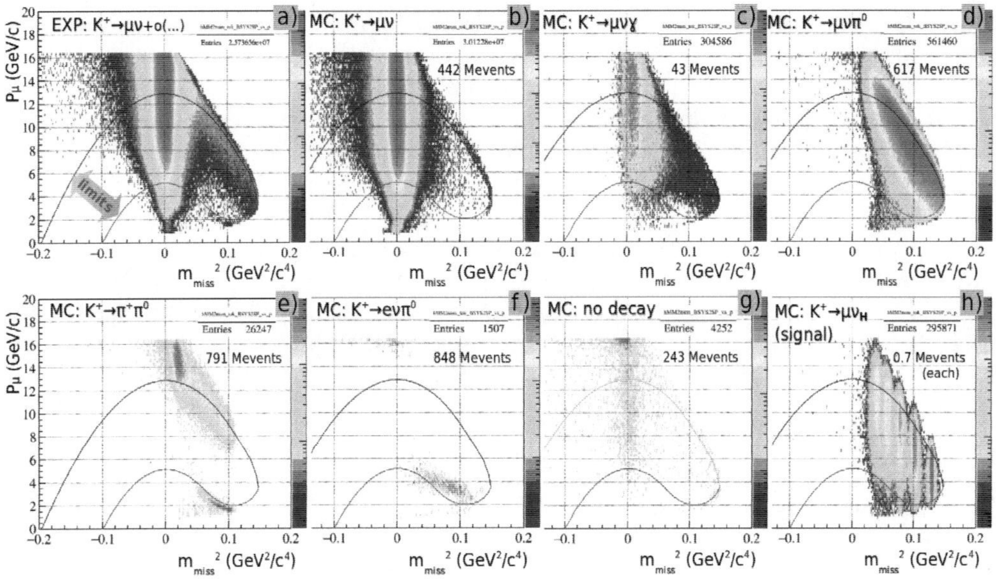

Figure 2. The distribution of p_μ vs. m^2_{miss} obtained from the analysis of the experimental data (plot a) and from MC simulation of background channels (plots b, ..., g). Distributions (MC) from a set of possible signals for $m_{\nu_H} = \{200, 240, 280, 320, 360\}$ MeV/c^2 are also shown superimposed on each other (plot h). Note logarithmic scale for the third dimension. Event entries passing all selections are shown. In case of MC, the initial statistics is also indicated.

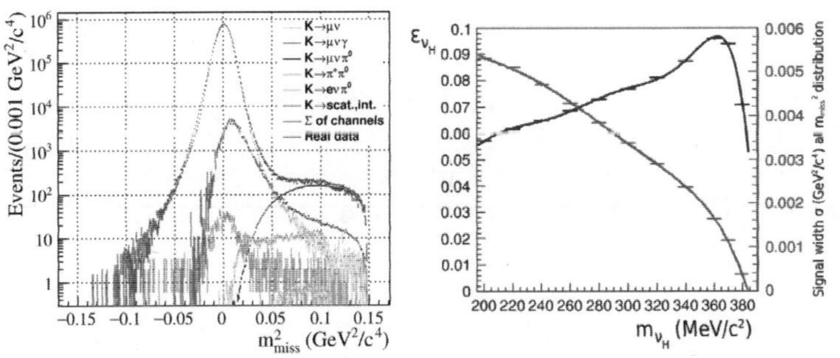

(a) (b)

Figure 3. (a): the m^2_{miss} distribution for the data and MC events inside kinematic limits, indicated in figure 2 (a). The normalization is relative to the experimental data. (b): efficiency and signal width parameterization for the set of HN masses, obtained with MC.

parameterization of the expected (HN) signal width only: a bin by bin scan is done within $0.05 < m^2_{miss} < 0.14$ GeV/c^2 interval in which a common fit with Gaussian shape for the signal of interest and polynomial approximation for the background, obtained from a wider region of $m^2_{miss} > 0.04$ GeV/c^2 from the data events (selected as already described above).

The second approach explicitly uses information about MC background processes and the signal width parameterization. It does bin by bin data scan (within the same range of interest $0.05 < m^2_{miss} < 0.14$ GeV/c^2) in which for every assumed m^2_{miss} we do a common fit of signal (with known width) and each background process (except for the $K^+ \to \mu^+\nu_\mu$) with a free scaling parameter. This multiplier is introduced for suppressed background channels because of very strong background suppression applied, hence one may assume that the obtained efficiency can be inaccurate by some factor of ~ 2 (e.g. relative suppressions for two most pronounced channels, $K^+ \to \mu^+\nu_\mu\gamma$ and $K^+ \to \mu^+\nu_\mu\pi^0$, in comparison to the main decay channel $K^+ \to \mu^+\nu_\mu$ are in the order of 10^{-1} and 10^{-2} correspondingly). I.e. in this approach the MC shape of background processes is used explicitly (within the high mass range $m^2_{miss} > 0.05$ GeV/c^2), while their magnitude is adjusted during fit procedure at each mass of interest.

The third approach is a reduced version of the second one. It uses distributions for each background process obtained from MC. On the first stage the efficiency for each background shape (except for the $K^+ \to \mu^+\nu_\mu$) are refined by the fit of (sum of) the MC distributions to the experimental distribution, hence the best background model is obtained with refined efficiencies for each background process in the mass range of interest (ignoring possible local contributions from NH signal). After that the bin by bin fit of signal with the Gaussian shape and known width is done in order to find possible local excess in data yield compared to the MC distribution (which is fixed at the first stage before the mass scan).

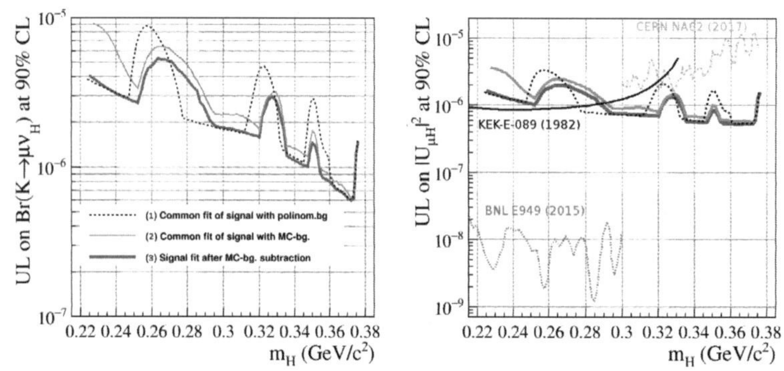

(a) (b)

Figure 4. Upper limit on branching (a) and on the mixing matrix element $|U_{\mu H}|^2$ (b) at 90% CL as a function of a heavy neutrino mass for three procedures of the fit. $|U_{\mu H}|^2$ is shown in comparison with preceding experiments, see text for details.

The fit procedure for the signal number estimate in all three approaches is not bound to positive values, the bin content is $\sim 10^2$, hence, the Wald approximation [15, 16] for the upper limit estimate can be used. No (statistically significant) indications of signal from ν_H are found in the considered mass region [14]. The upper limit at 90% CL is obtained for on branching of $K^+ \to \mu^+\nu_H$ and on the mixing parameter $|U_{\mu H}|^2$ between muon neutrino and one of heavy sterile neutrinos ν_H, see figure 4. The systematical error of around 10% to be added. We extend the result [7] (red dashed curve) obtained with stopped kaons toward higher masses and we update upper limit from [8] (dash-dotted green curve), also the old limit from [9] is shown (black solid curve).

References

[1] Shrock R E 1981 *Phys. Rev.* D **24** 1232
[2] Vuilleumier J L 1986 *Rep. Prog. Phys.* **49** 1293
[3] Kusenko A, Pascoli S and Semikoz D 2005 *JHEP* **11** 028
[4] Asaka T, Blanchet S and Shaposhnikov M 2005 *Phys. Lett.* B **631** 151-6
[5] Gorbunov D and Shaposhnikov M 2007 *JHEP* **10** 015
[6] Boyarsky A, Ruchayskiy O and Shaposhnikov M 2009 *Ann. Rev. Nucl. Part. Sci.* **59** 191-214
[7] Artamonov A *et al.* 2015 *Phys. Rev.* D **91** 052001
[8] Lazzeroni C *et al.* 2017 *Phys. Lett.* B **772** 712-18
[9] Hayano R S *et al.* 1982 *Phys. Rev. Lett.* **49** 1305
[10] Garkusha V I *et al.* 2003 *Preprint* IHEP-2003-4
[11] Citron A *et al.* 1979 *Nucl. Instrum. Meth.* **164** 31-5.
[12] Kurshetsov V *et al.* 2009 *Proc. of Sci.* **083** 051
[13] Yushchenko O *et al.* 2017 *arXiv*:1708.09587
[14] Sadovsky A *et al.* 2017 *arXiv*:1709.01473
[15] Cowan G *et al.* 2011 *Eur. Phys. J.* C **71** 1554
[16] Wald A 1943 *Trans. Amer. Math. Society.* 54 **3** 426-82

IOP Conf. Series: Journal of Physics: Conf. Series **934** (2017) 012010 doi:10.1088/1742-6596/934/1/012010

Current Results of NEUTRINO-4 Experiment

A Serebrov[1], V Ivochkin[1], R Samoilov[1], A Fomin[1], A Polyushkin[1], V Zinoviev[1], P Neustroev[1], V Golovtsov[1], A Chernyj[1], O Zherebtsov[1], V Martemyanov[2], V Tarasenkov[2], V Aleshin[2], A Petelin[3], A Izhutov[3], A Tuzov[3], S Sazontov[3], D Ryazanov[4], M Gromov[3], V Afanasiev[3], M Zaytsev[1, 4], M Chaikovskii[1]

[1] Petersburg Nuclear Physics Institute NRC KI, Gatchina, 188300 Russia
[2] NRC "Kurchatov institute", Moscow, 123182 Russia
[3] JSC "SSC RIAR", Dimitrovgrad, 433510 Russia
[4] DETI MEPhI, Dimitrovgrad, 433511 Russia

E-mail: serebrov_ap@pnpi.nrcki.ru

Abstract. The main goal of experiment "Neutrino-4" is to search for the oscillation of reactor antineutrino to a sterile state. Experiment is conducted on SM-3 research reactor (Dimitrovgrad, Russia). Data collection with full-scale detector with liquid scintillator volume of $3m^3$ was started in June 2016. We present the results of measurements of reactor antineutrino flux dependence on the distance in range 6– 12 meters from the center of the reactor. At that distance range, the fit of experimental dependence has good agreement with the law $1/L^2$. Which means, at achieved during the data collecting accuracy level oscillations to sterile state are not observed. In addition, the spectrum of prompt signals of neutrino-like events at different distances have been presented.

1. Introduction

At present there is a widely spread discussion about possible existence of a sterile neutrino having much less cross-section of interaction with matter, compared, for instance, with that of a reactor electron antineutrino [1, 2].

To search for neutrino oscillation in sterile state one has to observe the deviation of reactor antineutrino flux. Oscillation process can be described by the following equation:

$$P(\tilde{\nu}_e \to \tilde{\nu}_e) = 1 - \sin^2 2\theta_{14} \ \sin^2(1.27 \frac{\Delta m_{14}^2 [eV^2] L[m]}{E_{\tilde{\nu}}[MeV]}) \quad (1)$$

where $E_{\tilde{\nu}}$ is antineutrino energy, Δm_{14}^2 and $\sin^2 2\theta_{14}$ are the unknown oscillation parameters.

To carry out the model independent experiment it is required to perform measurements of antineutrino flux and spectrum at short reactor distances, e.g. 6 – 12 meters from almost point-like antineutrino source. Due to some peculiar characteristics of its construction, research reactor SM-3 provides the most favorable conditions for conducting the experiment [3]. However, SM-3 reactor is at the Earth surface, hence cosmic background is the main difficulty with this experiment. Here we present last results of measurements of flux distance dependence and neutrino spectrum at different distances within range 6 –12 m from reactor obtained with full-scale detector.

2. Full-scale antineutrino detector

Scheme of full-scale detector is presented in figure 1. The full-scale detector with liquid scintillator has volume of 3 m^3 (5×10 sections). The scintillation type detector is based on IBD (inverse beta decay) reaction: $\tilde{\nu}_e + p \rightarrow e^+ + n$. Detector model was designed multi-sectional to distinguish between positron emitted in inverse beta decay reaction and recoil proton from fast neutron elastic scattering. The internal active shielding is located on the top of detector and under it.

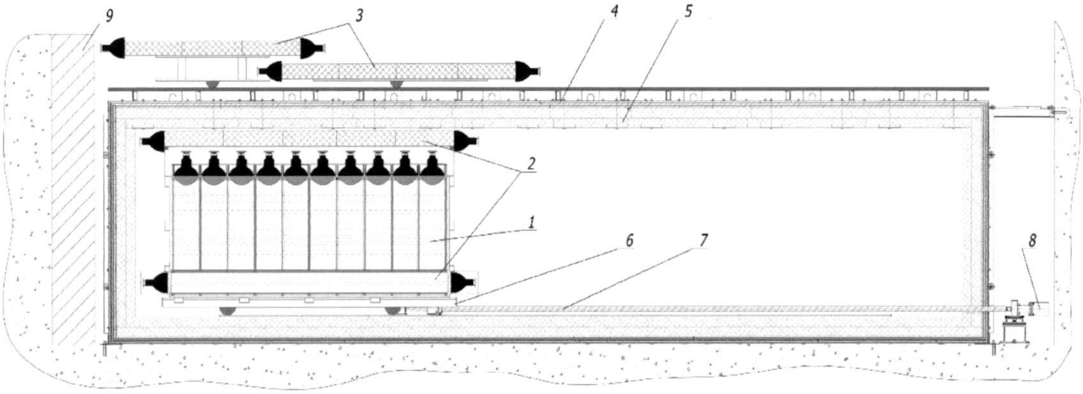

Figure 1. General scheme of experimental setup. 1 – detector of reactor antineutrino, 2 – internal active shielding, 3 – external active shielding (umbrella), 4 – steel and lead passive shielding, 5 – borated polyethylene passive shielding, 6 – moveable platform, 7 – feed screw, 8 – step motor, 9 – iron shot.

Detector can be placed in different positions for measurement with 0.47 m distance between them to avoid influence of section efficiency distinction. Shift could be made from one position to any other. The sectioned detector structure allows us to present the distance dependence with 0.47-meter step. The technique of making measurements was to move detector for 0.94 m starting with the end position. On the second stage the measurements were repeated with translation of starting position for 0.47 meters. Thus, practically each row of the detector measured the same point, averaging in this way somewhat different recording efficiency of detector's rows. For a full-scale detector, we use the first and the last rows of cells not for detecting a positron signal itself, but for annihilation gamma quanta at 511 keV.

Multi-section model was designed especially for detecting positron emitted in inverse beta decay reaction. Fast neutrons from cosmic rays are the main problem for Earth-surface experiments. Fast neutron scattering imitates neutrino reaction. The recoil proton mimics the prompt signal from positron. The delayed signal emits during neutron capturing by Gd in both reactions. The difference in prompt signals is that in neutrino process two γ-quanta are emitted due to positron annihilation. The recoil proton path with high probability lies in single section. 511 keV γ-quanta can be detected in adjacent sections. However, Monte-Carlo modelling of full-scale detector shows that only ~37% (on the average for detector considering the peculiarities of detection) of neutrino events are multi-section due to γ-quanta detection in neighbouring sections with respect to section where positron annihilated. The fact that event statistics is almost 3 times less if we consider only multi-section events is hardly acceptable, so we considered data analysis model with both multi and single section events. But we use ratio for multi and single section as additional examination. Hence, if the signal difference for reactor turning on (off) is in ~37% to ~63% ratio for multi and single section events then we consider them to be neutrino-like signals.

Difference in count rates in reactor on and off regimes for double and single starts integrated over all distances for full-scale detector was (37±4)% and (63±7)%. With considered precision this ratio allows us to regard registrated events as neutrino-like events.

IOP Conf. Series: Journal of Physics: Conf. Series **934** (2017) 012010 doi:10.1088/1742-6596/934/1/012010

Multi-sectional structure of full-scale detector allow us to improve statistical accuracy by increasing signal/background ratio from 0.3 to 0.6 and by suppressing accidental coincidences background.

3. Results

Results of measurements of difference in counting rate of neutrino-like events for full-scale detector is shown in figure 2 (left) as dependence of antineutrino on distance from reactor center. To increase statistical accuracy data was averaged in the range of 1 meter (last point in the range of 2 meters). At the same time, there is an experimental result of DANSS, kindly provided by M. V. Danilov, and presented at the 52nd Rencontres de Moriond [4], which points out that in the range of 10 – 13 meters the law $1/L^2$ holds true with rather good accuracy. We can make use of this information and combine results of measurements in the range of their overlapping of 10 – 12 meters. Results of association of the Neutrino-4 and DANSS experiments data are presented in figure 2 (right).

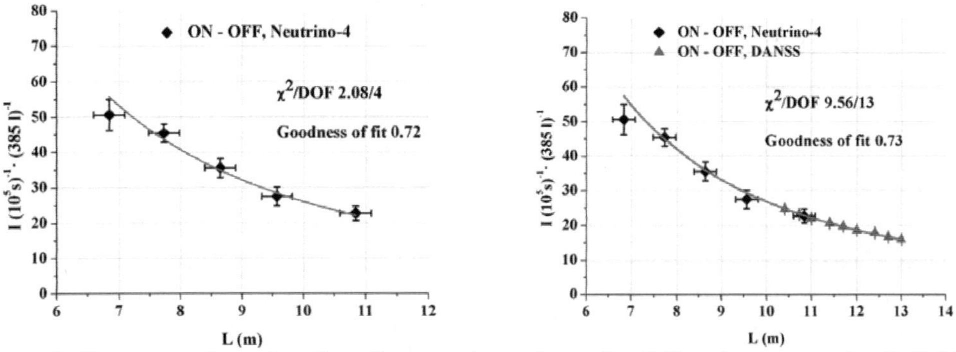

Figure 2. Reactor antineutrino flux distance dependence for full-scale detector itself (left) and after combining with DANSS experimental data (right). Solid line is the fit for dependence $1/L^2$, where L is the distance from the center of reactor core.

The difference spectra (reactor ON - reactor OFF) of prompt signals with 6 distance points (normalized to unity) are presented in figure 3.

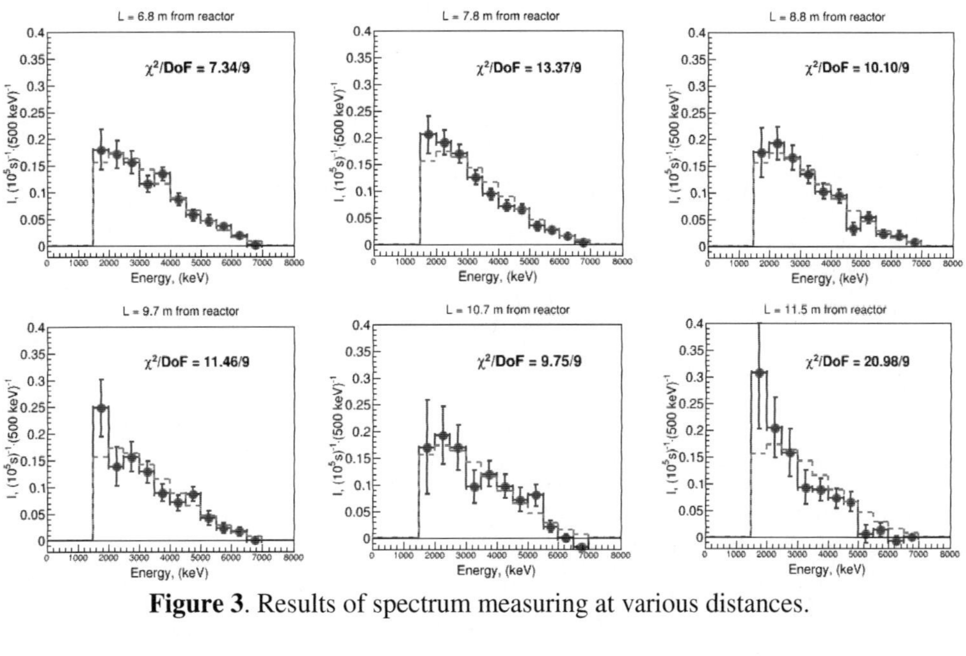

Figure 3. Results of spectrum measuring at various distances.

As one can see from figure 3, statistical accuracy is not enough for the further analysis of ratio of spectra at different distance. Figure 4 illustrates the spectrum of prompt signals summarized on all distances for increasing statistical accuracy (averaged distance 8.6 meters). This spectrum contains statistics of all measurements.

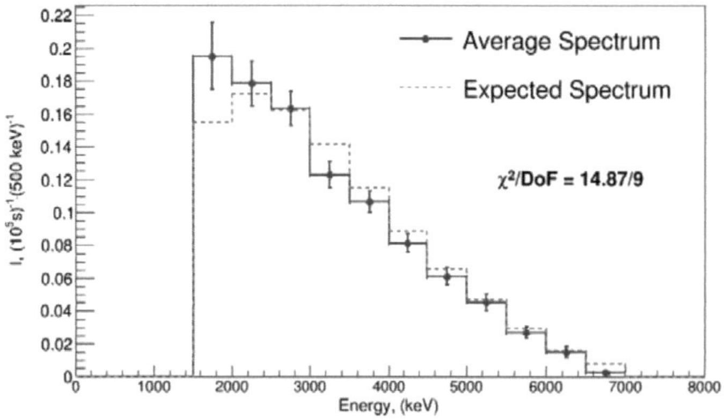

Figure 4. Spectrum of prompt signals for a total cycle of measurements summarized on all distances (average distance 8.6 meters). The red dotted line shows Monte Carlo simulation with neutrino spectrum for ^{235}U, as the SM-3 reactor works on highly enriched uranium.

It is of great interest to use the obtained experimental data of the Neutrino-4 and DANSS experiments for global fit of sterile neutrino model parameters. At the same time, it must be kept in mind that at long distances results of fit should satisfy the results of measurements at these long distances. As both experiments in DANSS and in Neutrino-4 use relative measurements, it is necessary to mention, how points of DANSS and Neutrino-4 together applied on an overall picture with results of other experiments [5]. In the picture of global fit, they can be fixed either with attaching to average value, which define "reactor anomaly", or to unity. In second case, data of other experiments are not used. It is reasonable to assume that the errors of DANSS points must have been determined by an error of binding procedure (i.e. error of average value of ratio observed/predicted of other reactor experiments [6,7]). Analysis of confidence level for parameters of oscillations Δm_{14}^2 and $\sin^2 2\theta_{14}$ is presented in figure 5 for spatial distribution for both cases. It is necessary to mention these areas on the $(\sin^2 2\theta_{14}, \Delta m_{14}^2)$ - plane of confidence level are mainly determined by high accuracy of DANSS experiment data.

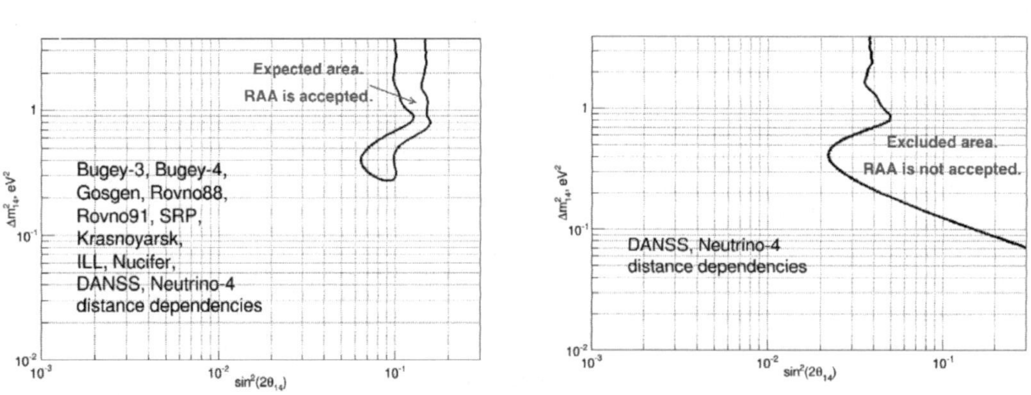

Figure 5. Confidence Level (CL): left – for spatial distribution of data for experiments of the Neutrino-4, DANSS (bound to RAA) and other reactor experiments from [2]; right – for spatial distribution of data for experiments of the Neutrino-4, DANSS attached to unity. Black curve corresponds to 99% CL.

Global picture with Neutrino-4 and DANSS data attached to average value 0.936 ± 0.021 with curve corresponding to minimum of χ^2 shown on figure 6.

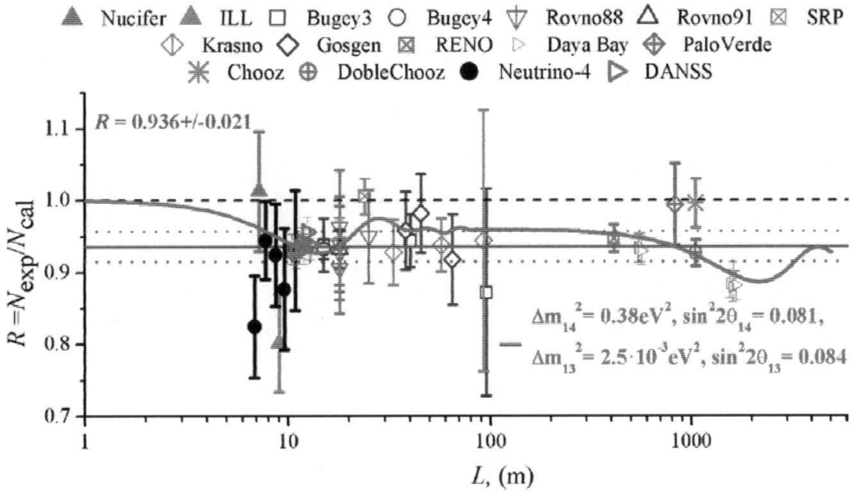

Figure 6. Global fit of the sterile neutrino model parameters with experimental data of the Neutrino-4, DANSS and data of the known experiments at long distances [6, 7].

4. Conclusion

Measurements of the flux of an antineutrino from the reactor at small distances of 6 – 12 m by means of the moveable detector are carried out for the first time. In the frame of the available statistical accuracy it is not revealed if there are any reliable deviations of antineutrino flux distance dependence from the law $1/L^2$ where L – distance from the center of reactor core, which is in a good agreement with data of DANSS experiment.

Measurements with full-scale detector will be continued in the same way (movement, spectrum measurement) to reach better statistical accuracy, but it does not solve the main problem of cosmic background in general. Therefore, our plan is to create second neutrino laboratory with longer base (up to 15m) on reactor SM-3 with two identical detectors (one stationary and one moveable). New detectors will have sectional structure too, but with smaller sections and scintillator with higher gadolinium concentration and better PSD capability will be used. DANSS (Russia) and NEOS (Korea) collaborations will be participants in this project.

Acknowledgments

The authors are grateful to the Russian Foundation for Basic Research for support under Contract No. 14-22-03055-ofi_m. The delivery of the scintillator from the laboratory leaded by Prof. Jun Cao (Institute of High Energy Physics, Beijing, China) has made a considerable contribution to this research.
The authors also express their gratitude to DANSS collaboration and personally to prof. M. V. Danilov for the DANSS experiment data they kindly provided.

References

[1] Mueller T *et al.* 2011 *Phys. Rev.* C **83** 054615
[2] Mention G *et al.* 2011 *Phys. Rev.* D **83** 073006
[3] Serebrov A *et al.* 2014 *Tech. Phys. Lett.* **40** 456 (*Preprint* 1310.5521)
[4] Danilov M 2017 *The 52nd Rencontres de Moriond Electroweak Interactions and Unified Theories,* Thuile *https://goo.gl/c8WUsV*
[5] Serebrov A *et al.* 2017 *Preprint* 1708.00421

[6] Giunti C 2016 *Neutrino 2016 27th Int. Conf. on Neutrino Physics and Astrophysics,* London *https://goo.gl/v1WrUi*
[7] Li Yu-Feng 2016 *Applied Anti-neutrino Physics 2016,* Liverpool *https://goo.gl/m2OJIj*

IOP Conf. Series: Journal of Physics: Conf. Series **934** (2017) 012011 doi:10.1088/1742-6596/934/1/012011

How Geoneutrinos can help in understanding of the Earth heat flux

L B Bezrukov[1], A S Kurlovich[1], B K Lubsandorzhiev[1], A K Mezhokh[1], V P Morgalyuk[2], V V Sinev[1,*] and V P Zavarzina[1]

[1]Institute for Nuclear Researches of Russian Academy of Sciences, Prospekt 60-letia Oktyabrya 7a, 115409 Moscow, Russia
[2]A. N. Nesmeyanov Institute of Organoelement Compounds of Russian Academy of Sciences, Vavilova 10, 115409 Moscow, Russia

E-mail: *vsinev@inr.ac.ru

Abstract. The Hydride Earth model predictions of geoneutrino flux and intrinsic Earth heat flux are discussed. The geoneutrino flux predicted by the model can be adjusted to the experimental one. The predicted intrinsic Earth heat flux is significantly larger than model dependent experimental value obtained under assumption that the main heat transfer mechanism is a thermal conductivity. We introduce an additional mechanism of heat transfer in the Earth's crust, namely the energy transfer by hot gases produced in the Earth crust at great depths. The experimental data supporting this idea, in particular the temperature profiles measured in the Kola super deep borehole, are discussed.

1. Introduction

So far there are only two detectors, Borexino [1] and KamLAND [2], which reported registration of geoneutrino signals. Geoneutrinos are electron antineutrinos produced in beta-decays of radioactive elements in natural families of ^{238}U, ^{232}Th and ^{40}K, accumulated inside the Earth. Geoneutrino flux on the Earth surface depends on the amounts of ^{238}U, ^{232}Th and ^{40}K in the Earth and their distribution with the depth.

Amounts of ^{238}U, ^{232}Th and ^{40}K in the Earth and their distribution are different in existing models of the Earth. The theory most popular at the moment is called the Bulk Silicate Earth (BSE) [3, 4]. Its main idea is that element abundances are the same as in meteorites. Based on this idea the amounts of ^{238}U, ^{232}Th and ^{40}K were obtained for the Earth:

$$M_{\rm BSE}(^{238}{\rm U}) = 0{,}81\cdot10^{17}\ {\rm kg},\ M_{\rm BSE}(^{232}{\rm Th}) = 3{,}16\cdot10^{17}\ {\rm kg\ and}\ M_{\rm BSE}(^{40}{\rm K}) = 5{,}73\cdot10^{16}\ {\rm kg}. \qquad (1)$$

These masses are distributed mainly in the Earth crust and partially in its mantle, but they are absent in its core. Experimentally observed antineutrino flux is in agreement with ^{238}U and ^{232}Th amounts from (1) under the assumption that they are distributed in the crust and in the upper mantle [1] only.

Each radioactive decay accompanied by a definite thermal energy emission. If we know the total amounts of ^{238}U, ^{232}Th and ^{40}K in the Earth, the value of radiogenic heat flux can be predicted and it can be compared with experimentally measured one.

The Earth thermal flux on continents in boreholes is explored by the measurement of the temperature gradient at depths of ~500 meters. The thermal flux in oceans is measured by dedicated apparatus which penetrates to the bottom floor by several meters and measures the temperature

IOP Conf. Series: Journal of Physics: Conf. Series **934** (2017) 012011 doi:10.1088/1742-6596/934/1/012011

gradient. To calculate the value of thermal flux one uses the idea that the main thermo-transfer mechanism is a thermal conductivity. Presently the value obtained from experimental measurements assuming the foregoing point of view is 47±2 TW.

The calculated inner Earth thermal flux appears to be equal to 17.5 TW. In this calculation the total masses of ^{238}U, ^{232}Th and ^{40}K in the Earth were taken from (1). Comparison of this value with experimental one did not confirm correctness of the BSE basic assumptions. Researches started to look for additional heat sources.

In this paper we will use an alternative Earth model, so called the Hydride Earth model (HE) [5], and discuss how predictions of geoneutrino and thermal fluxes are correspond to the observed data.

2. Hydride Earth model and geoneutrinos

The main idea of the HE model is the following: a planet element composition depends on the distance from the Sun. That is why the Earth's element composition should differ from composition of asteroids in the asteroid belt which is the source of meteorites. Vladimir Larin [5] proposed and used for element composition calculations the following equation:

$$(X_M/X_{Si})_{Earth} = (X_M/X_{Si})_{Sun} \cdot F(E_{IP}(M)), \quad (2)$$

where X_M – the share of element M in the Earth mass, X_{Si} – the share of Silicon in the Earth mass, $F(E_{IP}(M))$ – the mass share of chemical element M from the mass of this element in the Sun that existed at the Earth's orbit at its formation, $E_{IP}(M)$ – the ionization potential of element M.

In [6] the function F was proposed as an exponential one:

$$F(E_{IP}(M)) = A\cdot \exp\{-B\cdot E_{IP}(M)\}, \quad (3)$$

where coefficients A and B were chosen from the known element composition of the Sun and Earth's crust.

We can estimate the total masses of ^{238}U, ^{232}Th and ^{40}K in the Earth using a method of calculation proposed in [7] and quoting the value of $E_{IP}(M)$ and function $F(E_{IP}(M))$ from [8] and [6] respectively:

$$M_{HE1}\,(U) = 3{,}18\cdot 10^{17}\,kg,\ M_{HE1}\,(Th) = 1{,}0\cdot 10^{18}\,kg \text{ and } M_{HE1}(K) = 2{,}6\cdot 10^{19}\,kg. \quad (4)$$

It is noteworthy that the use of function $F(E_{IP}(M))$ as in [6] results in the values of ^{238}U and ^{232}Th masses roughly three times less than in (4) – M_{HE} (U) = $1.1\cdot 10^{17}$ kg, M_{HE} (Th) = $3.6\cdot 10^{17}$ kg.

As it is shown above, predictions of the HE model depends strongly on the $F(E_{IP}(M))$ function, which is known in its turn rather crudely. It is of fundamental importance that the HE model leaves room for larger values of the total masses of ^{238}U, ^{232}Th and ^{40}K in the Earth as well as allows considering ^{238}U, ^{232}Th and ^{40}K present in the Earth core in the primordial amount, i.e. at the moment of the Earth formation. In particular, the total mass of potassium is much more than predicted by the BSE model because of its small ionization potential.

In our previous paper [9] an attempt was done to estimate the maximal values of ^{238}U, ^{232}Th and ^{40}K masses in the Earth which are allowed by the modern experimental data on geoneutrino fluxes and the Earth thermal flux. Our analysis provided the following results:

$$M_{HE2}\,(U) = 1{,}7\cdot 10^{17}\,kg,\ M_{HE2}\,(Th) = 6{,}7\cdot 10^{17}\,kg \text{ and } M_{HE2}(K) = 1{,}2\cdot 10^{19}\,kg. \quad (5)$$

In the estimates we expanded potassium abundance known for the Earth's crust to the Earth whole volume.

Comparing estimates made by the HE and BSE models, (5) and (1) respectively, one can conclude that they are rather close, although the HE values are slightly larger.

Taking into account that the HE model predicts the total masses of ^{238}U, ^{232}Th and ^{40}K in the Earth's core in primordial abundances experimental arrays for ^{238}U and ^{232}Th geoneutrino flux detection should have substantially larger events statistics in comparison with the present arrays in order to have capabilities to distinguish those two models.

The total mass of ^{40}K predicted by the HE model is more than two orders of magnitude larger than the mass predicted by BSE model. In [10] the capability to detect the large geoneutrino flux due to ^{40}K flux by modern detectors was analysed. The conclusion is done that sensitivity of presently operating detectors is not sufficient for such a purpose. But for next generation detectors with significantly lower background and larger target possibility to measure geoneutrino flux due to ^{40}K should be conspicuously high. It is also important to have independent measurements of CNO cycle neutrinos spectrum, which is rather close to the spectrum shape of ^{40}K neutrinos.

3. Hydride Earth model and thermal flux of the Earth

One can calculate steady thermal flux of the inner Earth heat knowing the value of the heat energy released per radioactive decay. Thermal flux corresponding to the foregoing mass values of ^{238}U, ^{232}Th and ^{40}K in case of the HE model (5) turns out to be 304 TW. This huge amount of heat is the consequence of the larger amount of potassium in the Earth predicted by the HE model.

At first glance, comparison of the HE value with the experimentally measured one of 47 TW leads to the idea that the HE model is wrong. Indeed, the large amount of potassium follows from the basic HE postulates claiming that ionization potential determines a share of chemical element mass which existed in a planet's orbit at the moment of its formation. If to assume that the HE model is correct than it is necessary to bring forward arguments allowing understanding of so large difference between values of thermal fluxes predicted by the HE model and measured experimentally in boreholes by the temperature gradient method.

Furthermore let's consider here the experimental data of the Earth thermal flux different from the value in 47 TW.

The results of the temperature profile measured in super deep boreholes are rather easy to understand in the framework of the HE model. But, on contrary, the results were not predicted by the BSE model at all.

Scientific results of the Cola super deep borehole were published in [11]. The depth of 12 262 meters was achieved there. In fact all ideas of the Earth's core structure obtained from the measurements in the borehole were inconsistent with expectations. With the deepening the rocks do not get denser and their porosity does not decrease as it was expected earlier. On the contrary, the rocks at the multi-kilometer depth are penetrated by multiple pores and broken by cracks.

It was a sensation that there existed hydrogen and hydrogen-containing gases, in particular water, at high depth where the pressure reaches hundreds of atmospheres.

Finally, it appeared that the Earth is significantly hotter than it was assumed earlier. At the depth of 5 km the temperature exceeds 70°C and passes 120°C and 220°C at 7 km and 12 km depths respectively. The latter is 100°C more than the temperature extrapolated from the value measured at 1 km depth.

The HE model predicts element composition of the primordial Earth. Hydrogen should constitute about 18% of mass and it could be accumulated in the hydride form. Radiogenic heating of the Earth's core results in the production of free hydrogen at the border between the core and mantle. The HE model predicts existence of degasification processes from the Earth's surface, basically by hydrogen containing gases, e.g. water vapor. Gases should be produced at high depth where appropriate temperatures could exist. The gases provide production of multiple cracks and pores in rocks at high depth.

These gases could be the main carrier of thermal energy. Cracks and pores allow gases to move up to the surface. The heat could bypass the thermometers measuring the temperature gradient at the depths of about 500 meters or at the ocean bottom by two ways. The first one is absorption of thermal energy during gas production process. Thermal energy is transferred to the internal energy of gas molecules which is released later by exothermic decomposition reactions at small depths. The second one is transfer of thermal energy to the Earth surface by narrow jets of hot gases. These jets should have substantial distances between them. Existence of such vertical jets leads at large depths to the emergence of horizontal component of gas velocity. Therefore, only a fraction of heat at the depths of about 500 meters takes part in the process of vertical heat transfer by thermo-conductivity mode. The most part of heat energy is carried out by hot gases. We note that boreholes do not drilled at outlets of gas jets because on continental platforms gas outlets look like lakes or swamps. Volcanoes are gas outlets too.

The existence of the forgoing heat transfer mechanism, additional to thermo-conductivity, allows an understanding of unexpected high temperatures at large depths. It also helps to resolve contradiction between the large heat flux predicted by the HE model and results of temperature gradient measurements.

The ARGO experiment published recently interesting results showing temperature increasing of the world ocean in 2005–2010 [12]. Despite that period has been the period of the minimum of solar activity, the world ocean has been warming up. To provide such heating there is a need to have additional energy flux of 0.58 ± 0.15 W·m^{-2} for 6 years. If to multiply this value by the Earth surface area we will get the Earth thermal flux of 300 ± 80 TW.

In the context of the HE model, this experimental fact could be explained by an increase of gases outcome into the ocean and existence of exothermic reactions with gases in the ocean volume. The model includes an idea that the gas flux is not stable but cyclic. We can consider the present ocean heating is observed due to an increase of the Earth's internal heat flux during of the Earth's exit from the small ice age. The value of the heat flux necessary for the observed ocean heating does not cause surprise in frame of the HE model.

It is noteworthy to mention studies done at the Lebedev Physical Institute of the Russian Academy of Sciences (LPI) [13]. In that study the Moon thermal flux was measured by detection of radio wave emission from the Moon. The authors of the paper consider that the heat flux comes from deep interior of the Moon and has the same radiogenic nature as on the Earth. Basing on the idea that the Moon and Earth have the same element composition we can calculate the Earth heat flux using some scaling factor:

$$H_{\mathrm{earth}} = H_{\mathrm{moon}} \cdot M_{\mathrm{earth}} / M_{\mathrm{moon}} = 170\ \mathrm{TW}. \qquad (6)$$

Where: H_{moon} – the measured heat flux at the Moon, M_{earth} and M_{moon} – the Earth and Moon masses respectively. It is necessary to note that on the Moon, according to the HE model, degasification process finished because of its small size and practically all radiogenic energy comes out of the Moon by thermo-conductivity way. The heat flux value calculated using (6) supports the correctness of the HE model. It is important to note that the Moon thermal flux measurements by the radio wave emission method average the Moon thermal flux on the visible Moon surface and gives us the mean value.

In two different points on the Moon surface measurements of thermal flux were done by drilling boreholes during Apollo 15 and Apollo 17 missions. Basing on the Apollos data and using equation (6) it is possible to recalculate the Earth thermal flux. The resulting values are 43 and 65 TW. It is believed that these data do not contradict to LPI measurements because in that Apollo missions landing places the thermal fluxes should be significantly less than the mean value. It is supported by the Moon pictures made by the infrared telescope installed on board of the NASA satellite Lunar Reconnaissance Orbiter in 1996 during the full lunar eclipse.

Finally, let's estimate the upper limit of the Earth thermal flux. There are certain places on the Moon surface which are never illuminated by the Sun. The places have always constant temperature. We can assume that only radiogenic internal heat determines their temperature. So, the energy flux emitted by the Moon surface unit can be estimated according to the Stephan-Boltzmann law. Then it is possible to calculate the total thermal flux of the Moon. The Earth thermal flux recalculated using (6) is 420 TW. The value is obtained taking into account the temperature of 40°K in places never illuminated by the Sun at the Moon's South Pole region. The value of 420 TW for the Earth thermal flux can be regarded as an upper limit. This upper limit shows that the ^{238}U, ^{232}Th and ^{40}K masses in the Earth presented in (4) are too large. Whereas the masses shown in (5) result in the Earth thermal flux of 304 TW. The last value is smaller than the upper limit.

4. Conclusion

1. The HE model is a convenient tool for analysis of events taking place on the Earth. We demonstrated in the framework of the HE model that many complicated phenomena could find explanations in a natural way basing on just one model. Those are the geoneutrino fluxes, temperature profiles of super deep boreholes and experimentally observed increase of the ocean temperature.

2. It was shown that the HE model introduces the large concentrations of ^{238}U, ^{232}Th and ^{40}K in the Earth and allows their existence in the Earth core in primordial amount with correction for the elements decay. The most prominent amount appeared to be for potassium in comparison with the Bulk Silicate Earth model prediction due to its small ionization potential.
3. An assumption was done that the modern value of the Earth thermal flux resulting from the temperature gradient measurements does not take into account significant amount of heat transferred by hot gases. The Hydride Earth model predicts production of gases at high depths. The existence of this kind of heat transfer mechanism, additional to thermal conductivity, allows understanding of high temperature existing at big depths in deep boreholes on continents and in oceans.

Acknowledgments

Authors wishing to thank Igor Tkachev (INR RAS) for fruitful discussion, Allen Caldwell (MPI) for the opportunity to talk at MPI seminar and valuable discussion, Leo Stodolsky (MPI) for inspiring discussion, and in particular for his question on temperature profiles in continental platforms at existence of thermal energy transfer by hot gases.

References

[1] Agostini M *et al.* 2015 *Phys. Rev.* D **92** 031101 (*Preprint* 1506.04610 [hep-ex])
[2] Gando A *et al.* 2013 *Phys. Rev.* D **88** 033001 (*Preprint* 1303.4667 [hep-ex])
[3] Bellini G, Ianni A, Ludhova L, Mantovani F, McDonnough W F 2013 *Progress in Part. and Nucl. Phys.* **73** 1 (*Preprint* 1310.3732 [physics.geo-ph])
[4] Huang Yu, Chubakov V, Mantovani F, Rudnik R, Ludhova L, McDonnough W F 2013 *Preprint* 1301.0365 [physics.geo-ph]
[5] Larin V N 1993 *Hydridic Earth: the New Geology of Our Primordially Hydrogen-Rich Planet* (Hunt. Calgary, Alberta, Canada: Polar Publishing)
[6] Toulhoat H, Beaumont V, Zgonnik V, Larin N, Larin V 2012 *Preprint* 1208.2909 [astro-ph.EP]
[7] Bezrukov L 2014 *Preprint INR-1378/2014* (Moscow) (*Preprint* 1308.4163[astro-ph.EP])
[8] 2003 *CRC Handbook of Chemistry and Physics 84th Edition Section 10 Atomic, Molecular and Optical Physics: Ionization Potentials of Atoms and Atomic Ions* ed D R Lide (Boca Raton, Florida: CRC Press)
[9] Bezrukov L, Kurlovich A, Lubsandorzhiev B, Morgalyuk V, Sinev V and Zavarzina V 2016 On geoneutrino *EPJ Web of Conference QUARKS-2016* **125** 02004
[10] Bezrukov L, Litvinovich E, Machulin I, Skorokhvatov M, Sinev V and Sukhotin S 2015 *EPAN Letters* **46** 339 (*Preprint* 1405.3140 [hep-ex])
[11] Orlov V P, Laverov N P 1998 *Cola super deep hole. Scientific results and research experience* ed Brodsky P A *et al.* (Moscow: MF Technoneftegaz)
[12] Hansen J, Sato M, Kharecha P, and Schuckmann K von 2011 *Atmos. Chem. Phys.* **11** 13421
[13] Krotikov V D, Troitskyi V S 1964 Radio emission and nature of the Moon *Soviet Phys. Uspekhi* **6** 84

Calculation and measurement of ^{144}Ce-^{144}Pr β-spectrum

V. Atroschenko[1], V. Kopeikin[1], E. Litvinovich[1,2], L. Lukyanchenko[1], I. Machulin[1,2], M. Skorokhvatov[1,2], O. Titov[1*]

[1] National Research Centre "Kurchatov Institute", 123182, Moscow, Russia
[2] National Research Nuclear University "MEPhI" (Moscow Engineering Physics Institute), Kashirskoe highway 31, Moscow, 115409, Russia

E-mail: *titov_oa@nrcki.ru

Abstract. We calculate beta spectrum of Ce-Pr-144 taking into account several types of corrections. The result is compared with the experimental data obtained at NRC Kurchatov Institute. Using this comparison we estimate the reliability of theoretical calculations for electron and antineutrino spectra from beta decay.

1. Introduction

Prospective neutrino experiments with intense radioactive sources (such as SOX [1]) require a precise knoweledge of the emitted neutrino spectrum. One of the most promising antineutrino sources for such experiments is ^{144}Ce-^{144}Pr [2]. Antineutrinos are usually detected via inverse beta decay (IBD) process

$$\bar{\nu}_e + p \rightarrow n + e^+ \quad (1)$$

with a threshold $E_{thr} = 1806$ keV. In case of ^{144}Ce-^{144}Pr, the actual antineutrino source is ^{144}Pr, because it has two decay branches with endpoint energies greater than the IBD threshold E_{thr}. The first (dominant) branch is a non-unique first-forbidden Gamow–Teller transition $0^- \rightarrow 0^+$ with endpoint energy 2997.5 keV and 97.9% branching. The second one is a unique first-forbidden Gamow–Teller transition $0^- \rightarrow 2^+$ with endpoint energy 2301.0 keV and 1.04% branching.

The electron and antineutrino β-spectra are described basically by the same formula. So, to verify the reliability of antineutrino spectrum calculation one can compare theoretical electron spectrum with experimental data. We provide such a comparison using the experimental data on ^{144}Ce-^{144}Pr electron spectrum obtained at NRC Kurchatov Institute. In our calculation, only the main decay branch (with the branching $I = 97.9\%$) of ^{144}Pr is considered.

2. Beta spectrum calculation

The formula for the electron spectrum is

$$N(W_e) = Kp_eW_e(W_e - W_0)^2F(Z,W_e)H(W_e)\times \times L(Z,W_e)C(Z,W_e)S(Z,W_e)G_\beta(Z,W_e)B(W_e). \quad (2)$$

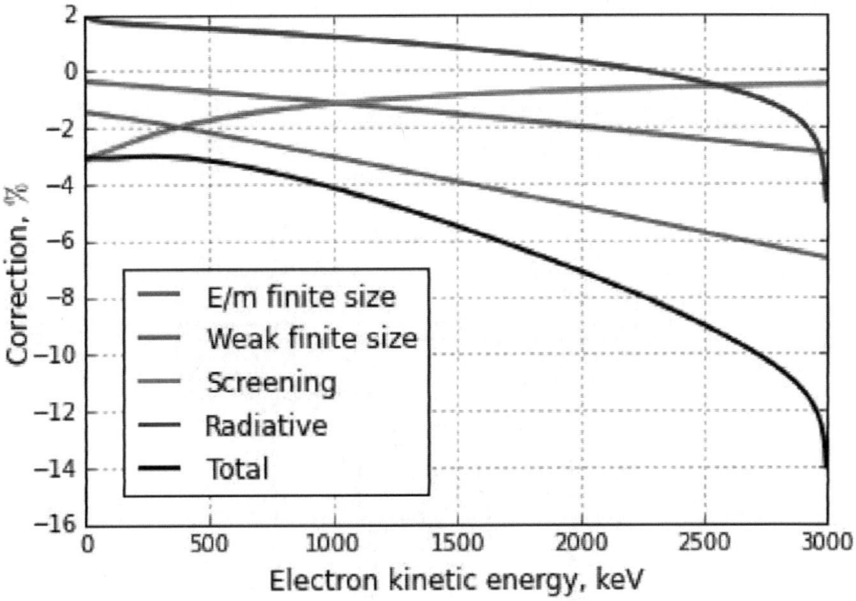

Figure 1. Corrections to electron spectrum for ^{144}Pr branch with endpoint energy 2.998 MeV.

Here W_e and W_0 are the total electron energy and the endpoint energy in units of m_ec^2, p_e is the electron momentum, K is a normalization constant. The antineutrino spectrum is obtained by replacing W with $W_0 - W$.

The Fermi function $F(Z, W)$ describes the effect of nuclear Coulomb field on the outgoing electrons [3]. In case of forbidden decays one has to take into account the shape factor $H(W)$. For the main decay branch of ^{144}Pr we take the shape factor in the following form [4] :

$$H(W_e, W_0) = p_e^2 + (W_e - W_0)^2 + 2p_e^2(W_e - W_0)/W_e.$$

Electromagnetic and weak finite-size corrections $L(Z, W)$ and $C(Z, W)$ are due to the distibutions of electric and weak isovector charges in the nucleus [5]. The effect of atomic electrons on the spectrum is given by the screening correction $S(Z, W)$ [6]. The first order QED radiative corrections are included in $G_\beta(Z, W)$ [7, 8]. Note that this is the only type of corrections that is different for electron and neutrino spectra. Finally, the weak magnetism correction $B(W)$ describes the interaction of the emitted electrons with the magnetic moment of the daughter nucleus. For the main decay branch of ^{144}Pr, which is a first-forbidden non-unique transition it equals zero [4, 9].

Figure 1 shows the correction values for the electron spectrum of the main ^{144}Pr decay branch. Only the energy dependent parts affecting the spectrum shape are presented.

3. Comparing theory with experiment

An experiment on ^{144}Ce-^{144}Pr beta spectrum measurement was conducted at NRC "Kurchatov Institute" with a beta-spectrometer. The experimental setup is shown on figure 2. The detector was calibrated using ^{207}Bi conversion spectrum. The detector resolution is $\sigma/E = 5\%$ at 1 MeV. To make a proper comparison between the theoretical calculations and the experimental data we introduced the response function $R(E, E')$. The measured spectrum is given by

$$M(E) = \int_0^\infty R(W_e, W_e')N(W_e')dW_e', \tag{3}$$

IOP Conf. Series: Journal of Physics: Conf. Series **934** (2017) 012012 doi:10.1088/1742-6596/934/1/012012

where $N(W'_e)$ is the theoretical spectrum (2).

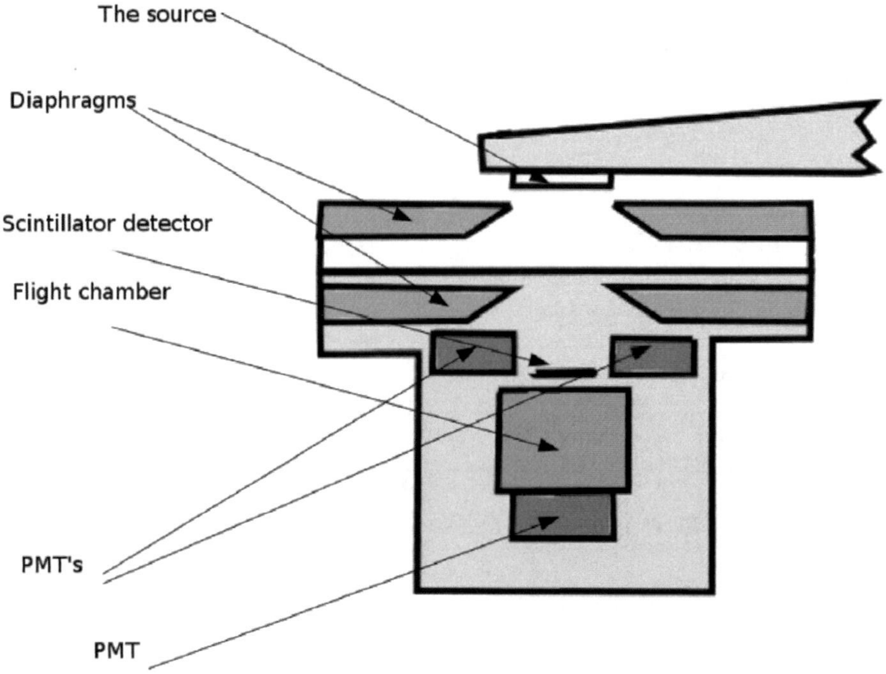

Figure 2. Beta-spectrometer scheme.

The response function $R(E, E')$ should take into account the detector resolution given by a Gaussian function and electron backscattering given by a step function (see [10]). According to our Monte-Carlo simulations, the step function accounts for about 3% of total area under the response curve. Note that usually one expects a stronger (about 10%) influence of electron backscattering [10], so further improvement of the Monte-Carlo calculations should be done.

Figure 3 shows the theoretical calculation and the experimental data compared (unconvoluted theoretical spectrum is also given to show the effect of the response function). The spectra are normalized; the total nomber of experimental events is $1.86 \cdot 10^5$. The theoretical and the experimental spectra are in general agreement with each other. In particular, the spectra match at low electron energies. These energies correspond to high antineutrino energies and are of interest for neutrino experiments.

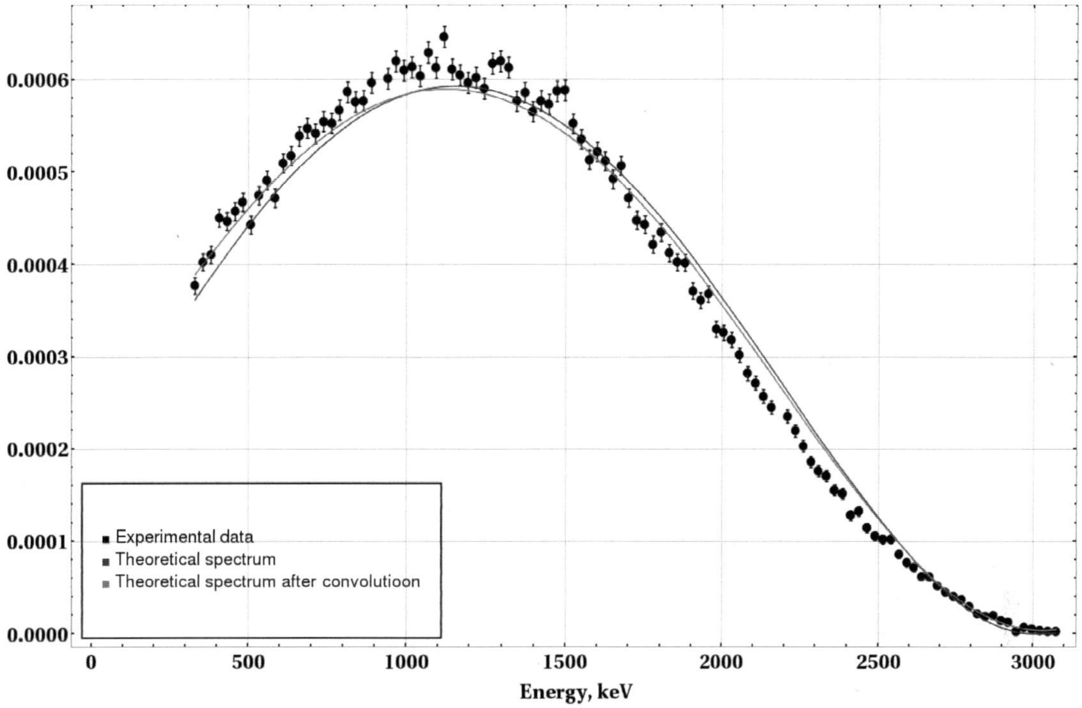

Figure 3. Comparison between calculated electron spectra for ^{144}Pr and experimental data. Total number of experimental events is $1.86 \cdot 10^5$.

4. Conclusion

We calculated the electron spectrum from the main branch of ^{144}Pr β-decay. The calculations are in general agreement with the experimental data. However, there is an uncertainty in the detector response function, so further data analysis is needed. A comparison with the data from other experiments is also desirable.

Acknowledgments

The work was partially supported by the Russian Foundation for Basic Research under grant 16-02-00616.

References

[1] Bellini G *et al.* (BoreXino collaboration) 2013 *JHEP* 8 p 38
[2] Gaffiot J *et al.* 2015 *Phys. Rev.* D **91** 072005
[3] Behrens H and Janecke J 1969 *Numerical Tables for Beta-Decay and Electron Capture* (Berlin: Landolt-Bornstein New Series, Springer Verlag)
[4] Hayes A C *et al.* 2014 *Phys. Rev. Lett.* **112** 202501
[5] Wilkinson D H 1990 *Nucl. Instr. and Meth.* A **290** p 509
[6] Behrens H and Buhring W 1982 *Electron Radial Wave Functions and Nuclear Beta Decay* (Oxford: Clarendon)
[7] Sirlin A 1967 *Phys. Rev.* **164** p 1767
[8] Raha U, Myhrer F and Kubodera K 2012 *Phys. Rev.* C **85** 045502
[9] Hayes A C and Vogel P 2016 *Annu. Rev. Nucl. Part. Sci.* **66** pp 219 – 244
[10] Tsoulfanidis N, Wehring B W and Wyman M E 1969 *Nucl. Instr. and Meth.* **73** pp 98 – 102

IOP Conf. Series: Journal of Physics: Conf. Series **934** (2017) 012013 doi:10.1088/1742-6596/934/1/012013

Nanotubes based neutron generator for calibration of neutrino and dark matter detectors

A S Chepurnov[1,2], V Y Ionidi[2], M A Kirsanov[3], E P Kitsyuk[4], A A Klenin[1], A S Kubankin[1,5*], A N Oleinik[1], A A Pavlov[4], A V Shchagin[1,6]

[1] Laboratory of Radiation Physics, Belgorod National Research University, Koroleva St. 2a, Belgorod, 308034, Russia
[2] Skobeltsyn Institute of Nuclear Physics, Lomonosov Moscow State University, Leninskie gory, GSP-1, Moscow, 119991, Russia
[3] National Research Nuclear University MEPhI (Moscow Engineering Physics Institute), Kashirskoe highway 31, Moscow, 115409, Russia
[4] Institute of Nanotechnology of Microelectronics of the Russian Academy of Science, Leninskiy Prospekt 32A, Moscow, 119991, Russia
[5] Lebedev Physical Institute, Leninskiy Prospekt 53, Moscow, 119333, Russia
[6] Kharkov Institute of Physics and Technology, Akademicheskaya St. 1, Kharkov, 61108, Ukraine

E-mail: kubankin@bsu.edu.ru*

Abstract. The compact 2.45 MeV fast neutron generator with a reduced supply voltage for calibration of low-background neutrino and dark matter detectors was tested. The generator is based on an array of carbon nanotubes. Neutron generation is carried out by applying a high voltage in the range of +10 to + 25 kV to a nanotube array, which cause an ionization of deuterium molecules with the following acceleration of ions in the direction of the grounded target covered by a deuterated polyethylene film. The d(d,n)^{3}He nuclear reaction happens as the result of ions collisions with the target. The dependences of the neutron yield as functions of the applied voltage were obtained for two different types of carbon nanotubes array. It is shown that the type of nanotubes array does not influence significantly on the neutron yield.

1. Introduction

The generation of neutrons via the d(d,n)^{3}He reaction is widely used in various of neutron sources [1, 2]. This reaction is very convenient because of the low energy threshold, which allows to use low energy ion accelerators and obtain neutron flux up to 10^{10} neutrons per second. Also, the useful advantage in some applications is the monochromaticity of the neutron flux. Such generators are widely used for such applications as a neutron diffraction, a neutron activation analysis, a neutron defectoscopy, etc.

The problem to have low-intensive neutron fluxes (about several hundreds of particles per second to the entire solid angle) for a calibration of low-background neutrino and Dark Matter detectors becomes actual recently [3, 4]. An elastic scattering of monochromatic neutrons on nuclei of the detector's target (usually liquid noble gases, such as LAr and LXe) causes the motion of nuclei and imitates such effects as coherent neutrino scattering or WIMP particles elastic scattering on baryonic matter. This effect permits to inspect all the detector systems and be sure that useful rare events will be fixed. Almost monochromatic and isotropic neutron yield could be reached via $d(d,n)^3He$ reaction within the neutron generator when the deuterium ions have energy range of 10-30 keV [4]. However, compactness of neutron generator dedicated for a detector calibration is strictly desirable in order to embed it into a low-background detector as close as possible to the sensitive ultra-low-background target. Such requirement makes it difficult to use standard neutron generators in the form of neutron tubes.

It was proposed to use a pyroelectric neutron generator [5, 6] for calibration of low-background detectors, but an instability of the electric potential generated by this source, low deuterium ion current, and a degradation of a deuterium ionizer - a single tungsten tip did not allow to obtain stable neutron flux level and limits life time of the device.

Another possible way to develop a compact neutron generator is to use many ionizers of deuterium molecules to increase an ion current. However, within the pyroelectric neutron generator these improvements will not give serious improvements due to a limited total charge, which is generated due to the a pyroelectric effect. The smaller amount of a charge per an individual ionizer causes consequently a lower efficiency of the ionization of deuterium [7]. Modern advances in semiconductor electronics make it possible to construct extremely compact 10-30 kV controllable high-voltage sources with output current of up to hundreds of microamperes with a primary power from a DC low voltage source or even from compact batteries. Such sources can be used to feed a high potential to initiate the $d(d,n)^3He$ reaction. To increase the efficiency of keeping the compactness we propose to use an array of carbon nanotubes (CNT) as an ionizer of deuterium.

2. The experimental results

Carbon nanotubes have been synthesized by plasma-enhanced chemical vapor deposition (PECVD) process on a n-type high-conductive silicon substrate in Institute of Nanotechnology of Microelectronics of the Russian Academy of Science. Aluminum layer (thickness – 0.5 μm) and catalyst (25 nm Ti and 5 nm Ni) were deposited on a silicone substrate. The CNT array is located in central part on a surface of substrate and forms a circle with a diameter of about 1.5 mm, the substrate itself have a size of 4*4*0.4 mm. The height of CNT array varied from 2 to 4 μm. Two types of CNT array were prepared: substrates with accidentally oriented nanotube bundles and substrates with vertically oriented nanotube bundles, (see Figure 1).

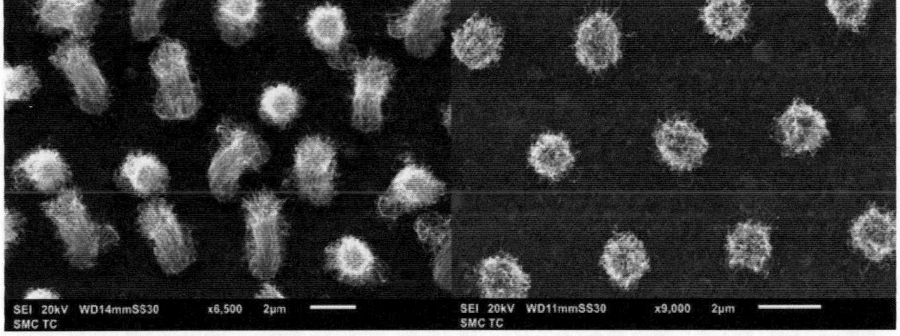

Figure 1. Image of fragments of the array of carbon nanotubes on Si substrate, obtained by scanning electron microscope (Performed in the NME RAS). Right - CNT with vertical orientation, left - with accidental orientation.

The scheme for testing the neutron generator on nanotubes is shown in Figure 2. The substrate with array of CNT (1) was fixed by electrically conductive epoxy glue in the duralumin holder (1), which in turn was fixed to a vacuum high-voltage input. The electric potential was applied to the CNT array via high-voltage positive-polarity source. The stainless-steel target (2), which is covered by deuterated film was mounted opposite the CNT array (dimensions of target: 50 * 30 * 0.8 mm) and grounded.

When high electric potential is applied to the CNT array (1), deuterium molecules are ionized by an electric field near tips of nanotubes and positive ions are accelerated to the target (2). The d(d,n)^{3}He reaction is becomes possible, when accelerated ions react with a deuterium atom in the deuterated film on a target surface. The neutron yield is almost isotropic relatively a velocity of incident ions, so a position of neutron detectors could be selected according a convenience of the general setup construction. The neutron yield was measured by two instruments in parallel. One detector (4) is the fast neutron spectrometer with an organic scintillator (para-terphenyl) and photomultiplier with implemented digital pulse shape discrimination algorithm for separation gamma/neutron events. The second detector of neutron was ^{3}He-counter (3). Both detectors were located outside the vacuum chamber.

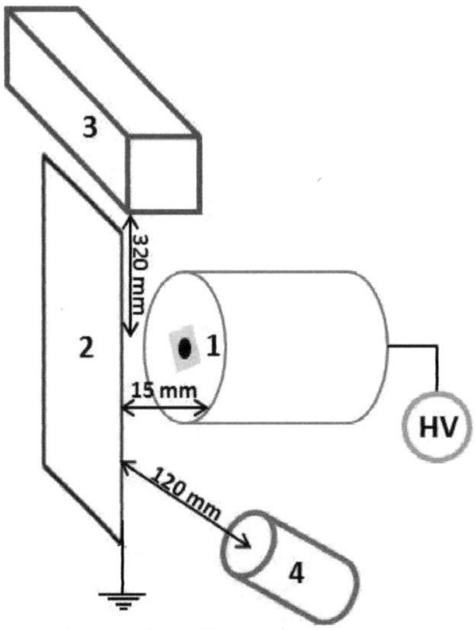

Figure 2. The scheme for testing the neutron generator on nanotubes. 1- duralumin holder with fixed silicon substrate with CNT array, 2 - target with deuterated film, 3 - ^{3}He counter, 4 - neutron scintillation spectrometer.

The neutron yield was measured as a function of a pressure of residual gas and the applied voltage for two different types of CNT arrays. As the residual gas pressure and the potential, which applied to the CNT array rise, the neutron yield increases and reaches about 40-50 neutrons per second at the pressure of 5 mTorr and the potential of +25 kV both for the array of vertically oriented CNT, and for the array with accidentally oriented CNT. There is no visible difference between two types of CNT array from the point of view of the possibility of neutron generation. Comparable amount of neutrons was observed at the same pressure and applied potential parameters with both types of CNT arrays. Figure 3 shows results of the experiment.

Thus, it was confirmed experimentally the possibility of neutron generation via d(d,n)^{3}He reaction when deuterium ions are created in an electric field near the tips of the carbon nanotube array by applying positive potential of 10-25 kV with the following acceleration by same potential to the grounded deuterated target.

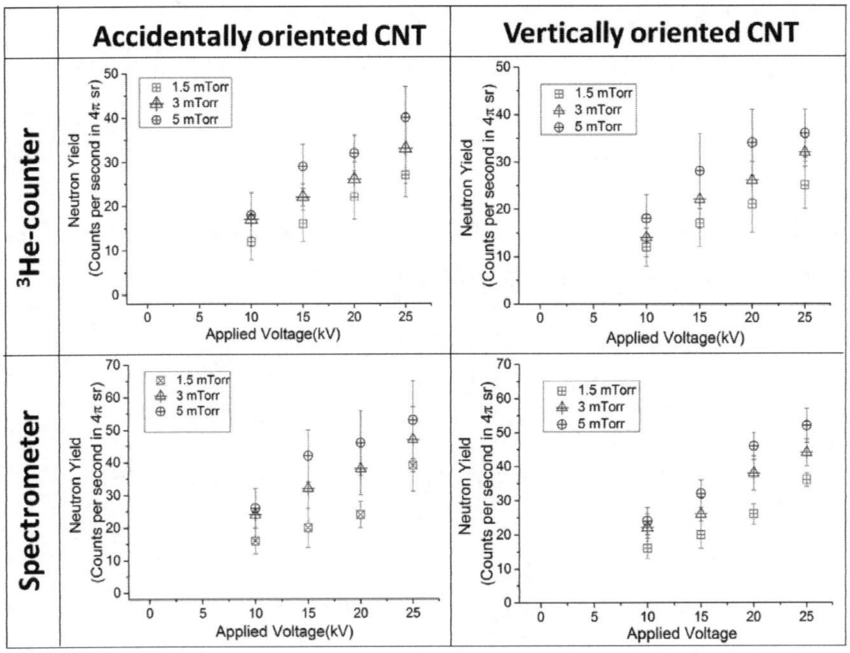

Figure 3. Dependence of the neutron yield on the residual gas pressure and the applied potential for two different types of CNT arrays, which measured using ^{3}He counter and scintillation spectrometer.

3. Conclusion

The experimental results extend an area of carbon nanotubes applications, and shows perspectives to use the phenomenon of atoms and molecules ionization by means of nanostructures. The possibility to construct a compact neutron generator based on the described construction is shown. Such a neutron source can be used, in particular, for a calibration of low-background neutrino and Dark Matter detectors.

Acknowledgements

The work was financially supported by a program of the ministry of education and science of the Russian Federation for higher education establishments, project №14.578.21.0192 (RFMEFI57816X0192).

References

[1] Reijonen J, Lawrence E 2005 *Proc. of Part. Acc. Conf.* (Knoxville: Tennessee 49)
[2] Leung K N *et al.*1991 *Rev. Sci. Instrum.* **62 (1)** 100
[3] The DarkSide project 2016 *JINST* **11** C02051
[4] Hagmann C, Bernstein A 2004 *IEEE Trans. Nucl. Sci.* **51**. 2151
[5] Chepurnov A S *et al.* 2016 *J. Phys.: Conf. Ser.* **675** 032031
[6] Chepurnov A S *et al.* 2017 *J. Phys.: Conf. Ser.* **798** 012119
[7] Geuther J A 2007 Ph.D. thesis

IOP Conf. Series: Journal of Physics: Conf. Series **934** (2017) 012014 doi:10.1088/1742-6596/934/1/012014

From the PAMELA mission to the GAMMA-400 project – the indirect search for signatures of dark matter

A M Galper[1, 2] and N P Topchiev[2,*]

[1] National Research Nuclear University MEPhI (Moscow Engineering Physics Institute), Kashirskoe highway 31, 115409 Moscow, Russia

[2] Lebedev Physical Institute, Russian Academy of Sciences, Leninskiy pr. 53, 119991 Moscow, Russia

E-mail: *tnp51@yandex.ru

Abstract. In 2008, the PAMELA magnetic spectrometer has discovered unpredicted abundance of the ratio of the galactic positron flux to the total positron and electron flux at high energies. It does not agree with the cosmic-ray fluxes calculated using the GALPROP code. This abundance was called the “PAMELA anomalous effect” and one of the explanations of this effect was the appearance of the additional electron and positron flux due to annihilation or decay of the dark matter particles. Later the precision PAMELA results were confirmed by the Fermi-LAT gamma-ray telescope and the AMS-02 magnetic spectrometer. Currently, the new GAMMA-400 project is being developed. The one of its main goals is to search for signatures of dark matter particles, which produce gamma rays. The GAMMA-400 gamma-ray telescope will have unprecedented angular and energy resolutions. PAMELA and GAMMA-400 are the instruments with the best characteristics for their time, which will improve our understanding of the nature of dark matter. At present, the problem of the nature of dark matter still remains the main challenge in high-energy astrophysics.

1. Searching for dark matter particles

The nature of dark matter is still unknown. Dark matter (DM) is ~23% of the Universe mass composition (figure 1). The presence of DM in the Universe is observed solely from its gravitational influence on the behavior of astrophysical objects. Among the huge number of possible candidates for the role of DM particles, supersymmetric particles, sterile neutrinos, axions, etc. are considered. The most commonly considered and studied are weakly interacting massive particles (WIMPs), whose mass can lie in the range from ~ GeV to ~ TeV.

Attempts to study the properties of DM particles are carried out by direct and indirect detection methods. The indirect detection method consists in recording not DM particles themselves, but the products of their annihilation or decay: charged particles of cosmic rays (the experiments PAMELA, AMS-02, ATIC, IACTs, Fermi-LAT, Auger, CTA, GAPS), cosmic gamma rays (the experiments Fermi-LAT, GAMMA-400, HESS-II, MAGIC, VERITAS, HAWC, CTA), neutrinos (the experiments IceCube/DeepCore/PINGU, ANTARES/KM3NET, BAIKAL-GVD, Super-Kamiokande/Hyper-Kamiokande) [1].

Possible channels for WIMP annihilation $W^+W^-, Z^0Z^0, b\bar{b}, t\bar{t}, \tau^+\tau^-$ with producing secondary electrons and positrons, protons and antiprotons, gamma rays and direct annihilation channels $\chi + \chi \rightarrow \gamma + \gamma$, which result in producing monoenergetic gamma rays are shown in figure 2 [2].

The PAMELA instrument had the best for its time characteristics allowing one to obtain the precision results, which can for the first time associate with dark matter. The upcoming GAMMA-400 space experiment also has the best characteristics at present, which can allow us to improve our knowledge on the DM nature.

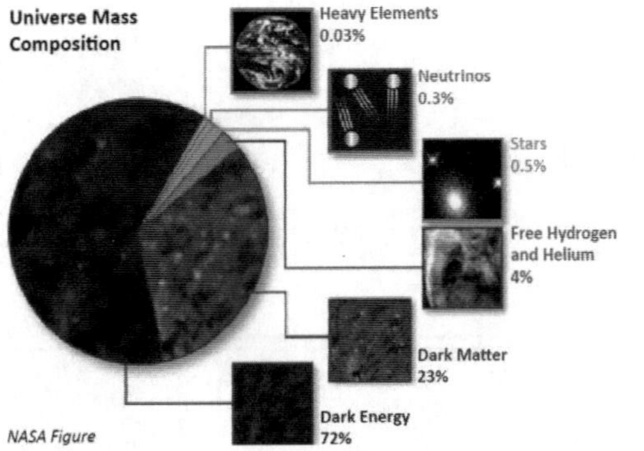

Figure 1. Universe mass composition.

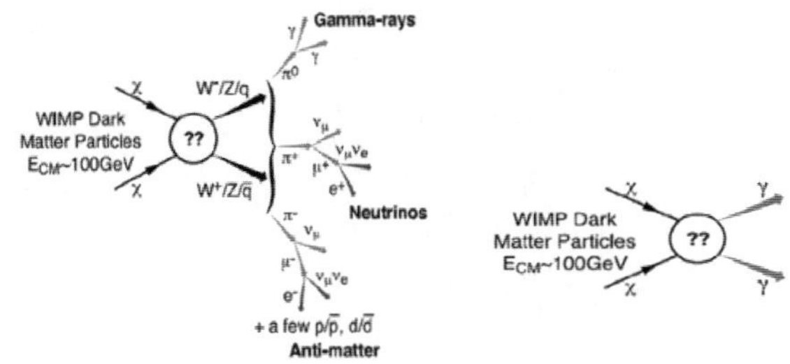

Figure 2. Possible channels of WIMP annihilation to produce electrons and positrons, protons and antiprotons, gamma rays, etc. [2].

2. The PAMELA magnetic spectrometer

The PAMELA (Payload for Antimatter-Matter Exploration and Light-nuclei Astrophysics) satellite experiment was designed to study the charged component of the cosmic radiation (especially antiparticles) and to search for structures in cosmic-ray spectra from new astrophysical sources or dark matter. PAMELA was launched onboard the Russian Resurs-DK1 satellite on June 15, 2006 from the Baikonur cosmodrome (Kazakhstan). The satellite was placed in a quasi-polar 70° inclination orbit at an altitude varying between 350 km and 600 km. The mission, which was planned to last 3 years, lasted almost 10 years.

Figure 3 shows the PAMELA instrument scheme [3]. PAMELA is a powerful particle identifier using a permanent magnetic spectrometer with a variety of specialized detectors for cosmic-ray direct measurements in space with unprecedented precision. A fully comprehensive description can be found in [4].

In 2008, PAMELA has discovered unpredicted abundance of the ratio of the galactic positron flux to the total positrons and electrons flux at high energies (figure 4, [5]). It does not agree with the cosmic-ray fluxes calculated using the GALPROP code [6]. This abundance was called the "PAMELA anomalous effect" and one of the explanations of this effect was the appearance of the additional electrons and positrons flux due to annihilation or decay of the DM particles.

Later the precision PAMELA results were confirmed by the Fermi-LAT gamma-ray telescope [7] and the AMS-02 [8] magnetic spectrometer (figure 5, [8]).

The antiproton intensity multiplied by $E3$ is shown in figure 6 [9] in comparison with the theoretical calculations [10] involving an additional source in the form of the annihilation of DM particles at high energies. Solid lines in figure 6 demonstrate uncertainty in the predicted flux of only the secondary component.

The PAMELA experimental results make it possible to study the nature of hypothetical massive DM particles, in particular, to set certain limits to the parameters of corresponding models. According to the PAMELA data the antiproton to proton ratio in the energy range > 10 GeV has some excess in comparison with theoretical calculations. This can be explained by the contribution of DM particles.

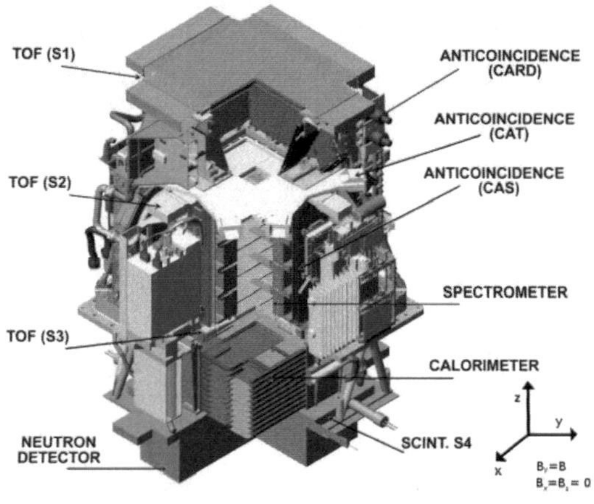

Figure 3. The PAMELA instrument scheme [3].

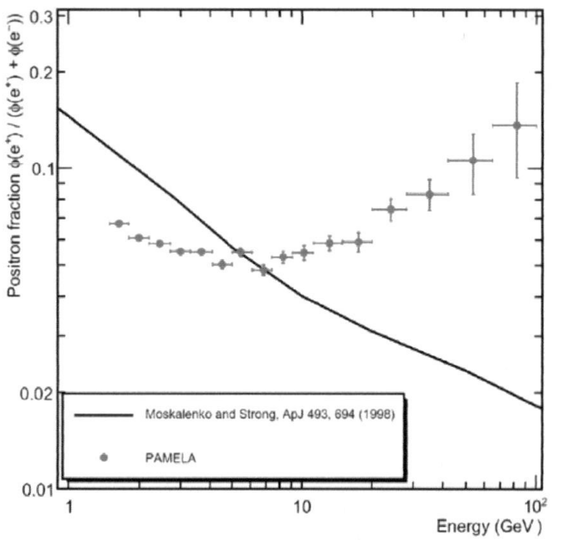

Figure 4. PAMELA positron fraction compared with a theoretical model [5]. The solid line shows a theoretical calculation [6] for pure secondary positron production during the propagation of cosmic rays in the Galaxy.

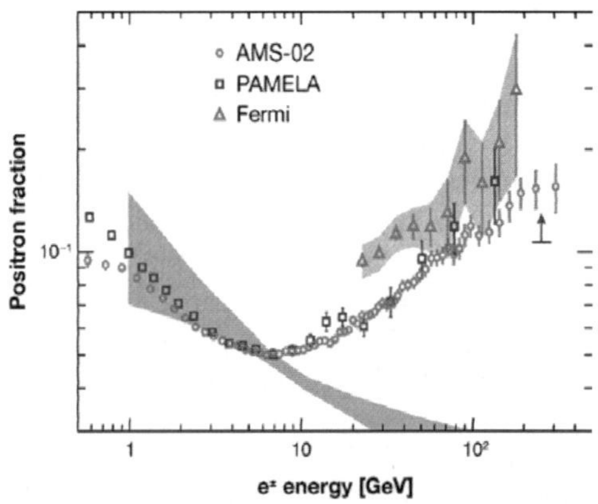

Figure 5. The positron fraction compared with the measurements from PAMELA, Fermi-LAT [7], and AMS-02 [8].

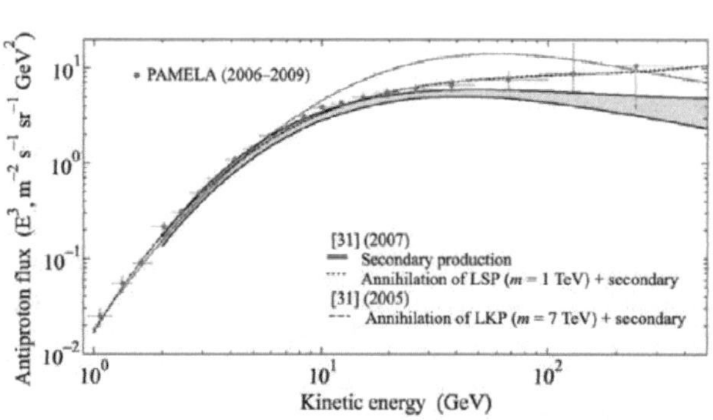

Figure 6. Antiproton intensity (multiplied by *E*3) vs the energy [9]. The yellow band shows uncertainty in the secondary antiproton flux for different diffusion models in the interstellar medium. Dashed and dotted lines correspond to the signal caused by the annihilation of different hypothetical DM particles [10].

3. The GAMMA-400 gamma-ray telescope

The GAMMA-400 gamma-ray telescope [11-15] is a bridge between the space-based and ground-based gamma-ray experiments, which will study gamma rays from low energy of ~20 MeV and up to high energy of several tens of TeV. The GAMMA-400 main scientific goals are: dark matter searching by means of gamma-ray astronomy; precise measurements of Galactic Center, Fermi Bubbles, Crab, Vela, Cygnus, Geminga, Sun, and other regions, extended and point gamma-ray sources, diffuse gamma rays with unprecedented angular (~0.01° at $E_\gamma = 100$ GeV) and energy (~1% at $E_\gamma = 100$ GeV) resolutions. GAMMA-400 will be installed onboard the Russian space observatory.

3.1 The GAMMA-400 physical scheme

The physical scheme of the GAMMA-400 gamma-ray telescope is shown in figure 7. The GAMMA-400 energy range for gamma-ray studies is from ~20 MeV to ~10000 GeV. The GAMMA-400 field of view (FoV) is ±45°.

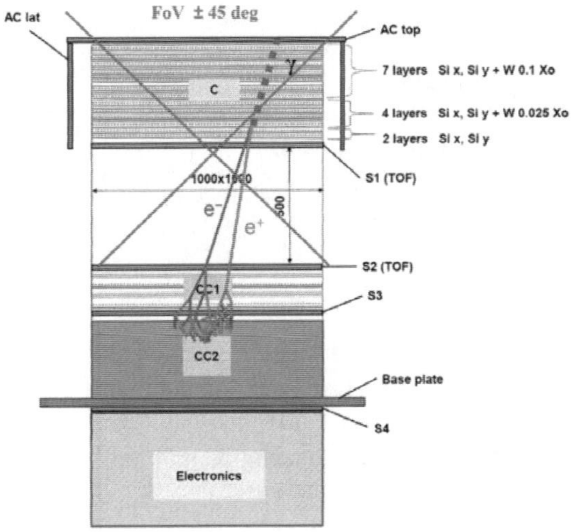

Figure 7. The GAMMA-400 physical scheme.

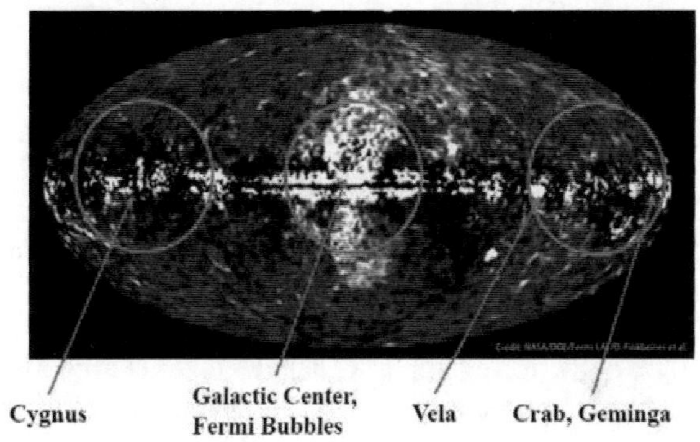

Figure 8. Galactic center, Fermi Bubbles, Crab, Cygnus, Vela, Geminga, and other regions will be observed with the GAMMA-400 FoV of ±45°.

GAMMA-400 consists of plastic scintillation anticoincidence top and lateral detectors (AC top and AC lat), converter-tracker (C), plastic scintillation detectors (S1 and S2) for the time-of-flight system (ToF), calorimeter (CC), and plastic scintillation detectors (S3 and S4).

The anticoincidence detectors surrounding the converter-tracker are used to distinguish gamma rays from significantly larger number of charged particles (e.g., in the region of 10-100 GeV, the flux ratios for gamma rays to electrons and protons are ~ $1{:}10^2{:}10^5$.

All scintillation detectors consist from two independent 1-cm layers. The time-of-flight system, where detectors S1 and S2 are separated by approximately 500 mm, determines the top-down direction of arriving particles. The additional scintillation detectors S3 and S4 improve hadron and electromagnetic shower separation.

The converter-tracker consists of 26 layers of single x and y silicon strip coordinate detectors with a pitch of 0.08 mm. Fourteen layers are interleaved with 0.1 X_0 tungsten conversion foils, next 8 layers with 0.025 X_0 tungsten conversion foils, and final four layers have no tungsten. This configuration allows us to measure gamma rays down to ~20 MeV and up to ~10000 with one trigger. The total converter-tracker thickness is ~1 X_0. The converter-tracker information is used to precisely determine the conversion point and the direction of each incident particle. Using the last four layers without tungsten will make it possible to measure the polarization of gamma rays in the energy range from ~20 to ~60 MeV.

The two-part calorimeter CC measures particle energy and consists of CsI(Tl) crystals. The total calorimeter thickness is ~23 X_0 or ~1.0 λ_0 when detecting vertical incident particles and ~43 X_0 or ~2.0 λ_0 when detecting laterally incident particles. Using the deep calorimeter allows us to extend the energy range up to several tens TeV for gamma rays, and to reach the energy resolution of ~1% at $E_\gamma =$ 100 GeV.

3.2 The GAMMA-400 gamma-ray observatory

The GAMMA-400 gamma-ray observatory will be installed onboard of the Navigator space platform, which is designed and manufactured by the Lavochkin Association.

Using the Navigator space platform gives the GAMMA-400 experiment a highly unique opportunity to install a scientific payload (mass of 4500 kg, power consumption of 2000 W, and telemetry downlink of 100 GB/day, with lifetime more than 7 years), which will also include X-ray telescope with the energy range of 5-30 keV together with the gamma-ray telescope. GAMMA-400 will provide with the means to significantly contribute as the next generation instrument for gamma-ray astronomy, X-ray, and cosmic-ray physics.

The GAMMA-400 experiment will be initially launched into a highly elliptical orbit (with an apogee of 300 000 km and a perigee of 500 km, with an inclination of 51.4°), with 7 days orbital period. Under the action of gravitational disturbances of the Sun, Moon, and the Earth after ~6 months the orbit will transform to about an approximately circular one with a radius of ~200 000 km and will not suffer from the Earth's occultation and shielding by the radiation belts. The launch of the GAMMA-400 space observatory is planned for the middle of the 2020s.

4. The anticipated GAMMA-400 scientific output

4.1. Galactic plane

GAMMA-400 will study continuously over a long-term period different regions of Galactic plane (figure 8), for example, Galactic center, Fermi Bubbles, Crab, Vela, Cygnus, Geminga with FoV of ±45°. In particular, using the gamma-ray fluxes obtained by Fermi-LAT, we can expect that GAMMA-400 when observing the Galactic center with aperture of ±45°during 1 year will detect: 57400 photons for $E_\gamma > 10$ GeV; 5240 photons for $E_\gamma > 50$ GeV; 1280 photons for $E_\gamma > 100$ GeV; 535 photons for $E_\gamma > 200$ GeV.

4.2 Dark matter searching

When detecting gamma rays from DM, the intensity from gamma rays is calculated by

$$\frac{d\Phi}{d\Omega dE} = \frac{1}{2}\frac{\langle\sigma v\rangle}{m_\chi^2}\frac{dN_\gamma}{dE} \times \frac{J}{4\pi}\,[\mathrm{erg\ s^{-1}\,cm^{-2}\,GeV^{-1}\,sr^{-1}}],\ J = \int_{l.o.s.} \rho^2(\vec{r})dl.$$

Here, $\langle\sigma v\rangle$ is particle annihilation cross section, m_χ is particle mass, dN_γ/dE is the spectrum of gamma rays from annihilation, $\rho(\vec{r})$ is the DM density of investigated object [16].

The Galactic center is, apparently, the best potential source of DM emission possessing the largest *J*-factor [17]. Moreover, recently, the anomaly excess of gamma-ray emission in the GeV energy range was revealed near the Galactic center (the region of about one degree) [18], which can be well described by DM with mass of several tens of GeV and annihilation cross section of about standard thermal 10^{-26} cm^3/s. However, this observed excess can have another interpretation - the presence a population of millisecond pulsars [19]. Therefore, the new GAMMA-400 observational data can help to solve this problem. Dwarf galaxies are considered for a long time as the strongest sources of constraints for DM, because they have sufficiently large *J*-factors and at the same time have considerably less gamma-ray background in comparison with the Galactic center. GAMMA-400 will able to specify the constrain area. Potentially interesting objects are other galaxies and their clusters, where DM presents and can emit gamma rays. GAMMA-400 with the highest energy resolution of ~1% will have unique sensitivity for detecting DM.

5. Conclusions

PAMELA mission has made a significant contribution to search for dark matter particles. GAMMA-400 mission represents a unique opportunity to continue the search for dark matter particles in the gamma-ray low- and high-energy range and X-ray range with unprecedented angular and energy resolutions. The best GAMMA-400 characteristics will lead to new best results. According the new approved Russian Federal Space Program 2016-2025 the GAMMA-400 space observatory is scheduled to launch in 2025-2026.

References

[1] Gaskins J M 2016 *arXiv*:1604.00014
[2] Baltz E A *et al.* 2008 JCAP **07** 013
[3] Adriani O *et al.* 2017 *Riv. del Nuo. Cim.* **40** 473
[4] Picozza P. *et al.* 2007 *Astropart. Phys.* **27** 296
[5] Boezio M *et al.* 2009 *N. J. of Phys.* **11** 105023
[6] Moskalenko I and Strong A 1998 *ApJ* **493** 694
[7] Ackermann M *et al.* 2012 *Phys. Rev. Lett.* **108** 011103
[8] Battiston R *et al.* 2014 *Phys. Dark Universe* **4** 6
[9] Adriani O *et al.* 2012 *JETP Let.* **96** 621
[10] Bringmann T and Salati P 2007 *arXiv*:0612514
[11] Dogiel V *et al.* 1988 *Sp. Sci. Rev.* **49** 215
[12] Galper A *et al.* 2013 *Adv. Space Res.* **51** 297
[13] Galper A *et al.* 2013 *AIP Conf. Proc.* **1516** 288
[14] Topchiev N *et al.* 2015 *Bull. Russ. Acad. Sci. Phys.* **79** 417
[15] Topchiev N *et al.* 2016 *J. Phys.: Conf. Series* **675** 032009
[16] Cirelli M *et al.* 2011 *JCAP* **03** 051
[17] Bertone G 2010 *Particle Dark Matter: Observations, Models and Searches* (Cambridge : Cambridge Univ. Press)
[18] Abazajian K and Kaplinghat M 2012 *Phys. Rev.* D **86** 083511
[19] Bartels R, Krishnamurthy S and Weniger C 2015 *arXiv*:1506.05104

The DAMPE experiment: 2 year in orbit

Fabio Gargano on behalf of DAMPE collaboration
INFN Bari - Via Orabona 4 Bari, Italy
E-mail: fabio.gargano@ba.infn.it

Abstract. The DArk Matter Particle Explorer (DAMPE) is a space mission within the strategic framework of the Chinese Academy of Sciences, resulting from a collaboration of Chinese, Italian, and Swiss institutions, and is a new addition to the growing number of particle detectors in space. It was successfully launched in December 2015 and has commenced nominal science operations since shortly after launch. Lending technologies from its predecessors such as AMS and Fermi-LAT, it features a powerful segmented electromagnetic calorimeter which thanks to its 31 radiation lengths enables the study of charged cosmic rays in the energy domain of up to 100 TeV and gamma rays of up to 10 TeV. The calorimeter is complemented with a silicon-tungsten tracker converter which yields a comparable angular resolution as current space-borne pair-conversion gamma-ray detectors. In addition, the detector features a top anti-coincidence shield made of segmented silicon plastic scintillators and a boron-doped plastic scintillator on the bottom of the instrument to detect delayed neutrons arising from cosmic ray protons showering in the calorimeter. An overview of the mission and a summary of the latest results in the domain of charged cosmic rays, gamma rays and heavy ions will be presented.

1. Introduction

The DArk Matter Particle Explorer (DAMPE) was successfully launched on December 17, 2015 and presently orbits sun-synchronously at an altitude of 500 km [1, 2]. The main scientific goals of DAMPE are the measurement of the fluxes of charged cosmic rays (electrons, protons and heavier nuclei), the measurement of high-energy gamma rays from various astrophysical sources and the search for possible signatures of dark-matter.

2. On-board instruments

The DAMPE detector is equipped with four different sub-detectors (Figure 1): a plastic scintillator detector array (PSD), a silicon-tungsten tracker (STK), a BGO calorimeter (BGO) and a neutron detector (NUD). The PSD is designed to work as a veto and to measure the charge (Z) of incident high-energy particles up to Z = 26. It has a double layer configuration and the directions of the scintillator bars in the two layers are perpendicular to provide information on the X and Y coordinates. Right below the PSD, the STK is used to the precise reconstruction of the particle tracks. It consists of 12 position-sensitive silicon detector planes (6 planes for the x-coordinate, 6 planes for the y-coordinate). Three layers of tungsten are inserted in between the silicon planes (2, 3, 4 and 5) to enhance the conversion probability of gamma rays in electron-positron pairs. The BGO calorimeter is used to measure the energy deposition of incident particles and to reconstruct the electromagnetic shower profile [3]. The calorimeter is composed of 308 BGO crystal bars optically isolated from each other and arranged horizontally in 14 layers, each one of 22 bars. The bars of a layer are orthogonal to those of the adjacent plane

IOP Conf. Series: Journal of Physics: Conf. Series **934** (2017) 012015 doi:10.1088/1742-6596/934/1/012015

in order to reconstruct the shower in both views (x-z and y-z). The total vertical depth of the calorimeter is 31 radiation lengths and 1.6 nuclear interaction lengths. The last sub-detector is the NUD, which consists of four boron-loaded plastic scintillators. All DAMPE sub-detectors are involved in photon selection with different roles. The BGO calorimeter is used to distinguish electromagnetic from hadronic showers in a wide energy range and its high separation power is further enhanced by the NUD, while the PSD and STK are mainly used for the separation between charged and neutral particles.

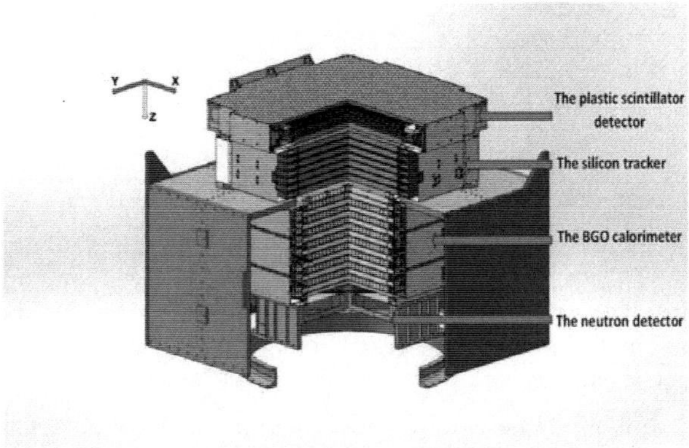

Figure 1. View of the DAMPE detector.

3. Trigger rate and data transfer

The trigger rate of DAMPE is closely related to the fluxes of the cosmic-ray particles on orbit. The effect of the high geomagnetic cutoff causes a significant reduction of the incident particles rate at low latitudes while in the pole regions where the energy cutoff value is lower the observed rate is higher.

Figure 2 (a) shows the overall trigger rate. The sudden change of the trigger rate at $\pm 20°$ latitude is due to different pre-scale factor applied to the triggers to reduce the triggered events near the poles. The acquisition rate registers values up to 200 Hz, but the rate of the "high-energy trigger" (the one used for the main physics analyses and not prescaled) can change from about 20 Hz in the equator region up to 70 Hz at the poles (Figure 2 (b)).

Data are sent to ground 4 times per day, when the satellite passes over the Chinese ground stations. An amount of 15 GB is transmitted to ground each day, structured in about 8 GB of raw data and the remaining of slow control and orbit information data. The off-line reconstructed data in ground computing facilities results in 85 GB/day, stored in ROOT files. A total amount of about 100 GB is stored each day with an average of 35 TB of data in each year of operations.

4. On-orbit absolute energy calibration

A method to determine the absolute energy scale of DAMPE has been developed within the collaboration [4] based on the measurement of the geomagnetic cutoff on cosmic-ray electrons and positrons. The rigidity cutoff on cosmic ray electrons and positrons was calculated using IGRF-12 [5] model and the cosmic-ray particle trajectory code of back-tracking in magnetic field. The interval of McIlwain L parameter ranges between 1.0 and 1.14 was selected to measure the electrons and positrons spectrum. A 1.25% higher energy scale has been found for DAMPE with respect to the expectation.

IOP Conf. Series: Journal of Physics: Conf. Series **934** (2017) 012015 doi:10.1088/1742-6596/934/1/012015

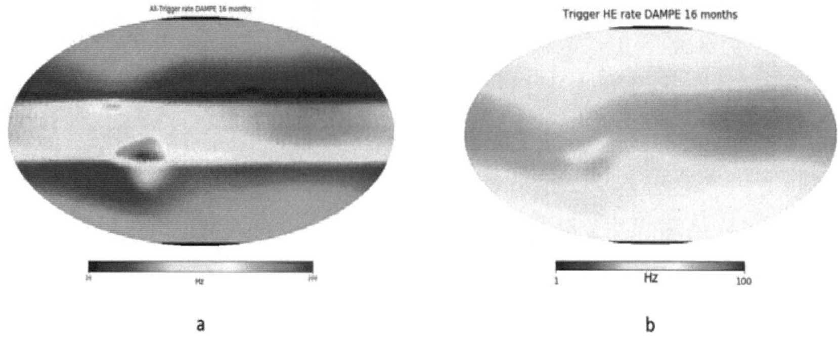

Figure 2. Trigger rate: (a) all the triggers (b) High Energy trigger

5. Electrons: identification and expected performance

The measurement of the electron and positron flux is one of the leading analysis within the DAMPE experiment. Measurements lead by other experiments are limited in the energy range and DAMPE has the capability to extend up to 10 TeV these measurements trying to answer to some of the most important questions about the behavior of this spectrum at very high energies. The electron acceptance has an almost constant value of 0.3 m^2sr for energies above 100 GeV. The energy resolution is is almost 1% for E $\geq$ 100 GeV [2].
The good energy resolution is due to the very thick BGO calorimeter. These characteristics allow to perform a good event selection for electrons, for which several PID strategies are being investigated within the collaboration, starting from classic cut-based analysis using *ad hoc* "shape parameters" to multivariate analysis involving machine learning algorithms and neural networks.

6. Protons and nuclei

The PSD and the STK play a major role in measuring the charge of incident high-energy particles. Good results in the measurement of ions nuclei charge have been achieved since the beam test campaign, where several tests were performed with Lead and Argon beams showing a dependency of the PSD charge resolution from Z ranging from 0.2 to 0.4. The charge measurement is done with the STK up to Oxigen while with the PSD it is possible to measure the charge from proton to iron.
Since DAMPE is a calorimetric-type, satellite-borne detector for high energy CR with a large geometric factor it is expected to improve significantly the direct measurement of the CR spectrum from several GeV up to 100 TeV, a region that is poorly investigated also by ground-based experiments due to their energy thresholds.
The first results on nuclei spectra in the first two years of on-orbit operations are on protons and helium spectra. Figure 3 (left) shows the preliminary proton flux as a function of the kinetic energy measured with DAMPE, compared with previous measurements by PAMELA, ATIC-2, AMS-02 and CREAM-III . The results show good agreement with the AMS-02 and PAMELA data for $E_k \lesssim 200$ GeV. The spectral hardening at $E_k \gtrsim 200$ GeV can be observed from DAMPE measurements, confirming the results from the other experiments.
Although very preliminary, also the helium flux measurement is in a good agreement with previous measurements done by other experiments and shows also an indication of a hardening of the spectrum. After only one year of data taking the number of helium candidates with a reconstructed BGO energy above 10 TeV is of about 150 events which correspond to already

few tens of candidates with an energy of about 100 TeV. The measured helium flux compared with the other experiment is shown in Figure 3 (right).

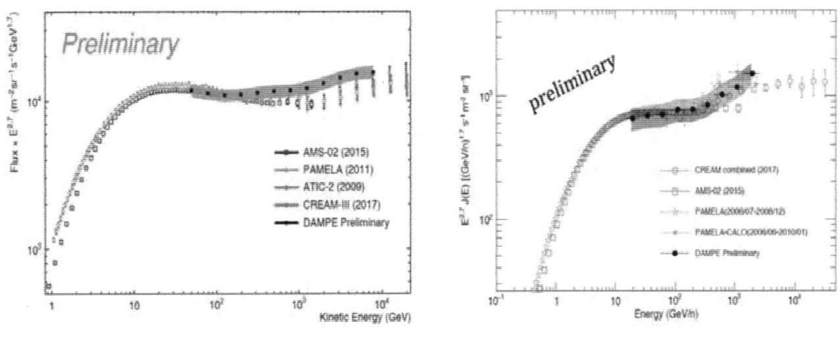

Figure 3. The preliminary $E_k^{2.7}$ weighted flux of CR protons (left) and helium (right) measured by DAMPE, compared with previous results by PAMELA [7], AMS-02 [8, 9], ATIC-2 [10], and CREAM-III [11]. The error bars represent statistical errors and the grey band shows the systematic errors.

7. Photon selection

The photon selection is one of the most challenging tasks of DAMPE, since the background events represented by the charged particles (electrons, protons and nuclei) have fluxes that are much higher compared to the main galactic gamma-ray emission. The minimum rejection power requested at 100 GeV is 10^5 for protons and 10^3 for electrons. Protons are mainly suppressed using the shower profile inside the BGO calorimeter, while electrons are mainly rejected using the PSD and the first layer of the STK. One of the main issues with charged particles is the back-scattering effect, that reflects on the acceptance for photons at high energies. Figure 4 shows the acceptances for photons and electrons after a selection based on the use of Convolutional Neural Networks and a Random Forest Classifier [6]. For energies above 500 GeV, the back-scattering affects the acceptance decreasing its value, while the decreas for energies below 5 GeV is due to the use of the high energy trigger in the events selection. Other particle identification (PID) algorithms are under study to further reduce the electrons contamination. An average number of about 150 photons is observed daily by DAMPE, and the count-maps after 16 months of data show a good resolution of the most bright gamma-ray sources, and a good agreement of their position with the ones observed by the Fermi-LAT.

8. Summary

The DAMPE detector is successfully operating after its launch on December 2015 with a large geometric factor of about 0.3 m^2sr for electrons, and its Si-W tracker provides for a 40 μm spatial resolution and a $0.15°$ angular resolution. Its thick BGO calorimeter is 31 X_0 deep and its energy resolution is about 1% for energies above 100 GeV for the electromagnetic component of the observed cosmic rays, while for hadrons it is almost 40%.
The on orbit operations still continue with high efficiencies, and an absolute energy calibration has been performed using the geomagnetic cut-off. In addition, a cross-check of DAMPE's pointing performance has been checked with bright sources in the observed photon map.

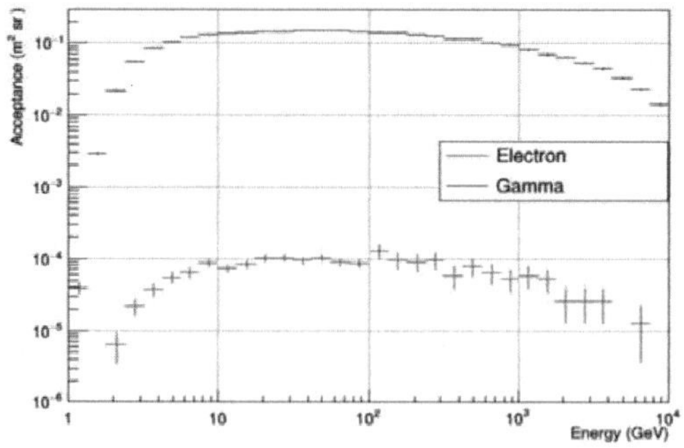

Figure 4. Photons acceptance and electrons contamination

9. Acknowledgments

The DAMPE mission was founded by the strategic priority science and technology projects in space science of the Chinese Academy of Sciences (No. XDA04040000 and No. XDA04040400). In China this work is supported in part by National Key Program for Re- search and Development (No. 2016YFA0400200), the National Basic Research Program (No. 2013CB837000), NSFC under grants No. 11525313 (i.e., Funds for Distinguished Young Scholars), No. 11622327 (i.e., Funds for Excellent Young Scholars), No. XDB23040000, No. 11273070, No. 11303096, No. 11303105, No. 11303106, No. 11303107, No. 11673075, U1531126, U1631111 and the 100 Talents program of Chinese Academy of Sciences. In Europe DAMPE activities are supported by the Italian National Institute for Nuclear Physics (INFN), the Italian University and Research Ministry (MIUR), the Swiss National Science Science Foundation and the University of Geneva.

References

[1] Chang J *et al.* 2014 *Chin. J. Space Sci.* **34**
[2] Chang J *et al.* 2017 *Astropart. Phys.* **95** 6
[3] Zhang Y L *et al.* 2012 *Chin. Phys.* C **36** 1
[4] Zang J *et al.* 2017 *Proc. of Sci.* PoS(ICRC2017)0637
[5] Thebault E *et al.* 2015 *Earth, Planets and Space* **67** 79
[6] Garrappa S *et al.* 2017 *Proc. of Sci.* PoS(ICRC2017)063
[7] Adriani O *et al.* 2011 *Science* **332** 69
[8] Aguilar M *et al.* 2015 *Phys. Rev. Lett.* **114** 171103
[9] Aguilar M *et al.* 2015 *PPhys. Rev. Lett.* **115** 211101
[10] Panov A D *et al.* 2009 *Bull. Russ. Acad. Sci. Phys.* **73** 564
[11] Yoon Y S *et al.* 2017 *Astrophys. J.* **839** 5

IOP Conf. Series: Journal of Physics: Conf. Series **934** (2017) 012016 doi:10.1088/1742-6596/934/1/012016

Measurements of the cosmic-ray electron and positron spectrum and anisotropies with the Fermi LAT

F. Loparco (on behalf of the Fermi LAT Collaboration)

Dipartimento di Fisica "M. Merlin" dell'Università e del Politecnico di Bari, I-70126 Bari, Italy
Istituto Nazionale di Fisica Nucleare, Sezione di Bari, I-70126 Bari, Italy

E-mail: francesco.loparco@ba.infn.it

Abstract. The Large Area Telescope (LAT) onboard the *Fermi* satellite is a pair-conversion telescope for high-energy gamma rays of astrophysical origin. Although it was designed to be a high-sensitivity gamma-ray telescope, the LAT has proved to be an excellent electron/positron detector. It has been operating in low Earth orbit since June 2008 and has collected more than 16 million cosmic-ray electron and positron (CRE) events in its first seven years of operation. The huge data sample collected by the LAT enables a precise measurement of the CRE energy spectrum up to the TeV region. A search for anisotropies in the arrival directions of CREs was also performed. The upper limits on the dipole anisotropy probe the presence of nearby young and middle-aged CRE sources.

1. Introduction

During their journey throughout the Galaxy, cosmic-ray electrons and positrons (CREs) rapidly lose energy by interacting through inverse Compton scattering with the interstellar radiation field and through synchrotron radiation with the Galactic magnetic field. The diffusion distances of CREs with energies above 100 GeV are of a few hundred pc, much shorter than the radius of the Galaxy, which is of the order of 10 kpc. A combined measurement of the spectrum and of the anisotropies of high-energy CREs can, therefore, provide evidence for local CRE sources of astrophysical (supernova remnants and pulsar wind nebulae) or exotic (dark matter) nature.

2. The Fermi LAT as an electron/positron detector

The Large Area Telescope (LAT) is the main instrument onboard the *Fermi* satellite and is a pair-conversion telescope designed to detect gamma rays with energies above 20 MeV. The LAT has a modular structure, consisting of 16 identical towers, each composed of a precision tracker-converter (TKR) and a calorimeter (CAL). The TKR consists of 18 x-y tracking planes of silicon strip detectors interleaved with tungsten converter foils, for a total on-axis thickness of 1.5 radiation lengths. The CAL consists of 96 CsI (Tl) crystals, hodoscopically arranged in 8 layers, for a total on-axis thickness if 8.6. radiation lengths. The towers are surrounded by a segmented anticoincidence detector (ACD) consisting of 89 plastic scintillators tiles, which is used for rejecting the charged cosmic-ray background. A full description of the LAT is given in ref. [1].

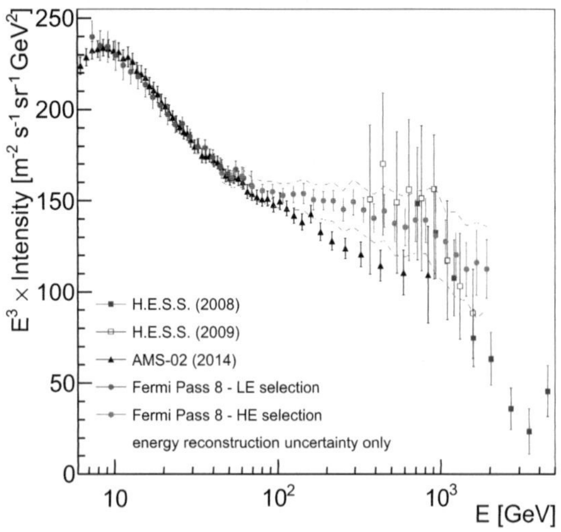

Figure 1. CRE spectrum measured by the *Fermi* LAT in its first seven years of operation. The results from the present analysis are compared with those from the AMS-02 [3] and H.E.S.S. [4, 5] experiments. The plot is taken from ref. [2].

Although it is designed to detect gamma rays, the LAT is able to collect and identify CREs with a large acceptance. In fact, like a gamma-ray, a CRE entering in the LAT will produce an electromagnetic shower which will start in the TKR and will develop in the CAL. The main difference between CRE and gamma-ray events resides in the response of the ACD: while in a gamma-ray event no signal is expected from the ACD tile intersected by the shower axis (since gamma-ray conversion must take place in the TKR), in a CRE event a signal from a singly-charged particle is expected.

The LAT trigger system takes inputs from the ACD, TKR and CAL detectors. After an event passes the hardware trigger, it is inspected by the onboard software filters, that are configured to identify events likely to be useful for scientific or calibration purposes. Four onboard filters are implemented:

- Gamma filter, designed to select gamma-ray candidates and events that deposit at least 20 GeV in the CAL;
- Heavy-ion filter, designed to perform calibrations on high energy scales;
- MIP filter, designed to select minimum ionizing charged particles;
- Diagnostic filter, designed to select an unbiased event sample for filter performance and background studies. Events selected by this filter are prescaled of a factor 250.

If an event is accepted by any filter, all its information is recorded and transmitted to the ground.

Two independent CRE analyses have been implemented: the high-energy (HE) analysis, above 42 GeV, which uses a sample of events passing the onboard gamma filter, and the low-energy (LE) analysis between 7 and 70 GeV, which uses a sample of events passing the diagnostic filter. In both cases, we first apply a set of preliminary cuts before performing a multivariate analysis in order to reduce the residual proton contamination as much as possible. For both the HE and the LE analyses, these cuts are combined with further selections based on multivariate classification analyses. A full description of the analysis is given in ref. [2].

3. The CRE energy spectrum

For the present analysis, we have used a sample of Pass 8 data collected by the LAT between August 4, 2008, and June 24, 2015. The overall live time for this data set is 4.68 years. In Fig. 1 the measured CRE spectrum is shown and is compared with the previous measurements performed by the AMS-02 [3] and by the H.E.S.S. [4, 5] experiments. The results from the HE and from the LE analyses match very well over the overlapping range from 42 GeV to 70 GeV.

The LAT results are in good agreement with the AMS-02 data up to about 200 GeV, while in the region above 400 GeV they are consistent with the H.E.S.S. data. The spectrum is well fitted by a broken power law, with a break energy of (53 ± 8) GeV and spectral indices 3.21 ± 0.02 and 3.07 ± 0.02 below and above the break respectively.

The H.E.S.S. data above 600 GeV reported in [4] are well reproduced by an exponentially cutoff power law with an index of 3.05 ± 0.02 and a cutoff at 2.1 ± 0.3 TeV. The LAT CRE spectrum above 50 GeV is compatible with a single power law with a spectral index of 3.07. A fit of the spectrum above 50 GeV with an exponentially cutoff power law does not yield statistically significant evidence for a cutoff below 1.8 TeV at 95% confidence level.

4. Search for anisotropies in the CRE arrival directions

The search for possible anisotropies in the CRE arrival directions has been performed using the HE sample in the range from 42 GeV to 2 TeV. The energy interval has been divided in 9 bins. To reduce geomagnetic effects (which could mimic a true anisotropy signal), we have performed the analysis with a reduced field of view. In particular, for each energy bin, we have set an allowed maximum off-axis angle θ_{max} (40° for 42 GeV $< E <$ 56 GeV, 50° for 56 GeV $< E <$ 75 GeV and 60° for $E >$ 75 GeV). In this way particles with large zenith angles, which might be affected by the geomagnetic field, have been discarded.

The strategy to search for possible anisotropies is to compare the observed sky maps in each energy bin with the corresponding reference maps that should be seen by the instrument if the CRE flux was isotropic, and represent the null hypothesis.

The starting point of the analysis is therefore the construction of the reference (isotropic) sky maps. Since the expected signal is tiny, in order to mitigate potential systematic uncertainties arising from the calculation of the detector exposure, we have built the reference maps using four different data-driven methods. The first method ("shuffling technique") consists of generating a set of simulated events starting from the real event sample and randomly associating detected event times and instrument angles. The sky direction of a simulated event is built combining the position and orientation of the LAT at the time of the i-th event with the angles in the LAT frame taken from the the j-th event, where (i, j) are two random numbers in the interval $[1, N]$ (N is the total number of observed events). In the second method ("event rate technique"), each simulated event is assigned a time generated from an exponential distribution with the same average rate as in the real data, and a direction in the LAT frame generated from the observed distribution $P(\theta, \phi)$. The other two methods combine the previous techniques, i.e. one can extract the event times sequence from an exponential distribution with given average rate and take the angles (θ, ϕ) from real events, or one can take the observed times from real events and draw the angles from the distribution $P(\theta, \phi)$. All these methods have been validated with dedicated Monte Carlo simulations [6].

Several methods have been implemented to compare the observed sky maps with the reference ones. The first option is to implement a direct pixel-to-pixel comparison. However, a more powerful approach is that of applying this method to integrated sky maps, in which each pixel contains the integrated number of events in a given circular region of fixed angular radius around the pixel itself. This approach increases the sensitivity to possible anisotropies with the same angular scale as the integration radius. Finally, another strategy is the spherical harmonic analysis of a fluctuation sky map. The fluctuation in each pixel is defined as $f_i = (n_i - \mu_i)/\mu_i$, where n_i and μ_i are the events of the real and of the reference map in the i-th pixel. The fluctuations map is expanded in the basis of spherical harmonics, producing a set of coefficients a_{lm}, which are used to build the multipole coefficients $\hat{C}_l$ of the angular power spectrum (APS). Each multipole coefficient is sensitive to anisotropies at an angular scale $180°/l$.

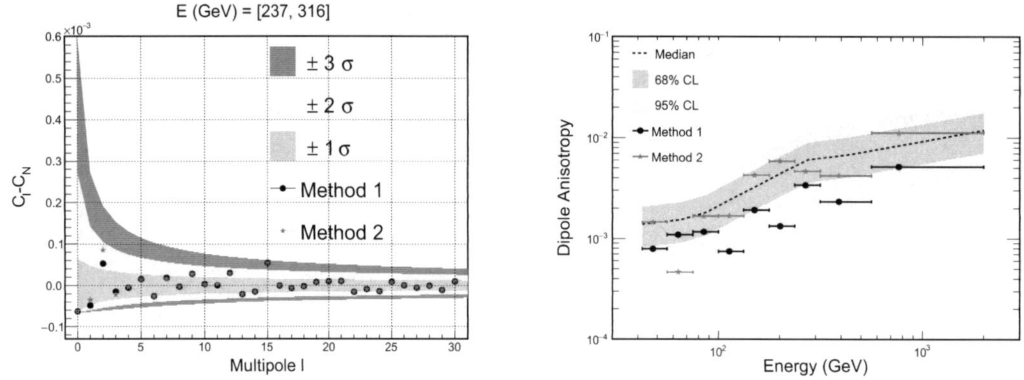

Figure 2. Left panel: measured APS in the energy bin from 237 GeV to 316 GeV. The difference $C_l - C_N$ is plotted as a function of the multipole l. Right panel: measured dipole anisotropy as a function of energy. The colored bands in both plots represent the expected central confidence intervals for an isotropic sky. The plots are taken from ref. [6].

Each multipole coefficient C_l can be viewed as a sum of two terms, i.e. $\hat{C}_l = C_N + \hat{C}_l^{ani}$, where C_N is the Poisson (white) noise term due to the finite number of events in the map, while $\hat{C}_l^{ani}$ is the anisotropy term. If the sky is observed with non-uniform exposure, the white noise is given by $C_N = \left(4\pi/N_{pixels}^2\right)\sum_i n_i/\mu_i^2$ where N_{pixels} is the number of pixels in the map [7]. To check whether the observed APS is statistically compatible with the Poisson noise, we have tested the null hypothesis $\hat{C}_l = C_N$ against the alternative hypothesis $\hat{C}_l > C_N$. The left plot of Fig. 2 shows an example of APS evaluated in the energy bin from 237 GeV to 316 GeV. The data lie within the expectation bands obtained assuming the null hypothesis. The same behavior is observed in all the energy bins.

The first multipole coefficient is directly related to the dipole anisotropy coefficient, which is given by $\delta = 3\sqrt{C_1/4\pi}$. The right plot of Fig. 2 shows the measured values of δ as a function of energy. The measured values are consistent with the expectations obtained assuming the null hypothesis. We can, therefore, conclude that no anisotropies are observed in the CRE data, and we can set upper limits on the value of the dipole anisotropy coefficient.

5. Discussion and conclusions

The measured CRE spectrum and anisotropies can be compared with the predictions from various astrophysical models. We have tested a model in which CREs arriving in the Solar System consists of a galactic diffuse component and of an additional component originated from a local nearby source.

To evaluate the spectrum of the galactic CRE component we have used the two-dimensional version of the DRAGON propagation code [8] with the secondary particle production cross sections of ref. [9], assuming that the scalar diffusion coefficient depends on the particle rigidity R and on the distance from the galactic plane z according to the parametrization $D = D_0(R/R_0)^{0.33}e^{|z|/z_t}$, where $D_0 = 4.25 \times 10^{28}\,\mathrm{cm^2s^{-1}}$, $R_0 = 4\,\mathrm{GV}$ and $z_t = 4\,\mathrm{kpc}$. The Alfvén velocity is set to $v_A = 33\,\mathrm{km\ s^{-1}}$

We have then considered the contributions from two possible sources located either in the Vela (distance 290 pc, age 1.1×10^4 yrs) or in the Monogem (distance 290 pc, age 1.1×10^5 yrs) positions. For both sources, we have adopted a burst like electron injection spectrum, in which the duration of the emission is much shorter than the travel time from the source, described by a

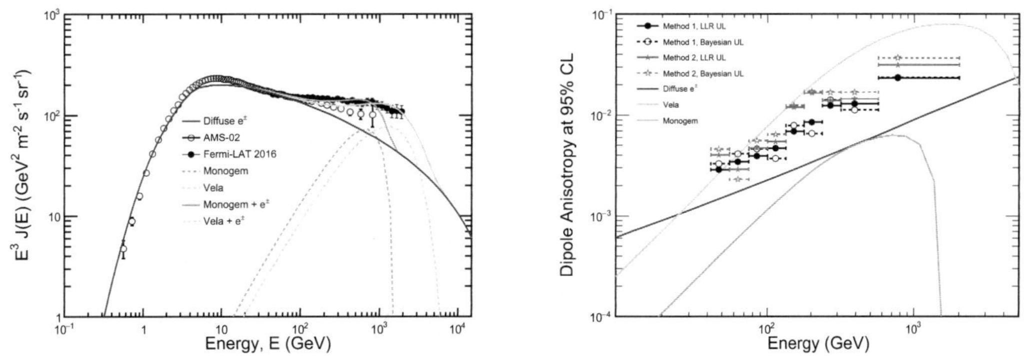

Figure 3. Left panel: CRE spectra measured by the *Fermi* LAT and AMS02 [3]. The blue line represents the galactic CRE contribution evaluated with DRAGON [8, 9]. The green (cyan) lines represent the contribution from Monogem (Vela) alone (dashed lines) and the sum of the Monogem (Vela) contribution with the galactic component (solid lines). Right panel: upper limits on the dipole anisotropy measured by the LAT compared with the predictions obtained assuming a CRE source in the Monogem (green line) or in the Vela (cyan line) position. The plots are taken from ref. [6].

power law with index $\Gamma = 1.7$ and with an exponential cutoff $E_{\rm cut} = 1.1\,{\rm TeV}$. For both sources, the normalization of the spectrum has been chosen to obtain a total flux not higher than that measured by the *Fermi* LAT and by AMS02 [3]. The left panel of Fig. 3 shows the CRE spectra measured by the *Fermi* LAT and by AMS02 compared with those predicted by the model.

In the right panel the upper limits on the CRE dipole anisotropy obtained by the LAT are compared with the predictions of the model. The LAT limits are consistent with the expected anisotropy in a scenario with a nearby middle-aged CRE source (Monogem), but are lower than the expected anisotropy in a scenario with a nearby young CRE source (Vela).

Acknowledgments

The *Fermi* LAT Collaboration acknowledges support for LAT development, operation and data analysis from NASA and DOE (United States), CEA/Irfu and IN2P3/CNRS (France), ASI and INFN (Italy), MEXT, KEK, and JAXA (Japan), and the K.A. Wallenberg Foundation, the Swedish Research Council and the National Space Board (Sweden). Science analysis support in the operations phase from INAF (Italy) and CNES (France) is also gratefully acknowledged. This work performed in part under DOE Contract DE-AC02-76SF00515.

References

[1] Atwood W B *et al.* 2009 *Astrophys. J.* **697** 1071
[2] Abdollahi S *et al.* 2017 *Phys. Rev.* D **95** 082007
[3] Aguilar M *et al.* 2014 *Phys. Rev. Lett.* **113** 221102
[4] Aharonian F *et al.* 2008 *Phys. Rev. Lett.* **101** 261104
[5] Aharonian F *et al.* 2009 *Astron. & Astrophys.* **508** 561
[6] Abdollahi S *et al.* 2017 *Phys. Rev. Lett.* **118** 091103
[7] Fornasa M *et al.* 2016 *Phys. Rev.* D **94** 123005
[8] Evoli C *et al.* 2008 *J. Cosmol. Astropart. Phys.* **10** 018
[9] Mazziotta M N *et al.* 2016 *Astrop. Phys.* **81** 21

IOP Conf. Series: Journal of Physics: Conf. Series **934** (2017) 012017 doi:10.1088/1742-6596/934/1/012017

Testing of the EPOS LHC, QGSJET01, QGSJETII-03 and QGSJETII-04 hadronic interaction models via help of the atmospheric vertical muon spectra

L G Dedenko[1,2], A V Lukyashin[3,4], T M Roganova[2] and G F Fedorova[2]

[1] Faculty of Physics M.V. Lomonosov Moscow State University, Leninskie Gory, 119991 Moscow, Russia
[2] Skobeltsyn Institute of Nuclear Physics, Lomonosov Moscow State University, 119234 Moscow, Russia
[3] Institute for Theoretical and Experimental Physics named by A.I. Alikhanov of National Research Center "Kurchatov Institute", 117218 Moscow, Russia
[4] National Research Nuclear University (MEPhI) Moscow Engineering Physics Institute, 115409 Moscow, Russia

E-mail: ddn@dec1.sinp.msu.ru, lukyashin.anton@physics.msu.ru

Abstract. The recent results of the very precise measurements of the primary cosmic protons and helium nuclei energy spectra by AMS-02 and some rather accurate estimates of these energy spectra generated in SNR allow us to elaborate the new approximation of the pimary nucleon energy spectra. As the acuracy of this approximation is rather high we can use it to test various models of hadronic interactions with the help of atmospheric muon energy spectra. The atmospheric vertical muon energy spectra have been calcullated in terms of the EPOS LHC, QGSJET01, QGSJETII-03 and QGSJETII-04 models in the energy range $10^2 \div 10^5$ GeV with help of the CORSIKA package and this new approximation of the primary nucleon spectrum. The comparison of calculations with the muon spectra observed by collaborations L3+Cosmic, LVD and MACRO has shown that all models predict approximately two times lower intensity of the muon energy spectra. As these muons are products of decays of the most energetic $\pi^{\pm}$ and $K^{\pm}$ mesons in the atmosphere, we can conclude that production of these $\pi^{\pm}$ and $K^{\pm}$ mesons is underestimated by EPOS LHC, QGSJET01, QGSJETII-03 and QGSJETII-04 models.

1. Introduction

The extensive air showers (EAS) are the only tool to understand the origin and composition of cosmic rays, their possible sources and a mechanism of acceleration.

All features of the energy spectrum, arrival directions and composition of the primary cosmic particles should be determined through an analysis of the EAS data. These data as some signals in the surface and underground detectors are usually interpreted in terms of various models of hadronic interactions [1–9]. It happened, that such interpretation leads to some inconsistency. As an example, energy of showers calculated in terms of the QGSJET II-03 [3] model with help of the surface detectors signals at Telescope Array [10] happened to be 1.27 times lager than this energy estimated with help of the fluorescence light. Usually these models are tested with the help of the accelerator data at small values (∼0) of the pseudorapidity η where most of secondary particles (mainly mesons) are produced [11–13]. However, calculations have shown that the maximal energy flow carried by secondary particles occurs at much larger values (∼8–10) of the pseudorapidity η [14]. So, it is of the primary importance to verify a production of the most energetic mesons simulated in terms of various models. This verification may be carried out by comparing model predictions of this muon fluxes with data of the classical experiments L3+Cosmic [15], MACRO [16] and LVD [17] in the energy interval of $10^2 - 10^5$ GeV. Showers induced by the primary protons and helium nuclei with different fixed energies have been simulated with help of the CORSIKA package [18] and the muon partial energy spectrum in each individual shower have been calculated. Then a convolution of these simulations for every type of the primary particles with intensities of these particles has been estimated. Inspired by new precision cosmic rays data base [19] (e.g.AMS-02 [20], PAMELA [21], ATIC-2 [22], CREAM [23], ARGO-YBJ [24], ARGO-YBJ & FWCTA [25], KASCADE [26], KASCADE-Grande [27], Tunka [28], IceCube [29], Telescope Array [30]), we suggested new approximations of cosmic ray energy spectra for primary protons and helium nuclei. Besides some calculations of spectra of the primary proton and helium nuclei in SNR [31] should also be used. Thus, with the help of any interaction models [1–9], the package CORSIKA and data on fluxes of the primary cosmic nuclei [20–31] one can predict the energy spectra of atmospheric vertical high energy muons at sea level. These predictions can be compared with data observed by collaborations L3+Cosmic, MACRO and LVD at energies above 100 GeV. In fact, some low energy models with the package FLUKA [32] have been tested in such a way. We are sorry that some our results of models testing in [33–35] are not correct. We do apologize for our mistake in input data for the atmosphere.

In this paper models EPOS LHC [7], QGSJET-01 [1], QGSJET II-03 and QGSJET II-04 [4] have been tested. A comparison of muon data observed in [15–17] with results of simulations allows to draw a conclusion that these models failed to be produce correctly the most energetic mesons.

2. Method

To estimate the energy spectra $D(E_\mu)$ of atmospheric vertical muons in the energy range of $10^2 - 10^5$ GeV we need to know the energy spectra dI_p/dE and dI_{He}/dE of the primary protons and helium nuclei within the energy interval of $10^2 - 10^7$ GeV and the partial energy spectra $S_p(E_\mu, E)$ and $S_{He}(E_\mu, E)$ of the vertical muons in EAS induced by the primary protons and helium nuclei with the various fixed energies E. Simulations of these partial muon spectra have been carried out in terms of the EPOS LHC, QGSJET-01, QGSJET II-03 and QGSJET II-04 hadronic interaction models in the same energy

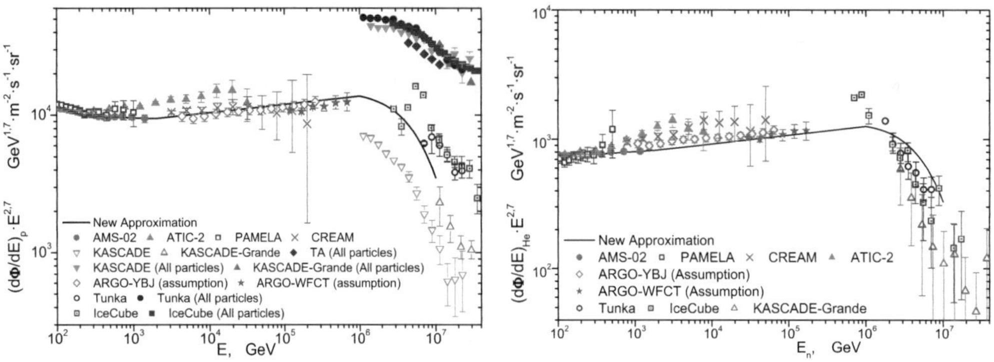

Figure 1. Energy spectra of primary nucleons: protons - left panel, helium - right panel. All particles spectra depends on energy per particle.

range of $10^2 - 10^7$ GeV.

The hypothesis of superposition [36] is also used. As direct results coincides with simulations in terms of the hypothesis of superposition, we will use this hypothesis.

We had used approximations the energy spectra of the primary protons and helium nuclei for (dI_p/dE) and (dI_{He}/dE) suggested in [37]. Figure 1 demonstrate how these approximations fit data [20-30].

The package CORSIKA 7.4 had been used to simulate the second important ingredients - the partial energy spectra $S_p(E_\mu, E)$ and $S_{He}(E_\mu, E)$ of vertical muons in showers induced by the primary protons and helium nuclei with the various fixed energies E.

The results of these calculations in the energy range of $10^2 - 10^7$ GeV were interpolated for 100 values of energies E with equal intervals in decimal logarithmic scale. The energy interval $10^2 - 10^5$ GeV of muons was divided into 60 equal bins also in decimal logarithmic scale. So, the width of the bin was equal to $h = \lg(E_{\mu,(i+1)}/E_{\mu,i}) = 0.05$.

The energy spectra $D_p(E_\mu)$ and $D_{He}(E_\mu)$ of muons for the primary protons and helium nuclei are calculated as integrals of products of functions $S_p(E_\mu, E)$ and $S_{He}(E_\mu, E)$ with corresponding intensities dI_p/dE and dI_{He}/dE of the primary protons, on energy E of primary nucleons.

$$D_p(E_\mu) = \int \left(\frac{dI_p}{dE}\right) \cdot S_p(E_\mu, E) \cdot dE \tag{1}$$

$$D_{He}(E_\mu) = \int \left(\frac{dI_{He}}{dE}\right) \cdot S_{He}(E_\mu, E) \cdot dE \tag{2}$$

Resulting energy spectrum of atmospheric muons is the sum of these energy spectra of muons produced by primary protons and helium nuclei.

$$D(E_\mu) = D_p(E_\mu) + D_{He}(E_\mu) \tag{3}$$

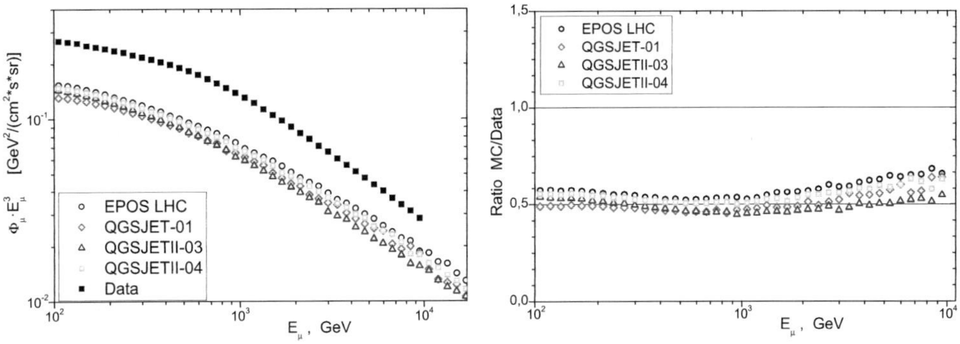

Figure 2. The energy spectra of near vertical muons - left panel. The ratio MC/DATA - right panel.

3. Results of simulations

The partial energy spectra $S_p(E_\mu, E)$ of the atmospheric vertical muons have been simulated for various fixed energies E of the primary protons in terms of the EPOS LHC, QGSJET-01, QGSJET II-03 and QGSJET II-04 models with statistics of $\sim 10^6$. It happened that this statistics end of the spectra is not enough and additional simulations with statistics 10^7 have been carried out.

We compared the total number of muons with energies above 10^2 and 10^3 GeV in showers induced by the primary protons with energies 10^5 and 10^6 GeV estimated in terms of the these models in our simulations and in [38]. The very reasonable agreement has been found.

The final results of the muon energy spectra $D(E_\mu)$ calculated in terms of the EPOS LHC, QGSJET-01, QGSJET II-03 and QGSJET II-04 models and data [15-17] are shown in figure 2 (left panel) and ratios of MC simulation to data - in the same figure (right panel). It is seen that calculated spectra are ~ 2 times below data in case of the QGSJET II-03 model and ~ 1.7 times below data for the EPOS LHC model. The result of the rest of models are in between of these limits. The main conclusion is quite clear. All considered models demonstrate the valuable deficit of muons.

4. Conclusion

Muons are produced in decays of the most energetic $\pi^\pm$-mesons and $K^\pm$-mesons generated in first interactions of the primary particles with nuclei in the atmosphere.

As calculated vertical muon energy spectra in case of the EPOS LHC, QGSJET-01, QGSJET II-03 and QGSJET II-04 models are $\sim 1.7 \div 2$ times below data we can conclude that production of the most energetic $\pi^\pm$-mesons and $K^\pm$-mesons in these models is considerably suppressed. This suppression may induce smaller values of signals in the surface scintillation detectors and will result in larger values of the calculated energy estimates. So, the coefficient 1.27 used by the TA collaboration [10] may be understood as a result of this suppression. The increased intensity of the primary particle flux observed at the Yakutsk array at super high energies [39] may be also a result of smaller values of calculated signals in surface scintillation detectors.

References

[1] Kalmikov N N and Ostapchenko S S 1993 *Phys. Atom. Nucl.* **56** 346
[2] Ostapchenko S S 2006 *Nucl. Phys.* B (Proc. Suppl.) **151** 143
[3] Ostapchenko S S 2006 *Phys. Rev.* D **74** 014026 (*Preprint* hep-ph/0505259)
[4] Ostapchenko S S 2011 *Phys. Rev.* D **83** 014018
[5] Ahn E-J *et al.* 2009 *Phys. Rev.* D **80** 094003
[6] Werner K 2008 *Nucl. Phys.* B (Proc. Suppl.) **175** 81
[7] Werner K, Liu F M and Pierog T 2006 *Phys. Rev.* C **74** 044902 (*Preprint* hep-ph/0506232)
[8] Ranft J 1999 *Phys. Rev.* D **51** 64 (*Preprint* hep-ph/9911213 and hep-ph/9911232)
[9] Werner K 1993 *Phys. Rep.* **232** 87
[10] Abu-Zayyad T *et al.* 2013 (Telescope Array Collab.) *Astrophys. J. Lett.* **768** L1
[11] Pierog T 2015 *EPJ Web of Conf.* **99** 09002
[12] Ostaphcenko S S 2012 *Progr. of Theor. Phys. Suppl.* **193** 204
[13] D'Enterria D *et al.* 2011 *Astropart. Phys.* **35** 98
[14] Engel R and Rebel H 2004 *Acta Phys. Polonica* B **35** 321
[15] Achard P *et al.* 2004 (L3 Collab.) *Phys. Lett.* B **598** 15-32
[16] Ambrosio M *et al.* 1995 (MACRO Collab.) *Phys. Rev.* D **52** 3793
[17] Aglietta M *et al.* 1998 (LVD Collab.) *Phys. Rev.* D **58** 092005
[18] Heck D *et al.* 1998 *Report FZKA 6019* Forschungszentrum Karlsruhe
[19] Maurin D, Melot F and Taillet R 2014 *Astron. Astrophys.* **569** A32
[20] Aguilar M *et al.* 2015 (AMS Collab.) Phys. Rev. Lett. **114** 171103
Aguilar M *et al.* 2015 (AMS Collab.) Phys. Rev. Lett. **115** 211101
[21] Adriani O *et al.* 2013 (PAMELA Collab.) *Adv. Space Res.* **51** 219–26
[22] Panov A D *et al.* 2007 (ATIC Collab.) *Bull. Russ. Acad. Sci. Phys.* **71** 494
Panov A D *et al.* 2009 (ATIC Collab.) *Bull. Russ. Acad. Sci. Phys.* **73** 564
[23] Ahn H S *et al.* 2009 (CREAM Collab.) *Astrophys.J.* **707** 593
[24] Bartoli B *et al.* 2015 (ARGO-YBJ Collab.) *Phys. Rev.* D **91** 112017
Di Sciascio G *et al.* 2015 (ARGO-YBJ Collab.) *J. of Physics: Conf. Series* **632** 012089
[25] Bartoli B *et al.* 2014 (ARGO-YBJ Collab.) *Chinese Physics* C **38** 045001
[26] Antoni T *et al.* 2005 (KASCADE Collab.) *Astropart. Phys.* **24** 125
[27] Apel W D *et al.* 2013 (KASCADE-Grande Collab.) *Astropart. Phys.* **47** 5466
[28] Prosin V V *et al.* 2014 (Tunka Collab.) *Nucl. Instr. Meth. Phys. Res.* A **756** 94101
[29] Rawlins K *et al.* 2015 The Hague (IceCube Collab.) *Proc. 34-th Int. Cosmic Ray Conf.*, PoS(ICRC2015)334
[30] Ivanov D *et al.* 2015 The Hague (Telescope Array Collab.) *Proc. 34-th Int. Cosmic Ray Conf.*, PoS(ICRC2015)349
[31] Berezhko E G 2014 *Nucl. Phys.* B (Proc. Suppl.) **256-257** 2335
Berezhko E G, Knurenko S P and Ksenofontov L T 2012 *Astropart. Phys.* **36** 3136
[32] Battistoni G *et al.* 2007 *Proc. Suppl. Nucl. Phys.* B **168** 286
[33] Dedenko L G, Roganova T M and Fedorova G F 2014 *JETP Lett.* **100** 223
[34] Dedenko L G, Roganova T M and Fedorova G F 2015 *Phys. Atom. Nucl.* **78** 840–48
[35] Dedenko L G *et al.* 2015 *EPJ Web of Conf.* **99** 10003
[36] Dedenko L G and Zatsepin G T 1960 Moscow *Proceedings of the 6-th ICRC* **II** *Extensive air showers and cascades process* 201–8
[37] Dedenko L G *et al.* 2017 Moscow *Proc. of International Symposium on Cosmic Rays and Astroparticles, http://iscra2017.mephi.ru/content/public/files/presentations/June20_Presentation_13_LukyashinAV.pdf*
[38] Lagutin A A, Tyumentsev A G and Yushkov A V 2004 *J. Phys. G.* **30** 573–96
[39] Glushkov A V *et al.* 2003 Tsukuba *Proc. of the 28-th ICRC* **1** 393

IOP Conf. Series: Journal of Physics: Conf. Series **934** (2017) 012018 doi:10.1088/1742-6596/934/1/012018

Recent Results of Search for Solar Axions Using Resonant Absorption by ^{83}Kr nuclei

A V Derbin[1], I S Drachnev[1], A M Gangapshev[2], Yu M Gavrilyuk[2], V V Kazalov[2], V V Kobychev[3], V V Kuzminov[2], V N Muratova[1], S I Panashenko[4], S S Ratkevich[4], D A Tekueva[2], E V Unzhakov[1], S P Yakimenko[2],

[1] Petersburg Nuclear Physics Institute, NRC "Kurchatov Institute", St. Petersburg, Russia
[2] Institute for Nuclear Research, RAS, Moscow, Russia
[3] Institute for Nuclear Research of NAS Ukraine, Kiev, Ukraine
[4] Kharkov National University, Kharkov, Ukraine

E-mail: derbin_av@pnpi.nrcki.ru

Abstract. A search for resonant absorption of the solar axion by ^{83}Kr nuclei was performed using the proportional counter installed inside the low-background setup at the Baksan Neutrino Observatory. The obtained model independent upper limit on the combination of isoscalar and isovector axion-nucleon couplings $|g_3 - g_0| \leq 8.4 \times 10^{-7}$ allowed us to set the new upper limit on the hadronic axion mass of $m_A \leq 65$ eV (95% C.L.) with the generally accepted values S=0.5 and z=0.56.

1. Introduction

If the axion exists, the Sun should be one of the most intense sources of these particles. The aim of this work is to search for monochromatic axions with an energy of 9.4 keV emitted in the M1 transition in the ^{83}Kr nuclei in the Sun [1]. Axions on the Earth can be detected in the inverse reaction of resonance absorption by detecting particles (γ- and X-ray photons, as well as conversion and Auger electrons) appearing at the decay of an excited nuclear level. The probability of the emission and subsequent absorption of axions depends only on the coupling constant with nucleons, which is minimally model dependent and is proportional to axion-nucleon coupling constant $(g_{AN})^4$.

The axion flux was calculated in [1] for the standard solar model BS05 [2] characterized by a high metallicity [3]. The differential flux at the maximum of the distribution is [1]:

$$\Phi_A(E_{M1}) = 5.97 \times 10^{23} \left(\frac{\omega_A}{\omega_\gamma}\right) \text{cm}^{-2}\text{s}^{-1}\text{keV}^{-1}. \quad (1)$$

where ω_A/ω_γ is the branching ratio of axions to photons emission. The cross section for resonance axion absorption is given by an expression similar to the expression for the photon-absorption cross section, the correction for the ratio ω_A/ω_γ being taken into account.

$$\sigma(E_A) = 2\sqrt{\pi}\sigma_{0\gamma} \exp\left[-\frac{4(E_A - E_M)^2}{\Gamma^2}\right] \left(\frac{\omega_A}{\omega_\gamma}\right), \quad (2)$$

IOP Conf. Series: Journal of Physics: Conf. Series **934** (2017) 012018 doi:10.1088/1742-6596/934/1/012018

where $\sigma_{0\gamma}$ is the maximum cross section of the γ -ray resonant absorption and $\Gamma = 1/\tau$. The total cross section for axion absorption can be obtained by integrating $\sigma(E_A)$ over the axion spectrum. The expected rate of resonance axion absorption by the ^{83}Kr nucleus as a function of the ratio ω_A/ω_γ, the combination of isoscalar and isovector coupling constants $|g_3 - g_0|$ and axion mass m_A can be represented in the form ($S = 0.5$, $z = 0.56$) [1]:

$$R_A[\mathrm{g}^{-1}\mathrm{day}^{-1}] = 4.23 \times 10^{21}(\omega_A/\omega_\gamma)^2 \quad (3)$$
$$= 8.53 \times 10^{21}(g_3 - g_0)^4(p_A/p_\gamma)^6 \quad (4)$$
$$= 2.41 \times 10^{-10}(m_A/1\ \mathrm{eV})^4(p_A/p_\gamma)^6. \quad (5)$$

2. Experimental setup

A large proportional counter filled with of ^{83}Kr (99.9 %) was used to detect X-rays and gammas, as well as conversion and Auger electrons, appearing in the decay of the excited level with an energy of 9.4 keV. The LPC is a cylinder with inner diameters of 137‘mm. A gold-plated tungsten wire of 10 μm in diameter is stretched along the LPC axis and is used as an anode. In order to reduce the influence of edge effects on the collection of the charge, the ends of the anode wire were surrounded by copper tubes (3 mm in diameter and 38 mm in length), which were at the anode potential and excluded gas amplification in this region. With the inclusion of Teflon insulators, the distance from the working area to the flanges of the chamber was 70 mm. The length of the working area of the chamber was 595 mm, which corresponded to a volume of 8.77 L. The chamber operated at a pressure of 1.8 bar. The mass of ^{83}Kr isotope in the working volume was 58 g.

Figure 1. A large proportional counter (LPC) with a casing of copper inside passive shielding.

IOP Conf. Series: Journal of Physics: Conf. Series **934** (2017) 012018 doi:10.1088/1742-6596/934/1/012018

The LPC is surrounded by passive shield made of copper (~20 cm), lead (~20 cm) and polyethylene (8 cm). The setup is located in the Deep Underground Low-Background Laboratory of Institute for Nuclear Research of Russian Academy of Sciences (BNO INR RAS) [4], at the depth of 4900 m w.e., where the cosmic muon flux is reduced by $\sim 10^7$ times in comparison to that above ground, and evaluated as $(2.6 \pm 0.09) \times 10^{-9}$ cm^{-2}s^{-1}.

A signal from the anode was supplied to a charge-sensitive preamplifier. The shape of the pulse was digitized in a time interval of 164 μs with a frequency of 12.5 MHz and was transmitted to a computer through a USB port. The rise time of the leading edge of the pulse and the ratio of amplitudes of secondary and primary pulses were determined for each event, because these parameter makes it possible to select events near the cathode and non-point events such as multiple Compton scattering. The procedure of the analysis of the shape of pulses was described in more detail in [5].

3. Results

The background spectra collected during 613.25 days and fit result curve are presented in figure 2. Two peaks are clear visible in the energy range (4 – 26) keV. The peak with energy 8.05 keV associates with the detection of $K_{\alpha 1,2}$ X-rays of copper. The structure of the second peak is more complicated, it is mixture of Kr and Br $K_{\alpha 1,2}$ X-rays and 13.5 keV from K-capture of cosmogenic ^{81}Kr. It is seen that the 9.4 keV peak is not manifested.

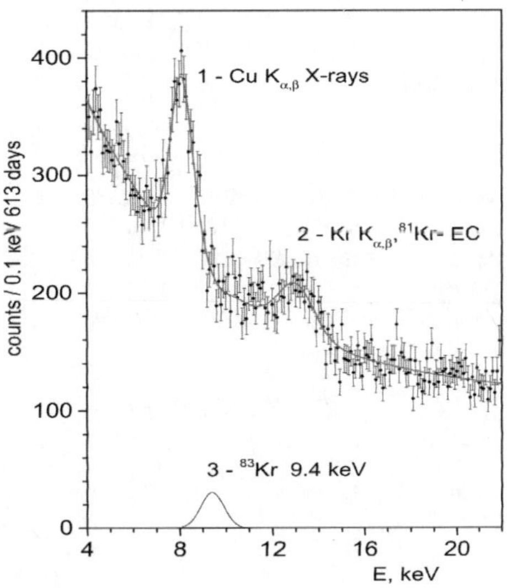

Figure 2. Energy spectra of the Kr LPC measured for 613 days, fitting results (red line) and expected axion peak for $3S_{\rm lim}$ (blue line).

The maximum likelihood method was used to determine the intensity of the peak. The fit of spectrum corresponding to the minimum χ^2 is shown by red solid line in figure 2. The minimum of χ^2 corresponds to the nonphysical value of the area of the 9.4 keV peak $S_A = -(102 \pm 92)$ events. The standard χ^2-profile method was used to determine the upper bound on the number of events in the peak. The upper bound thus determined for the number of events in the peak is $S_{\rm lim} = 127$ for 95% C.L.

The expected number of registered axions is:

$$S_A = RMT\epsilon \leq S_{\text{lim}}, \tag{6}$$

where $M = 58$ g is mass of ^{83}Kr isotope, $T = 613.25$ days is time of data taking, and $\epsilon = 0.825$ is the detection efficiency.

The upper limit on the excitation rate of ^{83}Kr by solar hadronic axions is defined as $R_{\text{exp}} = 4.29 \times 10^{-3}$ g^{-1}day^{-1}. The relation $R_A \leq R_{\text{exp}}$ limits the region of possible values of the coupling constants g_0, g_3 and axion mass m_A. In accordance with Eqs. (3-5), and on condition that $(p_A/p_\gamma) \cong 1$ provided for $m_A < 3$ keV one can obtain:

$$(\omega_A/\omega_\gamma) \leq 1.0 \times 10^{-12} \tag{7}$$

$$|g_3 - g_0| \leq 8.4 \times 10^{-7}, \text{ and} \tag{8}$$

$$m_A \leq 65 \text{ eV}, \text{ at } 95\% \text{ C.L.} \tag{9}$$

The limit (9) is stronger than the constrain obtained with 14.4 keV ^{57}Fe solar axions [6] and is stronger than our previous result obtained in ^{83}Kr experiment [1]. As in the case of ^{57}Fe nucleus the obtained limit on axion mass strongly depends on the exact values of the parameters S and z.

4. Acknowledgments

This work was supported by the Russian Foundation of Basic Research (grants 17-02-00305A, 16-29-13014ofi-m, arg16-29-13011ofi-m, 15-02-02117A, 14-02-00258A) and in part by the Russian Science Foundation (grant 17-12-01009).

References

[1] Gavrilyuk Yu M, Gangapshev A M, Derbin A V *et al.* 2015 *JETP Letters* **101** 664
[2] Bahcall J H, Serenelli A M and Basu S 2005 *Astrophys. J.* **621** L85
[3] Grevesse N and Sauval A J 1998 *Space Sci. Rev.* **85** 161
[4] Gavriljuk Ju M *et al.* 2013 *NIM* **A729** 576
[5] Gavrilyuk Ju M *et al.* 2010 *Instr. Exper. Techn.* **53** 57 (2010), 2013 *Phys. Rev.* C **87** 035501
[6] Derbin A V, Muratova V N, Semenov D A and Unzhakov E V 2011 *Phys. At. Nucl.* **74** 596

IOP Conf. Series: Journal of Physics: Conf. Series **934** (2017) 012019 doi:10.1088/1742-6596/934/1/012019

Calculation of total muon flux observed by Muon Monitor experiment

I Bandac[1], A Bayo[1], L Bezrukov[2], T Enqvist[3], A Fazliakhmetov[2,4], A Ianni[1], L Inzhechik[4], J Joutsenvaara[3], P Kuusiniemi[3], K Loo[5], B Lubsandarzhiev[2], M Nelubina[2,4], A Nozik[2,4], M Slupecki[5], W H Trzaska[5]

[1] Laboratorio Subterráneo de Canfranc
[2] Institute of Nuclear Research, Russian Academy of Sciences, Moscow, Russia
[3] Kerttu Saalasti Institute, University of Oulu, Finland
[4] Moscow Institute of Physics and Technology, Russia
[5] Department of Physics, University of Jyväskylä, Finland

E-mail: `fazliakhmetov@phystech.edu`

Abstract. An approach to calculate the flux of cosmicgenic muons detected by Muon Monitor experiment in lab LAB2400 of the Underground Laboratory in Canfranc (LSC) is described. The measuring apparatus consists of three layers of SC16 scintillation matrix detectors. The hardware function of the detector assembly was determined using computer simulation. Obtained value of the total muon flux turned out to be equal to $(4.35 \pm 0.2) \times 10^{-3}\ m^{-2}s^{-1}$.

1. Motivation

To eliminate background produced by cosmic rays, the experiments researching for rare processes are conducted in deep underground laboratories, where rock overburden attenuates flux of cosmogenic muons. Nevertheless, even such a small residual underground flux of cosmic muons produces background for physics experiments. The general goal of Muon Monitor project is to study muon background conditions in underground sites of Canfranc Underground Laboratory (LSC) and precisely measure muon flux taking into account considerable anisotropy of the flux because of complicated shape of mountains above the lab LAB2400 of LSC.

2. Experimental setup

The Muon Monitor consists of scintillation detecting assembly (SDA), data acquisition system (DAQ), and power supply system. The SDA has three layers consisting of the fast SC16 scintillation detectors: $3 \times 3 = 9$ detectors both in the top and in the bottom layers and $2 \times 2 = 4$ detectors in the middle layer (see figure 1). Effective square aperture is ~ 1 m^2. Maximum zenith angle of muon to be detected is approximately $80°$. Each SC16 detector contains a 4×4 matrix of 16 independent, square pixel detectors of $122 \times 122 \times 30$ mm^3 [1,2]. The internal structure of the pixel, schematic view of the SC16 and connection layout is presented in the figure 2.

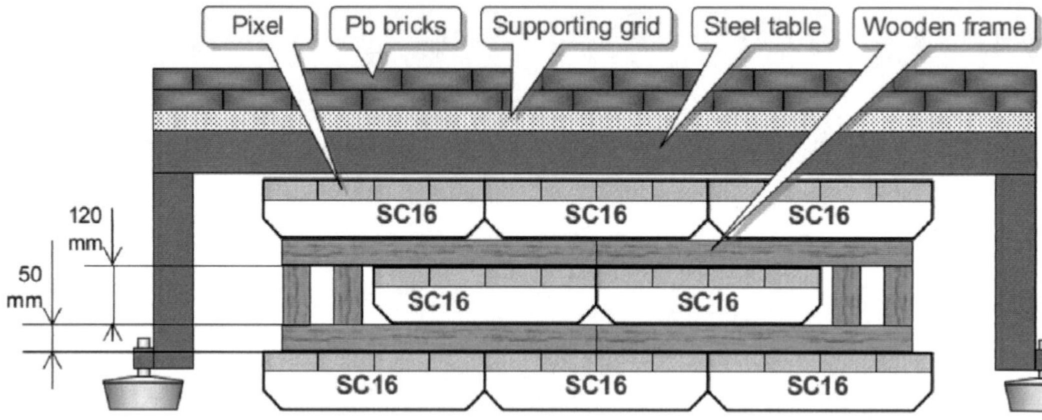

Figure 1. Schematic view of a detecting assembly

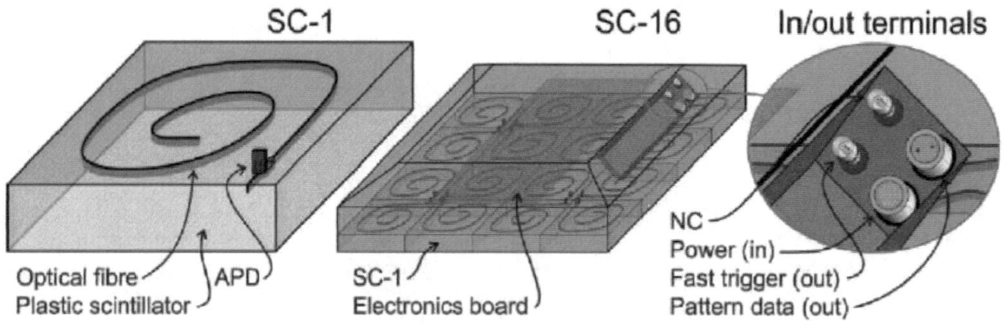

Figure 2. Schematic view of detector

3. Calculation of total muon flux

To calculate total muon flux with taking into account the geometrical response of SDA and individual efficiencies of subdetectors inside SC16 a Monte Carlo simulation was used. The simulation tool was written in Kotlin using Appache Commons Math for geometry description.

Only events that produce straight line tracks was selected to calculation. This was done using simulation program. Considering that simulation statistics of 10^7 is few orders of magnitude greater than the number of events in the experiment ($\sim 3 \times 10^5$), it is supposed that pixel combinations which do not appear in simulation, are not possible in the real data. Thus it was selected only events where set of pixels being hit exists in the simulation.

The efficiency for the given pixel set could be found from simulation with uniform initial distribution of muons in a following way:

$$R_i = \frac{N_{sim,i}}{N_{sim}}, \tag{1}$$

where $N_{sim,i}$ is the number of simulated muons that hit pixel set with index i and N_{sim} is the total number of simulated muons. The total flux for a given pixel set f_i can be found as:

$$f_i = \frac{g_i}{\int \varepsilon_i d(\Omega)} = \frac{g_i}{R_i}. \tag{2}$$

where g_i is experimental counts of given pixel combination. Each of f_i is not the estimate of the flux through the pixel set, but the flux through the monitor as a whole. If the initial distribution is uniform then all of f_i would give the same value and total flux could be found by simple averaging $f = \langle f_i \rangle$. In real experiment, the initial distribution is not uniform, but averaging still gives a consistent estimate of the total flux.

4. Results

This approach gives the total number of events $\langle f \rangle = \gamma \langle f \rangle_0 = 2.256 \cdot 10^5$ *events* observed by monitor during time of measurement equals to 51836385 seconds. The $\langle f \rangle_0$ has been calculated in assumption that average flux in unobservable large zenith angles region and detectable low ones is the same and γ is correction factor equals to the ratio between monitor angular coverage and total 2π solid angle. Uncertainties for calculation includes:

- uncertainties of pixels position: ~ 5 %
- statistical error: ~ 0.3 %
- uncertainties of total measurement time: less than 0.1%

With taking into account all uncertainties the total flux is:

$$\Phi = (4.35 \pm 0.2) \times 10^{-3}\, m^{-2}s^{-1}. \quad (3)$$

Paper containing overwiev of method, results and disciussion in detail is in progress[3].

Acknowledgments

This work was supported by the grant of the Ministry of Education and Science of the Russian Federation number 3.3008.2017.

References

[1] Akhrameev E V *et al* 2009 Multi-pixel geiger-mode avalanche photodiode and wavelength-shifting fibre-optics readout of plastic scintillator counters for the emma underground experiment *Nucl. Inst. Meth. in Phys. Res.* A **610** 419–22

[2] Volchenko V I *et al* 2010 Background and muon counting rates in underground muon measurements with a plastic scintillator counter based on a wavelength shifting fibre and a multi-pixel avalanche photodiode readout *Centr.Eur.J.Phys.* **8** 445–47

[3] Bandac I *et al* Measurement of muon flux in the Canfranc Underground Laboratory *Unpublished paper*

Constraints on the model of dark matter with Coulomb-like interaction explaining positron anomaly

K M Belotsky[1], **E A Esipova**[1] **and A A Kirillov**[1,2]

[1] National Research Nuclear University MEPhI (Moscow Engineering Physics Institute), Moscow, Russia
[2] Yaroslavl State P.G. Demidov University, Yaroslavl, Russia

E-mail: esipovaea@gmail.com

Abstract. The models of self-interacting dark matter (DM) are popular now, in particular dark matter with Coulomb-like interaction. It leads to enhancement of annihilation in Galaxy what helps to explain cosmic ray puzzles, e.g. positron anomaly. Though such models get constraints from cosmic microwave background (CMB), large scale structure. The constrained region of model parameters can be essentially enlarged if to take into account possibility of much stronger enhanced annihilation of dark matter particles due to their recombination accounted for by a classical dipole radiation. The given constraints are considered for dark disk model with self-interaction.

1. Introduction

The models of self-interacting dark matter (SIDM) became very popular in the last time (e.g. [1, 2, 3, 4, 5, 6, 7, 8, 9, 10, 11, 12, 13, 14, 15, 16]). Such models may avoid several problems of cold dark matter (CDM) scenario [1, 2, 12], such as predicted excess of dwarf galaxies and cuspy density profile in them, the problems with rotation curve and so called "too-big-to-fail". All these problems could be resolved due to collisions of SIDM particles during large scale structure formation. There can be alternative solution for aforementioned problems with SIDM when it becomes "warm" due to heating from a thermal background of the mediators (y-photons) of dark interaction (y-background). Warm DM may also avoid copious formation of dwarf galaxies and other CDM problems. But one should care here not to get heated so much to become "hot" (i.e. to prevent formation of greater size galaxies). It may restrict the model.

SIDM can have more advantages in implications to cosmic rays (CR). Positron anomaly in CR can be explained with annihilation of DM in Galaxy. If DM has a thermal origin in early Universe then annihilation cross section should be enhanced in Galaxy to provide observed CR excess. It can be done with the help of Gamow-Sommerfeld-Sakharov (GSS) enhancement [27, 28, 29, 30] accounted for by long-range interaction of DM particles. Also one can suppose that DM annihilation has a narrow resonance [31] with long-range ($U(1)$-gauge) interaction being involved, or some other model scenarios. However, such attempts face the problem with excess of gamma-flux over observed gamma-ray background, and model with GSS-effect are also constrained by Cosmic Microwave Background (CMB) data [32].

We suggested a dark disk model [17, 18, 19, 20, 21, 22] to avoid both constraints. It is based on assumption that some fraction of DM, which annihilate or decay, is concentrated in galactic disk. It allows to suppress gamma. Here we combine this hypothesis with that on Coulomb-like self-interaction (y-interaction), so test a self-interacting dark matter disk model. One more important ingredient of our consideration is a recombination of DM particles due to a classical dipole radiation [23, 24], which is found to be much stronger than that considered on common quantum level. It may change drastically description of such DM and, as a consequence, the constraints on it.

Our current consideration can be applied to other DM models with long-range interaction.

2. Building model and analysis

We need to get in Galaxy two components of DM: the dominant "passive" component and small "active" one concentrating in galactic disk. To realize this one can suppose that we have SIDM consisted of particles a, b, $\bar{a}$, $\bar{b}$ possessing Coulomb-like interaction with different amounts of particles and antiparticles. Particles a and b are assumed to have opposite y-charges.

Dark disk must make up $\lesssim 10^{-3}$ of all DM in order not to contradict observations. This limit follows from upper limit on the dark disk density ($\lesssim 0.4$ GeV/sm^3) and (our best fit) disk size ~ 0.4 kps [17].

We assume that a particles are heavier, have such an initial asymmetry between a and $\bar{a}$ that in modern epoch a constitutes dominant component of dark matter in the form of bound states and small fraction of $a\bar{a}$ pairs survived in free states. Particles b and $\bar{b}$ can be supposed to have the same amounts, i.e. $n_b = n_a$, $n_{\bar{b}} = n_{\bar{a}}$. Given so, we expect that the most of a compose bound states with b giving "passive" halo component; remained unbound a and $\bar{a}$ form dissipative ("active") component of dark disk. Modern number of free $\bar{b}$ should be strongly suppressed (negligible), since they are light and annihilate with b through recombination faster.

Particles a and $\bar{a}$ in dark disk annihilate producing fluxes in cosmic rays therefore mass of a is chosen ~ 400 GeV to explain positron anomaly. The mass of lighter particle b is to be adjusted to obtain sufficiently compact ab-atoms and prevent $\bar{a}$ annihilation with a due to higher ab-recombination cross section.

Important feature of evolution a, b, $\bar{a}$, $\bar{b}$-system is their recombination in two regimes in dependence which approximation works. If common quantum approximation is valid then recombination is slow, and it becomes fast at lower velocity (later in cosmological evolution) when classical approximation is switched on [23]. These two approximations are given by different cross sections of recombination. In quantum case it is (see, e.g. [23, 25])

$$\sigma = \frac{32\pi}{3\sqrt{3}} \frac{\alpha_y^3}{\mu^2} \frac{\ln(v^{-1})}{v^2}, \tag{1}$$

and in classical one

$$\sigma = (4\pi)^{2/5}\pi \frac{\alpha_y^2}{\mu^2} \frac{1}{v^{14/5}}. \tag{2}$$

Here μ is the reduced mass of recombining particles, v is their relative velocity, α_y is the y-interaction constant. Classical cross section is much bigger than the given quantum one. It is especially worth to note that quantum formula (1) includes enhancing GSS factor. So "classical" recombination is very fast. It is assumed to be on when [23, 25]

$$v \ll \alpha_y^{5/2}, \tag{3}$$

i.e. at low velocities.

The choice of model parameters (constant α_y, the mass of b particle and initial asymmetries) allows to regulate switching on “strong recombination” in different moments. One can obtain the case of “strong recombination” of $a\bar{a}$-pairs with immediate successive annihilation inside dark galactic disk, where their velocity can be very low ($\lesssim 50$ km/c).

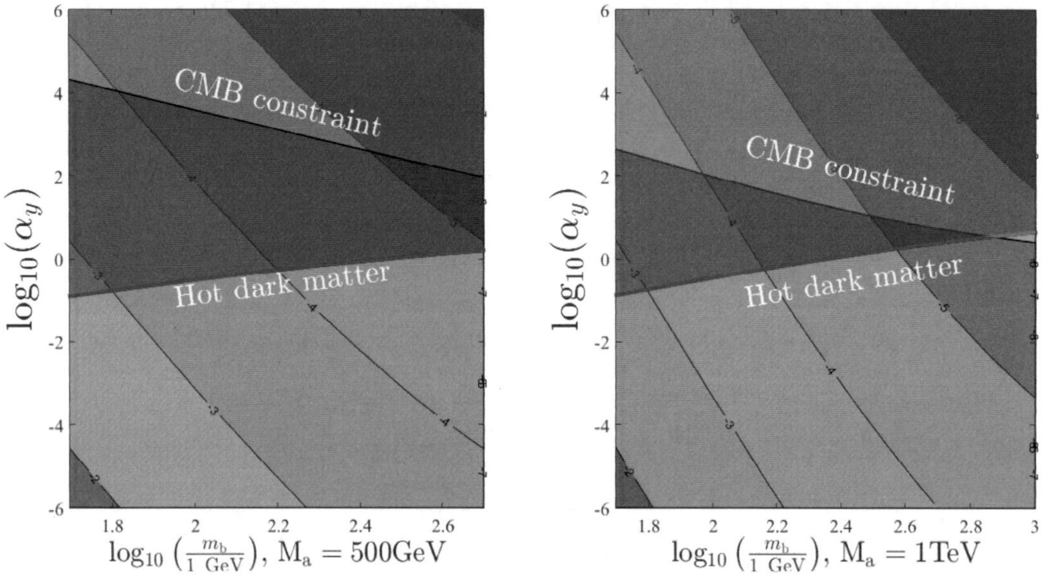

Figure 1. Contour plot for $\log_{10}(j/j_{\mathrm{teor}})$ in dependence of α_y and m_{b} for $M_{\mathrm{a}} = 500$ GeV (left) and 1 TeV(right) with the CMB constraints and HDM constraints. Excluded are the regions above line "HDM" and below line "CDM".

In all cases of interest classical approach becomes applicable after the moment when *ab*-particles decouple from y-background. Starting with this moment a, b-particles cool faster and condition (3) comes soon.

Calculations are done following work [23]. Aforementioned initial parameters are set here at the moment when condition (3) is reached. Modern fraction of free $a\bar{a}$ is not obtained to be bigger than 10^{-6}.

While such a self-interacting dark disk model is built, one needs to consider different constraints. We must not contradict observational data on the anisotropy of the CMB and prevent for DM to be hot (HDM). For the first, we can outline the region of parameters DM particle mass (of b) and α_y for which the value of σ/m satisfies the CMB constraint [26]. For the second, we require that the temperature of CMB, at which DM particles are decoupled from y-background, was $T_{\mathrm{dec}} > 5$ keV. Until DM particles are coupled to the relativistic background they wash out all the inhomogeneities which smaller the current horizon due to Silk dumping [33]. If it exceeds the galaxy scale, DM is hot and forbidden. When temperature $T_{\mathrm{dec}} \sim 5$ keV horizon corresponds to modern galaxy scale.

As to explanation of positron anomaly, we obtained [17] that in the framework of dark disk model it requires for emissivity value to be $j_{\mathrm{teor}} \equiv \frac{1}{4} n_a n_{\bar{a}} < \sigma v >= 1.4 \cdot 10^{-19} \mathrm{cm}^{-3} \mathrm{s}^{-1}$ within factor 3 (we assume that the probability of annihilation in the "visible" particles is 1).

In Fig. 1 you can see the contour plot for emissivity j in terms of the j_{teor} for two values of a particle mass M_a with the CMB and HDM constraints. Contour line with label "0" is of our interest. But as one can see, two constraints (CMB and HDM) exclude all j values (at any m_b and α_y) $\gtrsim 10^{-5} j_{\mathrm{teor}}$, what makes impossible explanation of positron anomaly. This conclusion is obtained when classical approximation (2) is valid. For galactic velocity it approximately corresponds to $\alpha_y \gtrsim 10^{-6}$, i.e. low borders of the plots in Fig. 1.

3. Conclusion

Dark disk model was suggested to explain positron anomaly. It could be realised assuming that DM possesses Coulomb-like interaction. In this case such model can avoid existing problems of standard CDM scenario. We built here some variant of such self-interacting dark matter disk model, introducing two species if DM particles (a and b) and their antiparticles. We have shown that two types of constraints, coming from CMB and HDM abundance, exclude parameter region where positron anomaly can be explained.

Acknowledgments

The work fulfilled in the framework of MEPhI Academic Excellence Project (contract № 02.a03.21.0005, 27.08.2013). The work of A. A. K. was supported by Russian Science Foundation (№ 15-12-10039). The work of K. M. B. was also funded by the Ministry of Education and Science of the Russia, Project № 3.6760.2017/BY.

References

[1] Tulin S and Yu H-B 2017 *Preprint* 1705.02358
[2] Kamada A, Kaplinghat M, Pace A B and Yu H-B 2017 *Phys. Rev. Lett.* **119** 111102
[3] Kaplan D E, Krnjaic G Z, Rehermann K R and Wells C M 2010 *JCAP* **05** 21
[4] Khlopov M Y 2011 *Mod. Phys. Lett.* A **26** 2823
[5] Cyr-Racine F-Y and Sigurdson K 2013 *Phys. Rev.* D **87** 103515
[6] Dudas E, Heurtier L, Mambrini Y and Zaldivar B 2013 *JHEP* **11** 083
[7] Petraki K, Pearce L and Kusenko A 2014 *JCAP* **07** 39
[8] Lindner M, Schmidt D and Watanabe A 2014 *Phys. Rev.* D **89** 013007
[9] Wise M B and Zhang Y 2014 *Phys. Rev.* D **90** 055030
[10] von Harling B and Petraki K 2014 *JCAP* **12** 33
[11] Foot R and Vagnozzi S 2015 *Phys. Rev.* D **91** 023512
[12] Foot R 2015 *Phys. Rev.* D **91** 123543
[13] Kouvaris C, Shoemaker I M and Tuominen K 2015 *Phys. Rev.* D **91** 043519
[14] Heikinheimo M, Spethmann C and Veermae H 2015 *Phys. Lett.* B **749** 236
[15] Boddy K K, Kaplinghat M, Kwa A and Peter A 2016 *Phys. Rev.* D **94** 123017
[16] Belotsky K M, Esipova E A, Khlopov M Yu and Laletin M N 2015 *Int. J. Mod. Phys.* D **24** 1545008
[17] Belotsky K, Budaev R, Kirillov A and Laletin M 2017 *JCAP* **01** 021
[18] Alekseev V V *et al.* 2015 *Phys. Proc.* **74** 48
[19] Alekseev V V *et al.* 2016 *J. Phys. Conf. Ser.* **675** 012023
[20] Alekseev V V *et al.* 2016 *J. Phys. Conf. Ser.* **675** 012026
[21] Alekseev V V *et al.* 2017 *Phys. Atom. Nucl.* **80** 713
[22] Belotsky K M, Budaev R I, Kirillov A A and Solovyov M L 2017 *J. Phys. Conf. Ser.* **798** 012084
[23] Belotsky K, Esipova E and Kirillov A 2016 *Phys. Lett.* B **716** 81
[24] Belotsky K M, Esipova E A and Kirillov A A 2015 *Phys. Proc.* **74** 24
[25] Elyutin P 1978 *Theor. and Math. Phys.* B **37** 112
[26] Galli S, Iocco F, Bertone G and Melchiorri A 2009 *Phys. Rev.* D **80** 023505
[27] Belotsky K M, Khlopov M Yu and Shibaev K I 2000 *Grav. Cosmol. Suppl.* **6** 140
[28] Belotsky K M, Khlopov M Yu, Legonkov S V and Shibaev K I 2005 *Grav.Cosmol.* **11** 27
[29] Belotsky K, Fargio D, Khlopov M and Konoplich R 2008 *Phys. Atom. Nucl.* **71** 147
[30] Arkani-Hamed N, Finkbeiner D P, Slatyer T R and Weiner N 2009 *Phys. Rev.* D **79** 015014
[31] Bai Y, Berger J and Lu S 2017 *Preprint* 1706.09974
[32] Planck collaboration 2016 *Astron. Astrophys.* **594** A13
[33] Bertschinger E 2006 *Phys. Rev.* D **74** 063509

The gamma-ray Moon seen by the Fermi LAT

F. Loparco (on behalf of the Fermi LAT Collaboration)

Dipartimento di Fisica "M. Merlin" dell'Università e del Politecnico di Bari, I-70126 Bari, Italy
Istituto Nazionale di Fisica Nucleare, Sezione di Bari, I-70126 Bari, Italy

E-mail: francesco.loparco@ba.infn.it

Abstract. When seen in gamma rays, the Moon appears brighter than the Sun. Gamma rays emitted by the Moon mostly originate from the decays of neutral pions produced by the interactions of cosmic rays with the lunar surface. Using the data collected by the *Fermi* Large Area Telescope (LAT) in its first seven years of operation, we measured the gamma-ray emission spectrum of the Moon in the energy range from 30 MeV up to a few GeV and we studied its time evolution, finding a correlation with the solar activity. We also developed a full Monte Carlo simulation based on the FLUKA code, which describes the production of gamma rays in the cosmic-ray interactions with the Moon. We used the simulation results to infer the cosmic-ray proton and helium spectra near the Earth from the lunar gamma-ray data.

1. Introduction

High energy gamma rays emitted from the Moon are produced in the inelastic collisions of cosmic-ray nuclei (CRs) with the lunar surface. The gamma-ray flux from the Moon will be therefore sensitive to the primary CR energy spectra, to the composition of the lunar surface and to the mechanisms of hadronic interactions of CR nuclei with the lunar regolith.

2. Data selection

This analysis uses the data collected by the *Fermi* LAT [1] from August 2008 to June 2015. It has been performed using the newest Pass 8 data [2] and selecting P8_SOURCE photon events starting from a minimum energy of 30 MeV.

In our analysis we have defined a signal region as a cone centered on the Moon position and a background region as a cone centered on a time-offset Moon position. Since the Moon orbits around the Earth with a period of $\sim$ 28 days, we chose a time offset of 14 days (i.e. at a given time, the center of the background region is in the position that the Moon will take 14 days later).

The angular radius of the two regions is given by:

$$\theta = \sqrt{[\theta_0(E/E_0)^{-\delta}]^2 + \theta_{min}^2} \tag{1}$$

where E is the photon energy, $E_0 = 100\,\mathrm{MeV}$, $\theta_{min} = 1°$, $\theta_0 = 5°$ and $\delta = 0.8$. The energy dependence of the angular radius follows the behavior of the 68% containment radius of the LAT point-spread function (PSF) [3]. This choice maximizes the signal-to-noise ratio. The value of θ_{min} in Eq. 1 has been chosen to account for the finite dimension of the Moon, which is seen from the Earth as an extended source of 0.25° angular radius. The position of the Moon is

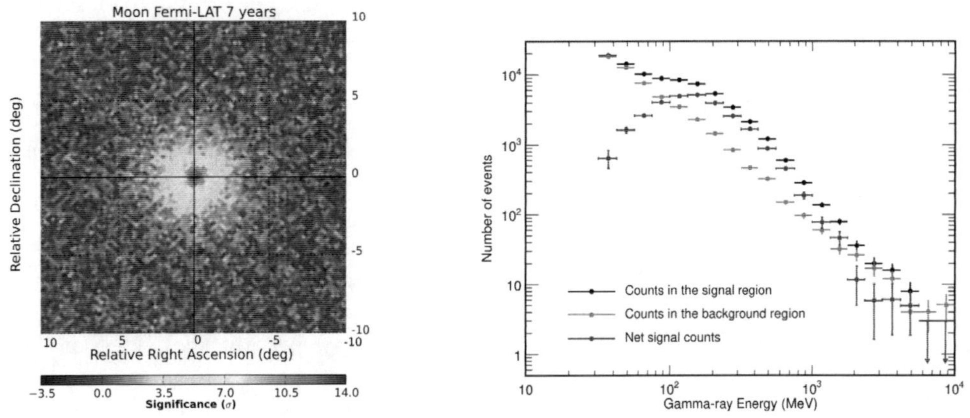

Figure 1. Left panel: significance map of the Moon as a function of right ascension and declination (the coordinates are centered on the Moon). The map is built with photons in the energy range from 30 MeV to 10 GeV. Right panel: photon count spectra in the signal (black circles) and background regions(red circles). The net signal count spectrum (blue circles) is superimposed. The plots are taken from ref. [10].

obtained from its ephemeris using a software interfaced to the JPL libraries [4] and correcting for *Fermi* orbital parallax.

For our analysis we selected the time intervals when the LAT was operating in its standard science operation configuration and was outside the South Atlantic Anomaly (SAA). To avoid contamination from the bright limb of the Earth we discarded the data taken during the times when the angular separation between a cone of angular radius $\theta_{max} = 15°$ centered on the Moon direction (or on the time-offset Moon direction) and the zenith direction exceeded 100°. We also discarded data taken during the times when the center of the signal (background) region was observed with off-axis angles θ larger than 66.4° (i.e. $\cos\theta < 0.4$). Furthermore we selected only the periods where the Moon (or the time-offset Moon) was at a Galactic latitude Finally, we required a minimum angular distance of 20° between the Moon (or the time-offset Moon) and the Sun and between the Moon (or the time-offset Moon) and any celestial source in the 2FGL *Fermi* LAT source catalog [5].

The left panel of Fig. 1 shows the significance map of the Moon, evaluated with photons in the energy range from 30 MeV to 10 GeV. The sky has been divided into 786432 pixels, each one covering a solid angle of $\approx 1.6 \times 10^{-5}$ sr using the HEALPix [6] pixelization tool. The significance of each pixel has been evaluated with the formula of ref. [7], starting from the counts in the signal and in the background regions and taking into account the lifetime ratio between the two regions. A clear peak is found in the center of the map, corresponding to the gamma-ray emission from the Moon.

The right panel of Fig. 1 shows the observed photon count spectra in the signal and background regions, and the net signal count spectrum, which has been calculated by applying in each energy bin the Bayesian procedure illustrated in ref. [8], taking into account the lifetimes of the signal and background regions and assuming uniform priors for the net signal counts in each energy bin.

3. Evaluation of the gamma-ray flux from the Moon

We have reconstructed the energy spectrum of gamma rays from the Moon starting from the observed count spectra in the signal and in the background regions and using an analysis

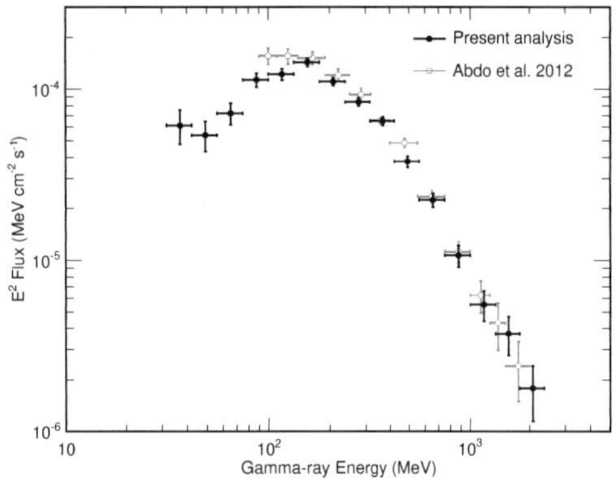

Figure 2. Gamma-ray flux from the Moon in the period from August 2008 to June 2015 compared with the results of ref. [11]. The central values of each bin represent the mean flux values, while the error bars represent the RMSs of the corresponding PDFs. The plot is taken from ref. [10].

method based on the software toolkit BAT [9], which allows to evaluate the posterior probability distribution functions (PDFs) for the parameters of a model. In our analysis we used BAT to evaluate, starting from the count spectra in the signal and background regions, the posterior PDFs for the signal and background gamma-ray fluxes in 15 energy bins. The calculation is performed by maximizing a Poisson likelihood function, in which the the expected counts are evaluated folding the fluxes with the exposure and taking energy dispersion into account. The full details of the analysis are discussed in ref. [10].

Figure 2 shows the reconstructed gamma-ray spectrum of the Moon, compared with the spectrum published in ref. [11], which was obtained from the analysis of the first 2 years of data taken by the *Fermi* LAT. At energies above 150 MeV the present results are consistent with those of ref. [11], while some discrepancies are observed at lower energies, due to the solar modulation effect on CRs. The data sample analyzed in ref. [11] was taken in the first 2 years of LAT operations, when the solar activity was at a minimum; on the other hand, the present data sample corresponds to a period of 7 years, covering more than half of a solar cycle.

4. Time evolution of the lunar gamma-ray emission

To study the time evolution of the lunar gamma-ray emission we have divided our dataset into smaller samples, each corresponding to a period of 6 months. The left panel of Fig. 3 shows the time evolution of the gamma-ray intensity from the Moon measured by the LAT above 56, 75, 100 and 178 MeV. As expected, the gamma-ray intensity from the Moon follows the evolution of the solar cycle. This feature is confirmed when looking at the correlations between the lunar gamma-ray intensity and the data from various neutron monitor stations installed in various locations on the Earth. As an example, in the right panel of Figure 3 it is shown a comparison of the lunar gamma-ray intensity measured by the LAT with the count rates of the McMurdo neutron monitor [12]. Furthermore, as the gamma-ray threshold energy is increased, the correlation with the solar cycle becomes weaker, as gamma rays of higher energies are produced by more energetic cosmic rays, which are not affected by the solar modulation.

5. Evaluation of the low-energy CR spectra and of the solar modulation potential

We have implemented a Monte Carlo simulation of the interactions of CR protons and ^{4}He nuclei with the lunar surface based on the FLUKA toolkit [13]. The Moon is described as a sphere of radius $R = 1737.1$ km, consisting of a mixture of different oxides (45% SiO_2, 22% FeO, 11% CaO, 10% Al_2O_3, 9% MgO, 3% FeO) with a density $\rho = 1.8$ g cm^{-3}.

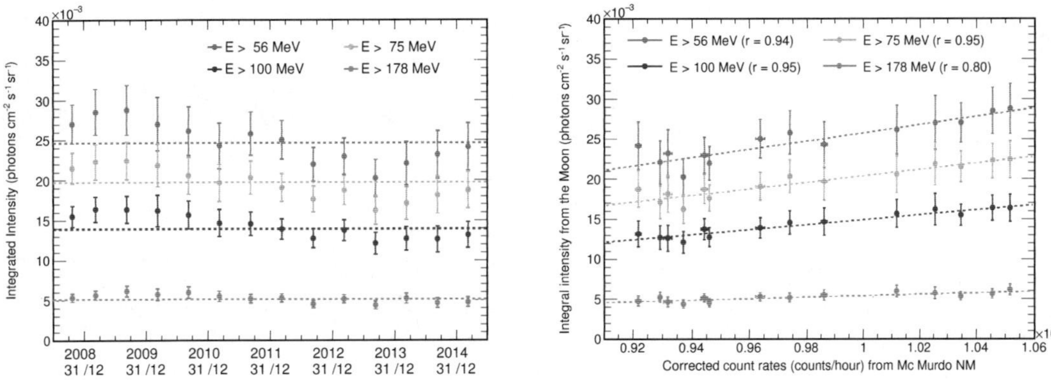

Figure 3. Left panel: Time evolution of the intensity of gamma rays emitted from the Moon above 56, 75, 100 and 178 MeV. The shaded areas indicate the average values over the whole data taking period. Right panel: Study of the correlations between the lunar gamma-ray intensity and the count rate of the McMurdo neutron monitor. The plots are taken from ref. [10].

The gamma-ray flux from the Moon can be expressed as.

$$\phi_\gamma(E_\gamma) = \frac{\pi R^2}{d^2} \sum_{i=p,He} \int Y(E_\gamma|T_i) I(T_i) dT_i \tag{2}$$

where d is the LAT-Moon distance, $I(T_i)$ is the intensity of CRs of the i-th species impinging on the Moon as a function of kinetic energy and $Y(E_\gamma|T_i)$ is the corresponding gamma-ray yield. To check our simulation, we have verified that the lunar gamma-ray flux calculated with eq. 2 using the CR proton and ^{4}He intensities measured by AMS-02 [16, 17] reproduces the LAT data taken in the same epoch.

The gamma-ray yields evaluated from the simulation have been used to fit the Moon data in order to evaluate the solar modulation potential. The fit procedure is based on BAT, and is similar to the one described in sec. 3 for the reconstruction of the gamma-ray fluxes. Here we have assumed a model for the local interstellar spectra (LIS) of CR protons and ^{4}He nuclei [14], and we have assumed that their spectra at the Solar System can be calculated from the LIS in the framework of the force field approximation [15] in terms of a single parameter, the solar modulation potential. In this analysis the parameters to be fitted are the background photon fluxes (which can be treated as nuisance parameters) and the solar modulation potential.

The fit procedure results into an average value of the solar modulation potential of (537 ± 12) MV during the whole data taking period. In the left panel of Fig. 4, the fitted gamma-ray spectrum from the Moon is compared with the measured fluxes in the individual energy bins. The fitted spectrum reproduces the data fairly well, although above 400 MeV it tends to overestimate the measured fluxes. The right panel of Fig. 4 show the modulated CR proton and ^{4}He spectra inferred from this analysis. Both the spectra are consistent with the results from direct measurements performed by PAMELA [18] and AMS-02 [16, 17].

Acknowledgments
The *Fermi* LAT Collaboration acknowledges support for LAT development, operation and data analysis from NASA and DOE (United States), CEA/Irfu and IN2P3/CNRS (France), ASI and INFN (Italy), MEXT, KEK, and JAXA (Japan), and the K.A. Wallenberg Foundation, the

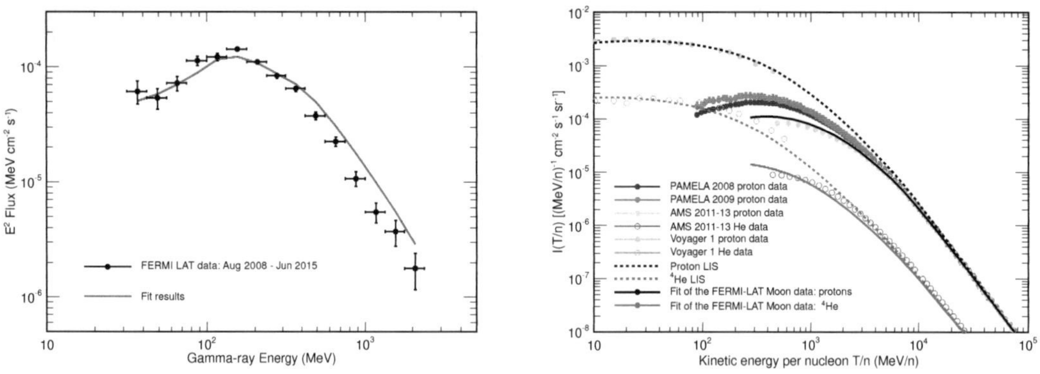

Figure 4. Left panel: gamma-ray flux from the Moon. The fitted spectrum (red line) is compared with the measured fluxes in the individual energy bins (black circles). Right panel: CR proton and ^{4}He spectra obtained from a fit of the *Fermi* LAT Moon gamma-ray data. The result is compared with the direct measurements performed by PAMELA [18] and AMS-02 [16, 17] in different epochs. The Voyager 1 proton and helium data [19] taken at ~ 120 AU from the Sun are also shown. The plots are taken from ref. [10].

Swedish Research Council and the National Space Board (Sweden). Science analysis support in the operations phase from INAF (Italy) and CNES (France) is also gratefully acknowledged. This work performed in part under DOE Contract DE-AC02-76SF00515.

References

[1] Atwood W B *et al.* 2009 *Astrophys. J.* **697** 1071
[2] Atwood W B *et al.* (Fermi-LAT Collaboration) 2013 Fermi Symposium proceedings *eConf C121028* (*Preprint* 1303.3514)
[3] http://fermi.gsfc.nasa.gov/ssc/data/analysis/documentation/Cicerone/Cicerone_LAT_IRFs/IRF_PSF.html
[4] Folkner W M *et al.* 2014 JPL Interplanetary Network Progress Report 42-196 http://ipnpr.jpl.nasa.gov/progress_report/42-196/196C.pdf
[5] Nolan P *et al.* 2012 *Astrophys. J. Supp.* **199** 31
[6] Gorski K M *et al.* 2005 *Astrophys. J.* **622** 759
[7] Li T and Ma Y 1983 *Astrophys. J.* **272** 317
[8] Loparco F and Mazziotta M N 2011 *Nucl. Inst. Meth.* A **646** 167
[9] Caldwell A, Kollar D and Krøninger K 2009 *Comput. Phys. Commun.* **180** 2197
[10] Ackermann M *et al.* 2016 *Phys. Rev.* D **93** 082001
[11] Abdo A A *et al.* *Astrophys. J.* **758** 140
[12] Home page of the University of Delaware Bartol Research Institute Neutron Monitor Program: neutronm.bartol.udel.edu
[13] http://www.fluka.org
[14] Mazziotta M N *et al.* 2016 *Astrop. Phys.* **81** 21
[15] Gleeson L J and Axford W I 1967 *Astrophys. J.* **149** L115
Gleeson L J and Axford W I 1968 *Astrophys. J.* **154** 1011
[16] Aguilar M *el al.* 2015 *Phys. Rev. Lett.* **114** 171103
[17] Aguilar M *et al.* 2015 *Phys. Rev. Lett.* **115** 211101
[18] Adriani O *et al.* 2013 *Astrophys. J.* **765** 91 (*Preprint* 1301.4108)
[19] Stone E C *et al.* 2013 *Science* **341** 150

The "Carpet-3" air shower array to search for diffuse gamma rays with energy E_γ>100TeV

D D Dzhappuev, V B Petkov V I , A U Kudzhaev*, A S Lidvansky, V I Volchenko, G V Volchenko, E A Gorbacheva, I M Dzaparova, N F Klimenko, A N Kurenya, O I Mikhilova, M M Khadzhiev and A F Yanin

Institute for Nuclear Research of the Russian Academy of Sciences, Pr. 60-letiya Oktyabrya 7a, Moscow 117312 Russia

E-mail: kudjaev@mail.ru*

Abstract. At present an experiment for measuring the flux of cosmic diffuse gamma rays with energy higher than 100 TeV (experiment "Carpet-3") is being prepared at the Baksan Neutrino Observatory of the Institute for Nuclear Research, Russian Academy of Sciences. The preparation of the experiment implies considerable enlargement of the area of both muon detector and surface part of the shower array. At the moment the plastic scintillation counters with a total continuous area of 410 m^2 are installed in the muon detector (MD) underground tunnels, and they are totally equipped with electronics. Adjusting of the counters and their electronic circuits is in progress. Six modules of shower detectors (out of twenty planned to be installed) have already been placed on the surface of the MD absorber. A new liquid scintillation detector is developed for modules of the ground –surface part of the array, whose characteristics are presented. It is shown that the "Carpet-3" air shower array will have the best sensitivity to the flux of primary gamma rays with energies in the range 100TeV – 1PeV, being quite competitive in gamma-ray astronomy at such energies.

1. Introduction

The interest to search of primary gamma rays with energies higher than 100 TeV has recently greatly increased in connection with the results of the IceCube experiment, where high energy neutrinos of astrophysical origin were found. In [1] it was suggested that such neutrinos are a result of decays of charge pions in the Galaxy. If so, the neutral pions of the same energy should exist, whose decays produce a considerable flux of gamma rays in the energy range 10^{14} - $5\cdot10^{17}$ eV. Investigation of diffuse gamma radiation at such energies is carried out by the EAS method in experiments in which one can separate the showers from primary protons and nuclei. Such a separation is possible due to the fact that showers produced by primary photons are essentially less abundant with hadrons (and, as result, they are muon-poor) in comparison with showers from primary protons and nuclei. Thus, if one selects hadron poor or muon –poor EAS, there is a hope to effectively distinguish between the showers produced by primary gamma rays and by nuclei. The "Carpet-3" multipurpose shower array which is constructed in the Baksan Neutrino Observatory of INR of RAS, already has the muon detector of an area of 410m^2 with possibility of increasing this area up to 615m^2. For increasing an area of detection for axes of EAS ground-surface modules of scintillation detectors will additionally be placed around the "Carpet". The preliminary estimates show that the new array will has the best sensitivity to the flux of primary gamma rays with energy in region 100TeV-1PeV [2-3].

2. "Carpet-3" array

The "Carpet-3" air shower array (Fig.1) is created on the bases of operating "Carpet-2" array. The Carpet-2 shower array of the Baksan Neutrino Observatory is located in the North Caucasus region near Mount Elbrus at an altitude of 1700m above see level (the atmospheric depth 840 g/cm^2) (Fig.1). Its coordinates are 43°25'N and 42°40'E, the respective vertical cutoff magnetic rigidity being 5.6 GV.

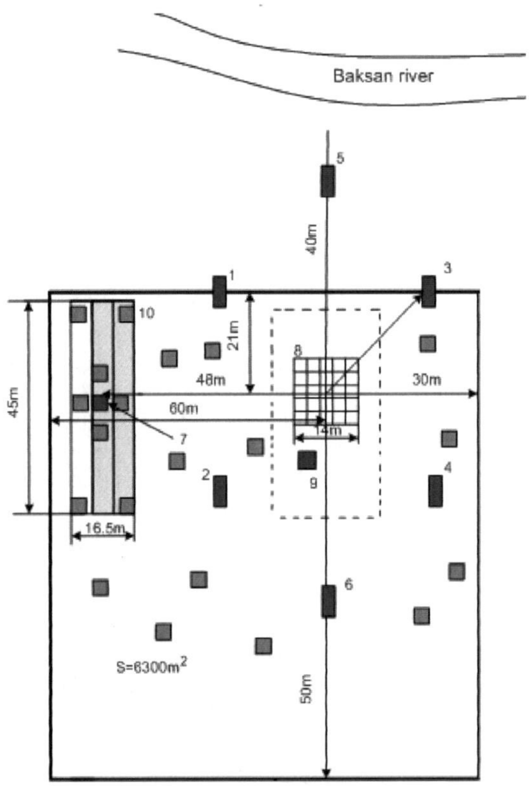

Figure 1. Layout of the "Carpet-3" air shower array: 1-7 are the outdoor huts with scintillators, 8 is the "Carpet", 9 is neutron monitor, 10 is the muon detector and red squares show new modules with scintillation detectors.

The "Carpet-3"array consists of a ground level detector called the "Carpet", six outdoor huts with 9m^{2}of scintillation detectors in each, an underground muon detector and neutron monitor. The twenty modules with liquid scintillation detectors (Fig1, red squares) will be installed on the ground around the "Carpet" array. At present six modules without scintillation detectors are installed on the surface above the MD absorber. The "Carpet" consists of 400 liquid scintillation detectors 0.7×0.7×0.3 m in size each, placed in the form of continuous square that has the side length of 14m and, accordingly, the total area of 196m^2 [4]. The measured range of energy release values in an individual liquid detector of the "Carpet" is 10-5000 relativistic particles (r.p.), where 1 r.p. is the most probable energy release produced by a cosmic ray particle crossing the detector, and it equals 50 MeV. Six outer stations (1-6 in Fig.1) intended for determining shower arrival directions are placed around the central part of the array. Four of them are positioned at a distance of 30m in such a way that square geometry is formed.

The remaining two are situated at a distance of 40m from the geometric center of the array. Each outer station has dimensions of 2.1×4.2 m and contains 18 liquid scintillation detectors, which are similar to those used in the "Carpet". The muon detector is arranged at a distance of 48m from the central part of the array; it is situated in underground tunnel at a depth of 500g/cm^2, which corresponds to the energy threshold of 1GeV[5]. The accuracy of determination of coordinates of EAS axes hitting the "Carpet" is no worse than 0.7m, while the arrival direction of showers are measured with an accuracy of better than ~ 3°.
The MD is an extended plane 5×35 m in size and it consists of 175 standard scintillation counters of 1m^2 area in each that are attached to the ceiling of the underground hall. Each counter comprises four plastic scintillators of size 0.5×0.5×0.05 m placed in a light proof casing. The plastic scintillators are viewed by FEU-49 PM tubes with 12 dynodes. The energy release upon the passage of one relativistic particle through the detector corresponds to about 280 photoelectrons. PM tubes are arranged in groups, five modules, each including 35 detectors.
A signal from the 12th dynode is recorded by a RC logarithmic converter with a threshold of 0.5Aμ generating a standard variable-duration signal delivered to hodoscope. The RC converter used is an amplitude logarithmic encoder whose performance relies on charging a C capacitor with a pulsed current from PM tubes and on its exponential discharge through a resistor R. The converter time constant τ=RC is 1μs. The filling of a converter signal with pulses of frequency 10 MHz ensures a 10% precision of respective measurements. The energy release values measured by the converter lie in the range between 6 and 20000 MeV. Two triggers of the "Carpet-3" array and the proper MD trigger developed by the coincidence scheme upon the actuation of any three of five MD modules are used to record information from the detectors. The view of the second tunnel of MD with the plastic detectors is presented in Fig.2.

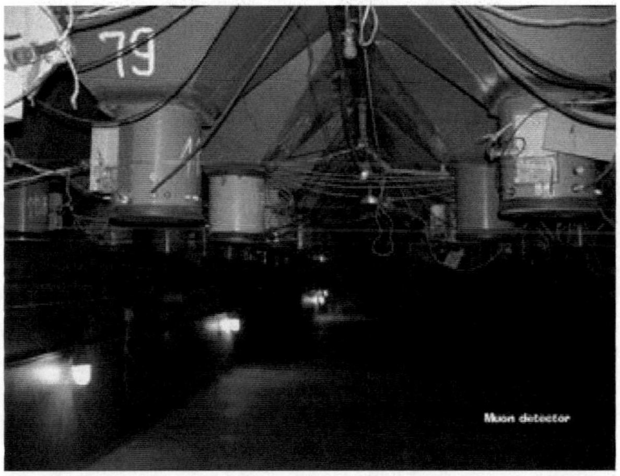

Figure 2. First stage of the Muon Detector, 175m2, in operation since 1999.

Preparation of the experiment suggested that continuous area of MD should be increased at first up to 410m^2 and then up to 615m^2 .In order to enlarge the detection area for EAS axes, additional 20 modules are to be installed, each including 18 liquid scintillation detectors of area of 0.5 m^2 each. At the present stage 410 scintillation counters with total continuous area of 410m^2 are installed in the MD underground tunnels, and they are totally equipped with electronics. The work on adjusting the electronics of scintillation counters and on constructing data acquisition system for the given configuration of MD is in progress. On the surface MD absorber six modules for scintillation counters are installed. They are made of heat-insulated material and supplied with heating elements. In each module the detectors will be installed in three rows and under each one three 200 W heating elements will be placed. The total power of heating elements for a module is 1.8 kW. The electric scheme of thermostatic system will provide stabilization of temperature for detectors at a level of ±0.25°C all the year round, which ensures stabilization of signals from new detectors at a level of 0.1%.

3. The new liquid scintillation detector for modules of "Carpet-3" shower array

For twenty modules of "Carpet-3" shower array a new liquid scintillation detector has been designed (Fig.3). The detector represents a container with dimensions 66x74x30cm, made from veneer with a thickness of 10mm. AT the bottom of the container 10 plastic bottles are placed. The plastic bottles with dimensions 14×14×31 cm each are filled with liquid scintillator using as a solvent the jet engine fuel T-6 with a density 0.82g/cm^3.The inner surface of the container is covered either by white paint (in one case) or by aluminium foil (in another) in order to choose the detector with the best light collection. The amplitudes spectra, taken with the digital oscilloscope LeCroy WaveJet , are shown in Fig.4. The high voltage for a detector is U=-1700V.

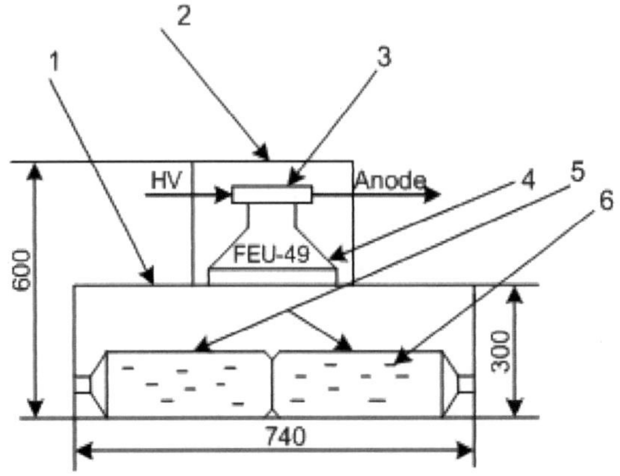

Figure 3. Scheme of the new liquid scintillation detector: 1 – container from veneer, 2 – light protective coat, 3 – divider of high voltage, 4 – PM tube FEU-49, 5 – plastic bottle, 6 – liquid scintillator.

As is seen in fig.4 the maximums of the first and second spectra correspond, respectively, to 8.5mv and 12.5 mV. This means that in the second case the efficiency of light collection is by 32% higher than in the first case. Hence, it follows that it is better to use for the new detector the inner surface covered by aluminium foil.

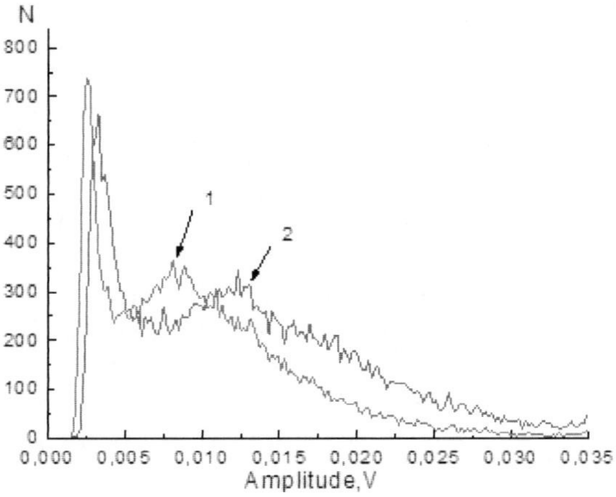

Figure 4. Comparison of amplitude spectra for single muons obtained for the new detector when coating its inner surface with 1 – white paint, 2 – aluminum foil.

The following characteristics were obtained for the new liquid scintillator detector:

- The amplitude of the anode signal is equal to12mV;
- Inhomogeneity of light collection of the detector is27%;
- The total number of photoelectrons at the photocathode is 360;
- Energy resolution equals 45%.

In addition, the cost of the new detector will be several times lower than that of the standard plastic scintillation detector, which is used in the MD of the "Carpet-3" air shower array.

4. Conclusion

Many years ago the experiment aimed at searching for local sources of gamma rays with energy higher than 100 TeV was made with the "Carpet" air shower array of the Baksan Neutrino Observatory of INR of RAS. A burst of gamma radiation with energy $E\gamma \geq 100$ TeV was detected from Crab Nebula [6], while for such possible sources of gamma rays as Cyg X-3, Her X-1, Cyg X-1, Geminga, and 4U115+63 the flux upper limits were obtained. The typical upper limit of the flux for them was $I_\gamma=(1\text{-}5)\times10^{-14}$ [$cm^{-2}\times s^{-1}$] [7]. After that the "Carpet" array was seriously modernized (the"Carpet-2" experiment). Now the new version of gamma - ray astronomy experiment is dedicated to searching for diffuse cosmic gamma ray emission. The "Carpet-3" array will have greatly increased area of the MD and enlarged area of surface detectors. This work is now in progress. A new type of liquid scintillation detector is designed to be used in modules of the "Carpet-3" array in order to increase the area of location of shower axes, thereby increasing the statistics of detected events and decreasing the energy threshold for primary cosmic radiation

Acknowledgments
The work was supported by the Russian Foundation for Basic Research, project numbers 16-02-0687-a and 16-29-13049.

References

[1] Kalashev O E and Troitsky S V 2016 *JETP Lett.* **100** 761(*Preprint* arXiv: 1410.2600)
[2] Dzhappuev D D *et al.*2017 *Izvestia RAS Ser. Fiz .* **81** 461
[3] Dzhappuev D D *et al.* 2017 *J.Phys. Conf. Ser.* **798** 012028
[4] Dzhappuev D D et al. 2007 *Bull. Rus. Acad. Sci.: Phys.* **71** 525
[5] Alekseev E N *et al.*, 1989 The big plastic scintillation detector. Modification of Baksan Underground telescope 8 *Preprint* P-0610 (Russian, INR of RAS, Moscow)
[6] Alekseenko V V *et al.* 1992 *J. Phys. G: Nucl. Part. Phys.* **18** 83
[7] Alekseenko V.V. *et al.* 1989 The "Carpet-2" array for search of local sources of gamma rays with energy higher 10^{14} eV (Russian, KSU, Alma-Ata: Materials of VCCR) part 1 pp 27 - 28

IOP Conf. Series: Journal of Physics: Conf. Series **934** (2017) 012023 doi:10.1088/1742-6596/934/1/012023

Study of cosmic ray sources using data on extragalactic diffuse gamma-ray emission

A Uryson

Lebedev Physical Institute RAS, Moscow

uryson@sci.lebedev.ru

Abstract. We discuss ultra-high energy cosmic rays from minor sources and their possible contribution to the extragalactic diffuse gamma-ray emission. As an illustration of minor sources we consider the possible specific type of active galactic nuclei in which supermassive black hole is surrounded by a super strong magnetic field of 10^{10}-10^{11} Gs. In this model we have calculated cosmic ray energy spectra at the Earth and intensity of cascade quanta produced by cosmic rays in extragalactic space. Proceeding from numerical results and the data by Pierre Auger Observatory and Telescope Array it is shown that these active galactic nuclei make a negligible contribution to the cosmic ray flux at the Earth. However cosmic rays from these active galactic nuclei produce significant gamma-ray flux as compared to extragalactic diffuse gamma-ray emission measured by Fermi LAT. We conclude that ultra-high energy cosmic rays from minor sources contribute noticeably to extragalactic diffuse gamma-ray emission.

1. Introduction

Sources of ultra-high energy (UHE) cosmic rays (CR) seem to be extragalactic. Evidently they are active galactic nuclei (AGN) but they are not still revealed. Source identification by particle arrival directions was not successful. A number of reasons make identification difficult. First, identification is carried out assuming that in extragalactic space particles propagate almost rectilinearly. However UHECRs are presumably deflected in magnetic fields. Next, error-boxes of $\sim 1^0$ around arrival directions result in a large number of astrophysical objects in the area near the particle arrival and this impedes source identification.

In extragalactic space UHECRs lose energy while interacting with the cosmic microwave background. This results in two effects: a suppression of the UHECR energy spectrum if UHECRs come from distances of more than approximately 50 Mpc – GZK-effect [1, 2] and in electromagnetic cascades in extragalactic space [3, 4]. Currently both UHECR spectra and cascade gamma-ray emission are studied to investigate UHECR sources. Data on spectra are obtained with CR giant arrays – Pierre Auger Observatory (PAO) and Telescope Array (TA). To analyse cascade emission Fermi LAT data [5] are used.

The common way to examine UHECR sources is to describe the particle energy spectrum measured and next to analyse emission produced by particles in extragalactic space. The intensity of

IOP Conf. Series: Journal of Physics: Conf. Series **934** (2017) 012023 doi:10.1088/1742-6596/934/1/012023

cascade gamma-quanta is required to be less than the measured intensity of extragalactic diffuse gamma-ray emission (excluding the contribution of unresolved gamma-ray sources). Models with parameters matching this requirement are selected for further analyse (see e.g. [6-8]). Other paragraphs are indented

In this work we consider the possible specific type of AGNs in which supermassive black hole is surrounded by a super strong magnetic field of 10^{10}-10^{11} Gs. Particles are accelerated there to ultra-high energies [9-12]. Here we show that these AGNs contribute negligibly to the UHECR flux at the Earth. However UHECRs generate noticeable diffuse gamma-ray flux that possibly amounts to few tens percent of the flux measured by Fermi LAT (excluding the contribution of unresolved gamma-ray sources). The same result was obtained previously in [13] using semi quantitative estimate. We conclude that UHECRs from minor sources can contribute noticeably to extragalactic diffuse gamma-ray emission. This is important when studying both UHECR source models and dark matter decays that add to diffuse gamma-ray emission.

Computing was performed with the code TransportCR [14].

2. The model

Model assumptions concern following points: injection spectra and cosmic evolution of CR sources, extragalactic background emission, and the extragalactic magnetic field.

We assume that UHECR sources are the points. Next we assume that UHECR sources are AGNs with super strong magnetic fields where protons are accelerated in induced electric fields to energies of 10^{21} eV. Due to acceleration mechanism it is reasonable to assume that the initial CR spectrum is monoenergetic with the energy $E= 10^{21}$ eV.

We assume also that UHECRs consist of protons.

Cosmic evolution of objects discussed is not clear. We analyze two possible cases of source evolution: as the evolution of Blue Lacertae objects (BL Lac's) [14, 15] and as the evolution of radio AGNs described in [16]. In the latter case only source density evolution is considered here (as it is not clear how luminosity of objects with super strong magnetic field varies with redshift).

In the model background emission is treated in the following way.

Microwave background emission has Planck distribution in energy with the mean value $\varepsilon_r = 6.7 \cdot 10^{-4}$ eV, the mean photon density is $n_r = 400$ cm^{-3}.

Extragalactic background light has characteristics described in [17].

For the radio background emission the model [18] with the pure luminosity evolution for radio galaxies is used.

Extragalactic magnetic fields are weak enough so cascade electrons lose negligible energy in synchrotron radiation ([14] and references therein).

3. Results

Calculated UHECR spectra along with the spectrum obtained at PAO are shown in figures 1, 2.
Calculated spectra are normalized to the PAO spectrum at the energy of $10^{19.5}$ eV.

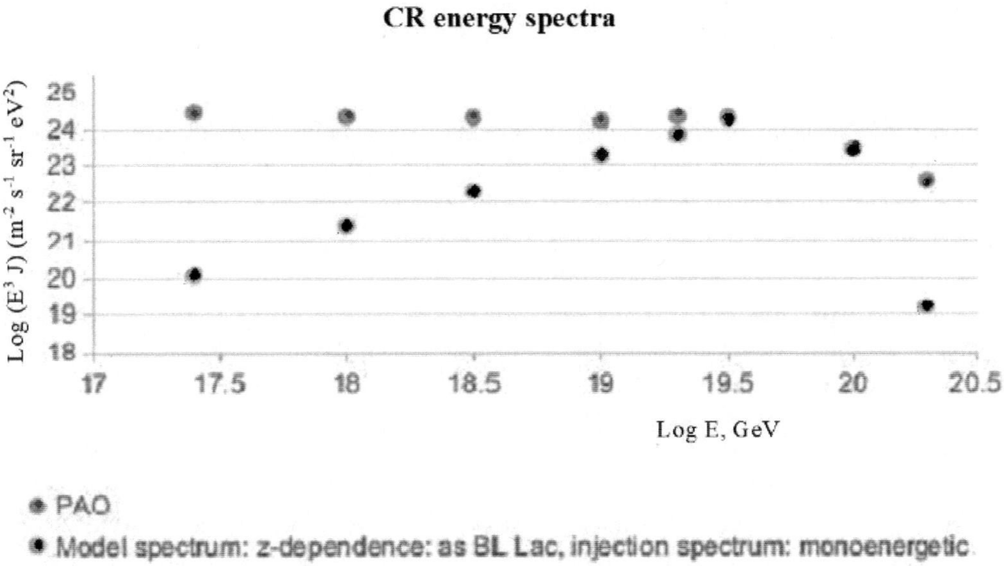

Figure 1. The fit of PAO CR energy spectrum [19] (blue points) and the calculated UHECR spectrum with source z-dependence as of BL LAC's (black points).

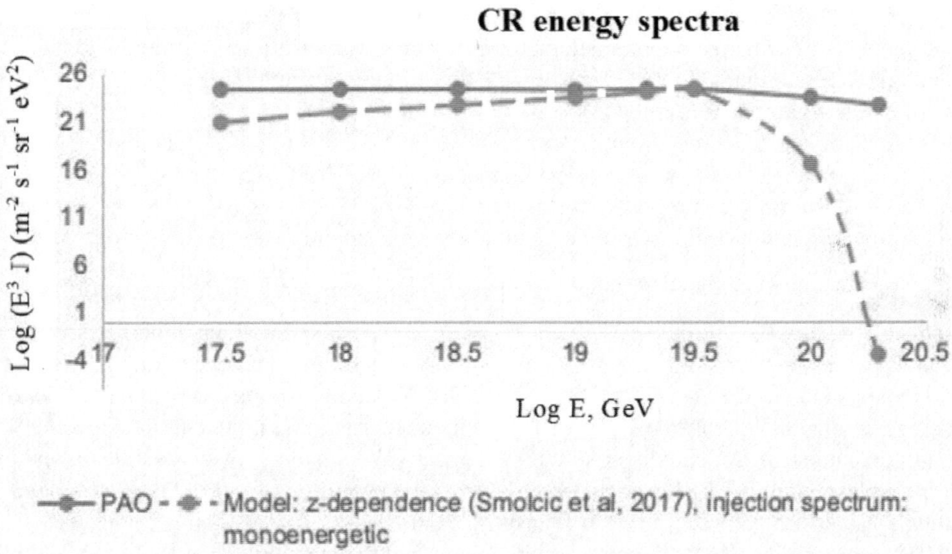

Figure 2. The fit of PAO CR energy spectrum [19] (blue solid line) and the calculated UHECR spectrum with source z-dependence as of radio AGNs with pure density evolution [16] (red dashed line).

In both cases of source evolution model spectra differ significantly in shape from the spectrum measured at PAO (as well as at TA). Spectra calculated are several orders lower than those measured.

We proceed now to the integral intensity of gamma rays above 50 GeV produced by UHECRs in extragalactic space. We choose this energy range as the contribution of unresolved gamma-ray sources is obtained in [20], and the contribution is taken into account when comparing with Fermi LAT data.

The integral intensity of cascade gamma rays is:

$$I_\gamma\ (E>50\ \text{GeV, source } z\text{-dependence: BL Lac's [14, 15]}) = 1.002\cdot10^{-9}\ (\text{cm}^{-2}\ \text{s}^{-1}\ \text{sr}^{-1}), \quad (3.1)$$

I_γ (E>50 GeV, source z-dependence: [16]) =1.641·10^{-9} (cm^{-2} s^{-1} sr^{-1}), (3.2)

The difference in values results from dissimilar z-dependence used in the calculation.

4. Discussion

Now we compare the integral intensity obtained with Fermi LAT data [5]. Truly diffuse extragalactic background – the isotropic diffuse gamma-ray background (IGRB) is produced be electromagnetic cascades which UHECRs and high-energy gamma rays initiate in extragalactic space.

IGRB obtained by Fermi LAT includes emission from unresolved individual extragalactic sources. At energies above 50 GeV their contribution equals to 86 percent [19].

From [5]

$$\text{IGRB} (>50 \text{ GeV}) = 1.325\cdot10^{-9} (\text{cm}^{-2}\ \text{s}^{-1}\ \text{sr}^{-1}) \quad (4.1)$$

and excluding contribution of unresolved sources we obtain

$$\text{IGRB}_{\text{without blazars}} (>50 \text{ GeV}) = 1.85\cdot10^{-10} (\text{cm}^{-2}\ \text{s}^{-1}\ \text{sr}^{-1}). \quad (4.2)$$

This value is much less than those calculated in the model.

Accounting for measurement errors, uncertainties in the model of diffuse galactic emission (which is used to obtain IGRB from Fermi LAT data), and uncertainty in the value of unresolved source contribution (the contribution can be 12 percent lesser) the value of IGRB $_{\text{without blazars}}$ (>50 GeV) lies in the band:

$$2.20\cdot10^{-10} (\text{cm}^{-2}\ \text{s}^{-1}\ \text{sr}^{-1}) \le \text{IGRB}_{\text{without blazars}} (>50 \text{ GeV}) \le 5.40\cdot10^{-10} (\text{cm}^{-2}\ \text{s}^{-1}\ \text{sr}^{-1}). \quad (4.3)$$

The values calculated in the model are several times higher as before.

In the model integral intensity of cascade gamma rays is derived normalizing UHECR spectra at the energy of 10$^{19.5}$ eV. However minor sources give negligible contribution to the UHECR flux. Thus it is mote natural to normalize spectra calculated to another value. As an example we choose *ad hoc* the normalizing value equals about 20 percent of the previous normalizing value. Then the integral intensity of cascade gamma rays equals:

I_γ (E>50 GeV, source z-dependence: BL Lac's [14, 15]) = 2.0·10^{-10} (cm^{-2} s^{-1} sr^{-1}), (4.4)

I_γ (E>50 GeV, source z-dependence: [16]) = 3.3·10^{-10} (cm^{-2} s^{-1} sr^{-1}), (4.5)

These values fall within the interval of measured IGRB (4.3) leaving room for the contribution of UHECRs from dominant sources to diffuse gamma-ray background.

5. Conclusion

We consider UHECRs from the possible specific type of AGNs in which supermassive black hole is surrounded by a super strong magnetic field of 10^{10}-10^{11} Gs. In sources protons are accelerated in induced electric fields to the energy of 10^{21} eV [9, 10, 12]. Due to the mechanism of acceleration it is reasonable to assume that the initial CR spectrum is monoenergetic with the energy E= 10^{21} eV.

Cosmic evolution of AGNs discussed is not clear. We analyze two possible cases of AGN evolution: as the evolution of BL Lac's [14, 15] and as the evolution of radio AGNs described in [16]. Computing was performed with the code TransportCR [14].

UHECR spectra calculated are several orders lower than those measured at PAO and TA. However CRs from AGNs discussed appear to produce significant gamma-ray flux – of several tens percent of IGRB measured by Fermi LAT [5].

AGNs discussed illustrate possible role of UHECRs from minor sources in contributing to diffuse gamma-ray emission. We conclude that UHECRs from minor sources can give noticeable part of the extragalactic diffuse gamma-ray background. This is important when studying both UHECR source models and dark matter decays that add to diffuse gamma-ray emission.

Acknowledgment
The author is grateful to O. Kalashev for discussion.

References

[1] Greisen K 1966 *Phys. Rev. Lett.* **16 748**
[2] Zatsepin G T and Kuzmin V A 1966 *J. Exp. Theor. Phys. Lett.* **4** 78-80

[3] Hayakawa S 1966 *Progr. Theor. Phys.* **37** 594
[4] Prilutsky O F and Rozental I L 1970 *Acta Phys. Hung. Suppl.* I **29** 51
[5] Ackermann M *et al.* 2015 *Astrophys. J.* **799** 1 (*Preprint* 1410.3696v1)
[6] Giacinty G *et al.* 2015 Unified model for cosmic rays above 10^{17} eV and the diffuse gamma-ray and neutrino backgrounds *Preprtint* 1507.07534v2
[7] Gavish E and Eichler D 2016 On ultra high energy cosmic rays and their resultant gamma rays *Preprint* 1603.04074
[8] Berezinsky V, Gazizov A and Kalashev O 2016 Cascade photons as test of protons in UHECR *Preprint* 1606.09293v2
[9] Kardashev N S 1995 *Mon. Not. R. Astron. Soc.* **276** 515
[10] Shatsky A A and Kardashev N S 2002 *Astronomy Rep.* **46** 639
[11] ZakharovA F, Kardashev N S, Lukash V N and Repin S V 2003 *Mon. Not. R. Astron. Soc.* **342** 1325
[12] Uryson A V 2004 *Astronomy Lett.* **30** 816
[13] Uryson A V 2017 *Astronomy Lett.* **43** 529
[14] Kalashev O E and Kido E 2015 *J. Exp. Theor. Phys.* **120** 790
[15] Di Mauro M *et al.* 2014 *Astrophys. J.* **786** 129 (*Preprint* 1311.5708)
[16] Smolcic v *et al.* 2017 The VLA-COSMOS 3 GHz Large Project: Cosmic evolution of radio AGN and implications for radio-mode feedback since z 5 *Preprint* 1705.07090
[17] Inoue Y *et al.* 2013 Extragalactic Background Light from Hierarchical Galaxy Formation: Gamma-ray Attenuation up to the Epoch of Cosmic Reionization and the First Stars *Preprint* 1212.1683
[18] Protheroe R J and Biermann P L 1996 *Astropart. Phys.* **6** 45
[19] Verzi V, Ivanov D and Tsunesada Y 2017 Measurement of Energy Spectrum of Ultra-High Energy Cosmic Rays *Preprint* 1705.09111v1
[20] Ackermann M *et al.* 2016 *Phys. Rev. Lett.* **116** 151105 (*Preprint* 1511.00693)

IOP Conf. Series: Journal of Physics: Conf. Series **934** (2017) 012024 doi:10.1088/1742-6596/934/1/012024

Search for dark matter particle candidates produced in association with a Z boson in pp collisions at a center-of-mass energy of 13 TeV with the ATLAS detector

A Basalaev, on behalf of the ATLAS Collaboration

National Research Centre "Kurchatov Institute" B.P.Konstantinov Petersburg Nuclear Physics Institute, St. Petersburg, Russia

E-mail: artem.basalaev@cern.ch

Abstract. A search for dark matter particle candidates produced in association with a Z boson in proton-proton collisions at the total center-of-mass energy of 13 TeV is presented. The search uses 36.1 inverse femtobarn of data collected by the ATLAS experiment at the Large Hadron Collider in 2015 and 2016. Events with large missing transverse momentum and consistent with the decay of a Z boson into oppositely charged electron or muon pairs were selected in the analysis. Background estimates and corresponding systematic uncertainties are shown. Exclusion limits on the dark matter candidate and mediator masses are reported.

1. Introduction

Astrophysical measurements from several independent sources indicate that the majority of the mass of the Universe exists in the form of non-baryonic dark matter (DM) [1]. However, its nature remains undiscovered and no evidence exists for its non-gravitational interactions [2]. DM is widely hypotheised to consist of weakly interacting massive particles (WIMPs) [3]. Experiments at the Large Hadron Collider (LHC) search for WIMPs production associated with a detectable final state, where a WIMP can be identified by a signature involving missing transverse momentum $E_{\mathrm{T}}^{\mathrm{miss}}$ [4]. Presented is a search for WIMP candidates produced in association with a Z boson with the ATLAS detector [5] at the LHC. Events with a large $E_{\mathrm{T}}^{\mathrm{miss}}$ and two oppositely charged electrons or muons consistent with a decay of a Z boson were selected in the analysis.

The results were interpreted in terms of simplified DM models, where WIMP production is mediated by a vector or an axial-vector particle (mediator). The exclusion limits on the WIMP mass (m_{χ}) and the mediator mass (m_{med}) were set for fixed coupling constants of the mediator to quarks (g_q) and to WIMPs (g_{χ}). The WIMP pair is produced through the s-channel exchange of an axial-vector or vector mediator. This choise provides a useful framework for comparison with direct detection experiments[1], furthermore, LHC searches can be more sensitive than direct

[1] Direct detection experiments aim to observe recoil of DM particles against nuclei, they typically contain large volumes of scintillator and are situated deep underground to reduce background contamination.

searches to WIMP production in this particular model with an axial-vector mediator [4]. The leading tree-level diagram for WIMP production is presented on figure 1.

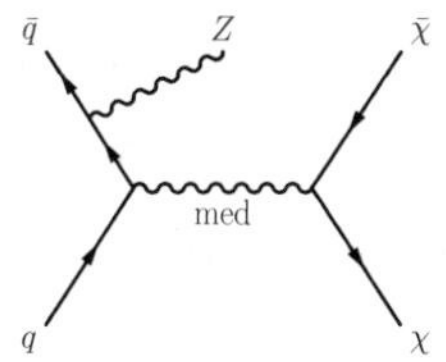

Figure 1. Leading tree-level diagram for WIMP production in simplified DM models [6].

Monte Carlo (MC) simulation of DM signal events with a vector and axial-vector mediator and fermionic WIMPs was performed for m_{med} and m_χ, ranging from 10 to 1000 GeV. Following recommendations in [7], the coupling constants were set to $g_q = 0.25$ and $g_\chi = 1$, and mediator width was set to be minimal, such that it decays only to DM or quarks.

Full description of the analysis and the results can be found in [6].

2. Event Selection and Background Estimation

Electron candidates are reconstructed from isolated energy deposits in the electromagnetic calorimeter, matched to inner detector tracks [8]. Muon candidates are reconstructed from a combined fit of tracks independently found in the muon spectrometer and inner detector [9]. Jets are reconstructed using the anti-k_t algorithm [10] with a radius parameter $R = 0.4$. Basic kinematic cuts along with some additional selections are also applied to reduce the probability of misidentification of the reconstructed objects [6].

The invariant mass of a lepton pair is required to be in the range 76 GeV$\leq \text{M}_{\ell\ell} \leq$ 106 GeV to reject background processes with two leptons that do not originate from the prompt decay of a Z boson (non-resonant-$\ell\ell$). Events are required to pass quality checks for errors in sub-detectors during recording. A combination of a lower p_{T} threshold trigger with an isolation requirement and a higher p_{T} threshold trigger without any isolation requirement is used, yielding the trigger efficiency above 98%. To suppress the non-resonant-$\ell\ell$ background further, candidate events are required to have $E_{\text{T}}^{\text{miss}} > 90$ GeV and $E_{\text{T}}^{\text{miss}}/H_{\text{T}} > 0.6$ where H_{T} is calculated as the scalar sum of the p_{T} of the selected leptons and jets. $E_{\text{T}}^{\text{miss}}$ is expected to be back-to-back with the Z boson. This is accounted for in the $\Delta\phi(Z, E_{\text{T}}^{\text{miss}})$ variable, which is required to be greater than 2.7 radians, and the selected leptons must be close to each other, which is ensured by $|\Delta R_{\ell\ell}| = \sqrt{(\Delta\phi_{\ell\ell})^2 + (\Delta\eta_{\ell\ell})^2} < 1.8$. Any possible imbalance, implying the presence of fake $E_{\text{T}}^{\text{miss}}$, is accounted for by an upper threshold on the fractional p_{T} difference variable, defined as $(p_T^{\ell\ell} - |\vec{E}_{\text{T}}^{\text{miss}} + \sum \vec{p}_{\text{T}}^{\text{jet}}|)/p_T^{\ell\ell} < 0.2$. Events are removed if they contain b-jets to suppress top-quark pair background. Events containing a third lepton with $p_{\text{T}} > 7$ GeV, satisfying looser identification requirements, are removed to suppress diboson background.

The dominant background process (59%) is $ZZ \to \ell^+\ell^-\nu\bar{\nu}$, it is estimated using MC simulation, main uncertainties on it originate from parton density function choice, perturbative calculation, and parton shower modelling. The other important background (25%) is $WZ \to \ell\nu\ell^+\ell^-$, which is contributing if a lepton from W boson decay is undetected, or a τ decays hadronically. It is estimated from data using 3-lepton control region; main uncertainties consist of systematic uncertainties on the ratio of events in control and signal region and statistical uncertainties on data. The non-resonant-$\ell\ell$ background (8%) is estimated from data using the absence of signal in the $e\mu$ channel and the relative production rate of 1 : 1 : 2 for the ee, $\mu\mu$, and $e\mu$ channels. An $e\mu$ control region similar to the signal region is defined, and the background estimate for the ee and $\mu\mu$ signal regions is obtained from the number of $e\mu$ events in the control region after correcting for different acceptances and efficiencies. Main uncertainties consist of systematic uncertainties on the bias of the method and statistical uncertainties on data. The Z + jets background (8%) is estimated from data using a set of two uncorrelated variables, constructing 4 regions (one signal region, A, and three Z-enriched control regions, B,C,D) and using a relation $N_A = N_B \times N_C/N_D$ to estimate the yield in the signal region. The W+jets (estimated using fake-factor method), VVV and $t\bar{t}V(V)$ (estimated using MC simulation) backgrounds have minor contributions ($\leq 1\%$).

3. Results and Conclusions

Table 1. Observed data yields and expectations for the signal and background contributions in the signal region with statistical and second systematic uncertainties. The DM signal contribution with $m_{\text{med}} = 500$ GeV and $m_\chi = 100$ GeVis scaled with a factor of 0.27 to the best-fit contribution. The background contributions from the W+jets, VVV and $t\bar{t}V(V)$ processes are summed up [6].

Final State	ee	$\mu\mu$
Observed Data	437	497
Signal		
DM (m_{med} = 500 GeV, m_χ = 100 GeV) × 0.27	10.8 ± 0.3 ± 0.8	11.1 ± 0.3 ± 0.8
Backgrounds		
$qqZZ$	212 ± 3 ± 15	221 ± 3 ± 17
$ggZZ$	18.9 ± 0.3 ± 11.2	19.3 ± 0.3 ± 11.4
WZ	106 ± 2 ± 6	113 ± 3 ± 5
Z+jets	30 ± 1 ± 28	37 ± 1 ± 19
Non-resonant-$\ell\ell$	30 ± 4 ± 2	33 ± 4 ± 2
W+jets, VVV and $t\bar{t}V(V)$	1.4 ± 0.1 ± 0.2	2.5 ± 2.0 ± 0.8
Total Background	399 ± 6 ± 34	426 ± 6 ± 28

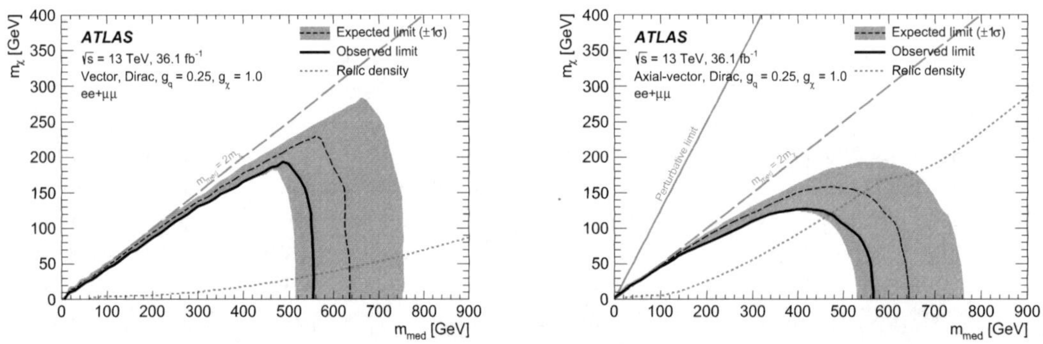

Figure 2. DM exclusion limits in the two-dimensional phase space of WIMP mass m_χ vs mediator mass m_{med} for a fermionic WIMP, vector (left) or axial-vector (right) mediator with coupling constants of mediator to quarks and to DM $g_q = 0.25$ and $g_\chi = 1$, correspondingly [6]. Regions bounded by the limit curves are excluded at the 95% CL.

The search was performed on 36.1 fb^{-1} of ATLAS dataset collected in 2015 and 2016 at $\sqrt{s} = 13$ TeV. Theoretical and experimental systematic uncertainties are estimated for all studied backgrounds, including uncertainties associated with pileup reweighting, luminosity calculation, object and $E_{\text{T}}^{\text{miss}}$ reconstruction algoritms and methods of background estimation. Data yields

IOP Conf. Series: Journal of Physics: Conf. Series **934** (2017) 012024 doi:10.1088/1742-6596/934/1/012024

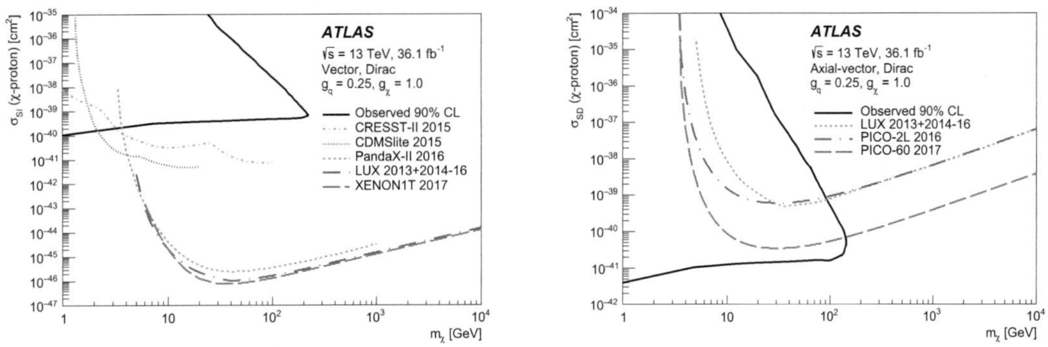

Figure 3. Limits on the WIMP and proton scattering cross section as a function of WIMP mass, obtained for simplified DM models with a vector mediator (left) and an axial-vector mediator (right) [6]. The solid black line shows the observed limit at the 90% CL from this search. Following results of direct-search experiments are shown: LUX [11], PICO-2L [12], PICO-60 [13], CRESST-II [14], CDMSlite [15], PandaX-II [16], and XENON1T [17].

with associated statistical and systematic uncertainties are shown in table 1. There is a small data excess in the $\mu\mu$ channel which corresponds to a significance of about 2.2σ. When combining the ee and $\mu\mu$ channels, the significance is 1.5σ. No significant deviations from the Standard Model predictions have been observed.

The 95% C.L. DM exclusion limits in the two-dimensional phase space of WIMP mass m_χ vs mediator mass m_{med} are presented on figure 2. Additionally the limits translated into χ-nucleon scattering cross section and compared with direct DM searches are shown on figure 3.

References

[1] Komatsu E *et al.* 2011 *Astrophys. J. Suppl. Ser.* **192** 18
[2] Olive K A *et al.* (Particle Data Group) 2014 *Chin. Phys.* **C38** 090001
[3] Lee B W and Weinberg S 1977 *Phys. Rev. Lett.* **39** 165
[4] Malik S A *et al.* 2015 *Phys. Dark Univ* **9-10** 51
[5] ATLAS Collaboration 2008 *JINST* **3** S08003
[6] ATLAS Collaboration 2017 Search for an invisibly decaying Higgs boson or dark matter candidates produced in association with a Z boson in pp collisions at $\sqrt{s} = 13$ TeV with the ATLAS detector *Preprint* arXiv:1708.09624 [hep-ex]
[7] Abercrombie D *et al.* 2015 Dark Matter Benchmark Models for Early LHC Run-2 Searches: Report of the ATLAS/CMS Dark Matter Forum *Preprint* arXiv:1507.00966 [hep-ex]
[8] ATLAS Collaboration 2015 Electron identification measurements in ATLAS using $\sqrt{s} = 13$ TeV data with 50 ns bunch spacing Tech. Rep. ATL-PHYS-PUB-2015-041 CERN Geneva
[9] ATLAS Collaboration 2016 *Eur. Phys. J. C* **76** 292
[10] Cacciari M, Salam G P and Soyez G 2008 *J. High Energy Phys.* **04** 063
[11] Akerib D S *et al.* 2017 *Phys. Rev. Lett.* **118**(2) 021303
[12] Amole C *et al.* 2015 *Phys. Rev. Lett.* **114**(23) 231302
[13] Amole C *et al.* 2017 *Phys. Rev. Lett.* **118** 251301
[14] Angloher G *et al.* 2016 *Eur. Phys. J. C* **76** 25
[15] Agnese R *et al.* 2016 *Phys. Rev. Lett.* **116**(7) 071301
[16] Tan A *et al.* 2016 *Phys. Rev. Lett.* **117**(12) 121303
[17] Aprile E *et al.* 2017 First Dark Matter Search Results from the XENON1T Experiment *Preprint* arXiv:1705.06655 [astro-ph.CO]

IOP Conf. Series: Journal of Physics: Conf. Series **934** (2017) 012025 doi:10.1088/1742-6596/934/1/012025

Measurements, status and plans of the TOTEM experiment at the LHC

F Ferro on behalf of the TOTEM Collaboration
INFN Genova, via Dodecaneso 33, 16146 Genova, Italy
E-mail: `fabrizio.ferro@ge.infn.it`

Abstract. The TOTEM collaboration at the LHC has measured the elastic, inelastic and total proton-proton cross sections at several center of mass energies and is carrying on a rich program of measurements of diffractive physics together with the CMS collaboration. The talk will review the TOTEM measurements mainly focusing on the newest and most significant ones. The status of the experimental apparatus, its latest changes and the current and future technological challenges will be discussed as well.

1. Introduction
The TOTEM experiment has been designed to measure the total proton-proton cross-section, elastic scattering and diffractive processes at the LHC [1]. The experimental apparatus of TOTEM is composed of three subdetectors: the Roman Pots (RP) and the T1, T2 inelastic forward telescopes. The detectors are placed symmetrically on both sides of the Interaction Point 5 (IP5), which is shared with the CMS experiment. The RPs are moveable beam-pipe insertions, hosting edgeless silicon detectors to detect leading protons scattered at very small angles. In order to maximize the acceptance of the experiment for elastically scattered protons, the RPs are able to approach the beam center to a transverse distance as small as 1 mm. Before the first LHC long shutdown (LS1) the RPs, used for measurements, were located at distances of 215−220 m from IP5, while the current layout is different in RP location and quantity. The RP stations previously installed at ±147 m from IP5 have been relocated at ±210 m. Moreover, two newly designed horizontal RPs have been installed between the two units of the ±220 m station [2].

2. Total cross section and elastic scattering
For each tagged elastic event the four-momentum transfer squared t is reconstructed using the LHC optical functions, characterized with the so-called betatron amplitude at IP5 β^*. The reconstruction of the kinematic parameters of the interacting protons is done using a novel experimental method that estimates the optical functions at the RP locations, using the measured elastically scattered protons [3, 4, 5, 6, 7]. The total inelastic rate N_{inel}, measured by the T1 and T2 telescopes, and the total nuclear elastic rate N_{el} with its extrapolation to zero four-momentum transfer squared $t = 0$ are combined with the optical theorem to obtain the total cross-section in a luminosity independent way

$$\sigma_{tot} = \frac{16\pi}{1+\rho^2} \frac{dN_{el}/dt(t=0)}{N_{el}+N_{inel}}$$

where ρ is the ratio of the real to the imaginary part of the forward hadronic scattering amplitude. The ρ value can be taken from data driven theoretical extrapolations by the COMPETE collaboration [8].

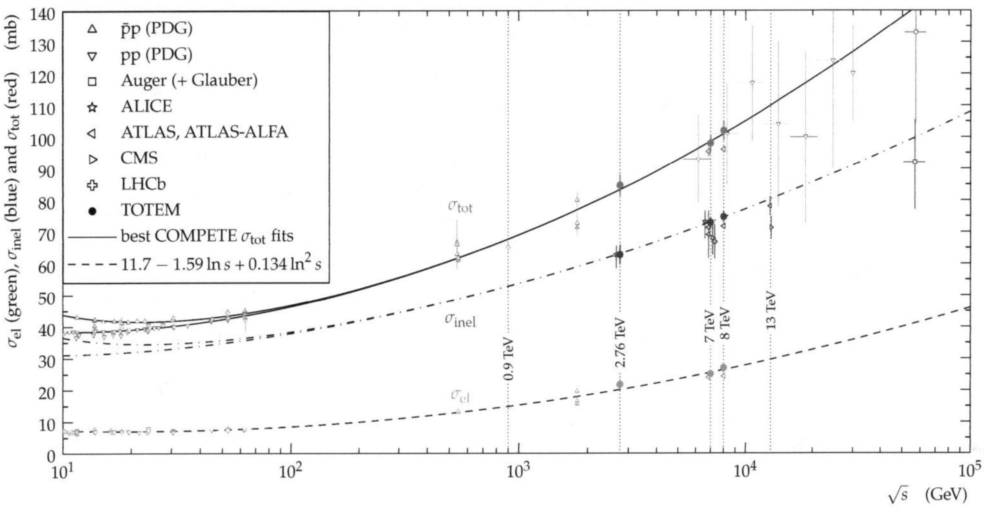

Figure 1. Total (red), inelastic (blue) and elastic (green) cross-section measurements plotted as a function of energy. The solid lines represent the pre-LHC fits by COMPETE [8]. The dashed line shows a $\sigma_{\rm el}$ fit by TOTEM, including its 7 and 8 TeV and PDG [9] points. The dashed-dotted lines corresponds to the difference between the $\sigma_{\rm tot}$ fits by COMPETE and the $\sigma_{\rm el}$ fit by TOTEM.

The total pp cross-section at $\sqrt{s} = 7$ TeV determined using the luminosity independent method [7] was shown to be consistent with the total cross-sections measured in independent ways, see table 1.

Table 1. Total cross section at $\sqrt{s} = 7$ TeV measured with different methods: luminosity independent [7], ρ independent [7], from elastic scattering only [3, 5].

	$\mathcal{L}$ ind.	ρ ind.	Elastic only
σ_{tot}	98.0±2.5 mb	99.1±4.3 mb	98.3±2.8 mb, 98.6±2.2 mb

The elastic and inelastic cross-sections were also separately measured: $\sigma_{el} = 25.1\pm1.1$ mb and $\sigma_{inel} = 72.9\pm1.5$ mb. The measurement was repeated at $\sqrt{s} = 8$ TeV, yielding to $\sigma_{tot} = 101.7\pm2.9$ mb, $\sigma_{el} = 27.1\pm1.4$ mb, $\sigma_{inel} = 74.7\pm1.7$ mb [10] and at $\sqrt{s} = 2.76$ TeV, yielding to $\sigma_{tot} = 84.7\pm3.3$ mb, $\sigma_{el} = 21.8\pm1.4$ mb, $\sigma_{inel} = 62.8\pm2.9$ mb. A compilation of the results is shown in figure 1, which also demonstrates that the observed cross-sections are in agreement with low-energy data and cosmic ray results as well. Moreover, an energy dependence of the total cross section like $\sigma_{tot} \propto ln^2 s$ seems to be favored.

The nuclear part of the differential elastic cross section at low t is usually described as an exponential function of t. Actually, the acquisition of a very-high statistics data set at $\sqrt{s} = 8$

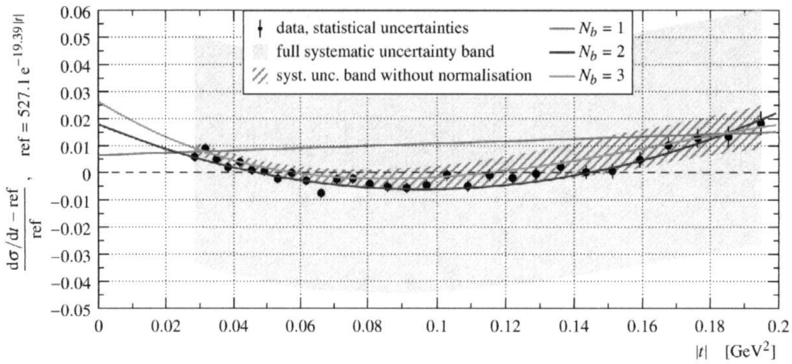

Figure 2. Differential cross-section measured at $\sqrt{s} = 8$ TeV LHC energy plotted as relative difference from a reference exponential. The black dots represent data points with statistical uncertainty bars. N_b is the degree of the t polynomial used for the fit.

TeV energy with $\beta^* = 90$ m allowed the TOTEM collaboration to exclude a purely exponential behavior [11]. The exclusion significance is greater than 7σ in the $|t|$ range from 0.027 to 0.2 GeV^2, see figure 2. Refined parametrizations for the extrapolation to the optical point ($t = 0$) yield total cross-section values $\sigma_{tot} = 101.5 \pm 2.1$ mb (with quadratic polynomial in the exponent) and $\sigma_{tot} = 101.9 \pm 2.1$ mb (with cubic polynomial in the exponent), compatible with the previous TOTEM measurement.

The TOTEM experiment performed its first measurement of elastic scattering in the Coulomb-nuclear interference (CNI) region [12]. The data have been collected at $\sqrt{s} = 8$ TeV with a special beam optics with $\beta^* = 1000$ m which allowed the detection of protons with t as low as $t_{min} = 6 \cdot 10^{-4}$ GeV^2. Using these data the ρ parameter was for the first time measured at the LHC: $\rho = 0.12 \pm 0.03$. The full historical series of ρ measurements fitted by the COMPETE collaboration show that the values measured by TOTEM fit the theoretical expectations. The measurements at $\sqrt{s} = 13$ TeV, expected to be published soon, might shed new light on the interpretation of the elastic scattering; for example, a ρ value significantly lower than expected might result in a first evidence of the Odderon at the LHC energies [14].

The measurement of the differential elastic cross section in a wide range of t confirmed the so-called shrinkage of the forward peak with energy, which consists in a decreasing of the t values corresponding to the dip position. It made also possible the comparison of the data with several theoretical models. None of the considered models was able to describe the data in the full t range, moreover, since no structure at large t was observed, some models were ruled out.

3. Charged particle multiplicity

The charged particle pseudorapidity density $dN_{ch}/d\eta$ has been measured by CMS and TOTEM at $\sqrt{s} = 7$ TeV and 8 TeV as shown in figure 3 [15, 16]. The TOTEM measurement was done in the pseudorapidity range $5.3 < |\eta| < 6.4$ using the T2 telescope. Moreover, some runs done with a displaced vertex with respect to the nominal position resulted in a modified acceptance for T2 ($3.7 < \eta < 4.8$ and $-7 < \eta < -6$) [17] widening the pseudorapidity range where data were available. All the analysis were done with data triggered with T2, requiring at least one reconstructed track in one of the two T2 arms. The main contribution to the errors is given by the systematics, such as, for T2, the subtraction of secondary particles, the track efficiency and the misalignment uncertainties.

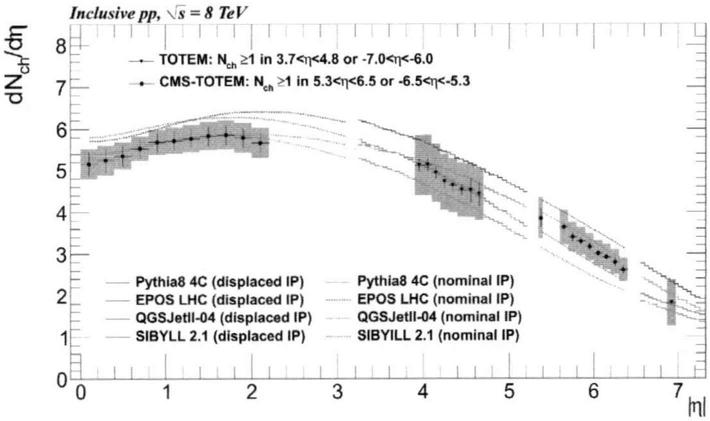

Figure 3. Charged particle distribution as a function of pseudorapidity η measured by CMS and TOTEM. The predictions made by some of the most used Monte Carlo generators are also shown as reference.

4. TOTEM and CMS combined analyses

The CMS and TOTEM experiments share the same interaction point, can take data independently but can also share the triggers and are capable of combining the data offline. The combined TOTEM and CMS detector system has an unprecedented coverage: the CMS tracker and calorimeters cover $|\eta| < 5.5$, while TOTEM T1 and T2 $3.1 < |\eta| < 4.7$ and $5.3 < |\eta| < 6.5$ rapidity ranges, respectively, and the TOTEM RP detectors provide the reconstruction of the leading protons.

The processes that mostly profit from this experimental setup are the so-called processes of central exclusive production (CEP) $pp \to pXp$. They are characterized by the production of a central system, which may be studied in the central (CMS+T1+T2) detectors, and by the two scattered protons that survived the interaction and that can be detected in the RP detectors.

A key ingredient for the study of this kind of processes is the measurement of the mass m_X of the central system X that can be determined both by the direct reconstruction in the central detector and by the reconstruction of the kinematic parameters of the scattered protons: $m_X = \sqrt{s\xi_1\xi_2}$ where $\xi_{1,2}$ are the fractional momentum losses of the two protons. Moreover, the proton survival imposes selection rules for the X system ($J_X^{PC} = 0^{++}, 2^{++}, ...$) that improves the QCD background suppression and would readily permit a clean determination of the quantum numbers of any new resonance.

Since at low momentum transfer LHC becomes a sort of gluon-gluon collider, CEP is also ideal to study low mass resonances and glueballs. Resonances in the mass range $m_X \sim 1-4$ GeV (e.g. f_0, f_2) can be produced very purely in CEP and their decay modes can be characterized quite accurately. CMS and TOTEM collaborations are pursuing these studies and publications are expected soon.

5. The CT-PPS project

The present and future of the TOTEM collaboration is tightly joint with that of CMS also for what concerns the measurements of low cross-section processes which need continuous data taking at nominal LHC luminosities. TOTEM baseline program has been carried on with short periods of data taking during dedicated LHC runs with special optics. In the last years TOTEM developed an upgrade program together with CMS in order to build a proton spectrometer that could operate also at high luminosities, the CT-PPS project.

The CMS-TOTEM Precision Proton Spectrometer [18] focuses on the study of rare central diffractive states, which can be produced at the LHC by $\gamma\gamma$ or Pomeron-Pomeron and $\gamma-$Pomeron interactions. In particular, it will put new constraints on the anomalous quartic gauge coupling of the Standard Model vector bosons and measure the exclusive production of $\gamma\gamma$, WW and ZZ bosons with a better precision with respect to the results obtained by CMS or ATLAS without the proton tagging. For each side of the interaction point (IP), the spectrometer consists of two tracking stations, one TOTEM silicon strip and one 3D pixels [19]. Moreover each side of the IP has also one timing station equipped with 3 scCVD diamonds planes [20] and one UFSD plane [21]. The stations will intercept protons produced in central diffractive events with masses from ~300 GeV up to ~2 TeV. For the first time, timing detectors are used at a collider to measure the time of flight of the protons, with an ultimate expected precision of 20 ps. Timing detectors are indeed necessary to reconstruct the longitudinal position of the diffractive protons, which can be associated unambiguously to one of the vertices reconstructed by the CMS central detectors even in high pileups runs. The experiment started to take data in June 2016 and collected about 15 fb^{-1} of data before the TS1 of the LHC. During 2017 the experiment collected additional 15 fb^{-1} both with the tracking and the timing detectors and it is still taking data. The collected data sample already constitutes the largest hard central diffraction sample ever collected in high energy physics. Several analyses are currently ongoing as the searches of exclusive dileptons or $\gamma\gamma$ and the measurements of central diffraction events with large missing masses.

6. Conclusions

TOTEM has taken data at the LHC at $\sqrt{s}$ =2.76, 7, 8 and 13 TeV accurately measuring the proton-proton total, inelastic and elastic cross section and the elastic scattering in a wide range of t. With CMS, TOTEM measured the charged multiplicity in a large pseudorapidity range and is studying the diffractive processes with a special focus on the Central Exclusive Production. An ambitious upgrade program has been carried on in the last years with CMS to build a precision proton spectrometer which could operate at nominal LHC luminosity. CT-PPS is currently taking data and the first physics results are expected to be published soon.

References

[1] The TOTEM Collaboration 2004 *TDR* CERN-LHCC-2004-02
[2] The TOTEM Collaboration 2013 *TOTEM Upgrade Proposal* CERN-LHCC-2013-009
[3] TOTEM Collaboration 2011 *EPL* **96** 21002
[4] TOTEM Collaboration 2011 *EPL* **95** 41001
[5] TOTEM Collaboration 2013 *EPL* **101** 21002
[6] TOTEM Collaboration 2013 *EPL* **101** 21003
[7] TOTEM Collaboration 2013 *EPL* **101** 21004
[8] COMPETE Collaboration 2002 *Phys. Rev. Lett.* **89** 201801
[9] Particle Data Group 2014 *Chin. Phys.* C **38** 090001
[10] TOTEM Collaboration 2013 *Phys. Rev. Lett.* **111** 012001
[11] TOTEM Collaboration 2015 *Nucl. Phys.* B **899** 527
[12] TOTEM Collaboration 2016 *Eur. Phys. J.* C **76** 661
[13] TOTEM Collaboration 2013 *Phys. Rev. Lett.* **111** 262001
[14] Nicolescu B *et al.* 2007 *12th International Conference on Elastic and Diffractive Scattering (Blois Workshop) — Forward Physics and QCD* DESY-PROC-2007-02 pp 173–180
[15] TOTEM Collaboration 2012 *EPL* **98** 31002
[16] The CMS and TOTEM Collaborations 2014 *Eur. Phys. J.* C **74** 3053
[17] TOTEM Collaboration 2015 *Eur. Phys. J.* C **75** 126
[18] The CMS and TOTEM Collaborations 2014 *CTPPS TDR* CERN-LHCC-2014-021
[19] Ravera F 2016 *JINST* **11** C11027
[20] TOTEM Collaboration 2017 *JINST* **12** P03007
[21] Berretti M *et al.* 2017 *JINST* **12** P03024

IOP Conf. Series: Journal of Physics: Conf. Series **934** (2017) 012026 doi:10.1088/1742-6596/934/1/012026

Search for Multi-quark Exotic States with Heavy Flavor at D0 Experiment

A Popov
NRC "Kurchatov Institute" — IHEP, Protvino
E-mail: Alexei.Popov@ihep.ru

Abstract. We present the results for multi-quark exotic states search from D0 Collaboration at the FNAL Tevatron. This includes an evidence for a $B_s^0\pi^\pm$ state ($X(5568)$) with hadronic decays of B_s meson, a confirmation of the $X(5568)$ state with semileptonic decays of B_s meson, and a search for exotic baryons decaying to $J/\psi\Lambda$ pairs.

1. Introduction

Since the creation of the quark model it was understood that exotic mesons containing more than one quark-antiquark pairs and exotic baryons containing more than three quarks are possible. It was also understood that multi-quark states that contain heavy quarks can be more recognizable owing to the distinctive decay structure of heavy quark hadrons. The 2003 discovery by the Belle experiment [1] of the $X(3872)$ in the channel $B^\pm \to K^\pm X(\to \pi^+\pi^- J/\psi)$ was the first accepted exotic meson in which heavy flavor quarks participate. From that time more than 20 multi-quark exotic states candidates have been discovered. We present the recent results for such search from D0 collaboration.

2. Evidence for a $B_s^0\pi^\pm$ state

In 2016 the D0 collaboration presented evidence for a possible four-quark state with the decay to $B_s^0\pi^\pm$ where $B_s^0 \to J/\psi\phi$ [2]. A study of the $B_s^0\pi^\pm$ invariant mass spectrum was performed using a data sample of 10.4 fb^{-1} collected with the D0 detector at the Fermilab Tevatron collider.

Events used in this analysis are collected both with single muon and dimuon triggers. They are required to include a pair of oppositely charged muons both with $p_T > 1.5$ GeV/c in the invariant mass range $2.92 < m(\mu^+\mu^-) < 3.25$ GeV/c^2 consistent with J/ψ decay, accompanied with two additional oppositely charged particles assumed to be kaons, each with $p_T > 0.7$ GeV/c, with an invariant mass $1.012 < m(K^+K^-) < 1.03$ GeV/c^2, consistent with ϕ decay, and a third charged particle with $p_T > 0.5$ GeV/c assumed to be a pion. To form a $B_s^0\pi^\pm$ combination, we select the B_s^0 candidates in the mass range $5.303 < m(J/\psi\phi) < 5.423$ GeV/c^2. To improve the resolution of the invariant mass of the $B_s^0\pi^\pm$ system we define this mass as $m(B_s^0\pi^\pm) = m(J/\psi\phi\pi^\pm) - m(J/\psi\phi) + m(B_s^0)$, where $m(B_s^0) = 5.3667$ GeV/c^2. We study events as a function of mass in the range $5.5 < m(B_s^0\pi^\pm) < 5.9$ GeV/c^2. To suppress background the $B_s^0\pi^\pm$ system is required to have $p_T > 10$ GeV/c. To further reduce background we impose a limit on the difference between the directions of the B_s^0 candidate and the pion to be $\Delta R = \sqrt{\Delta\eta^2 + \Delta\phi^2} < 0.3$, where η is the pseudorapidity and ϕ is the azimuthal angle. In addition to increasing the signal-to-background ratio this "cone cut" limits backgrounds that are not included in available simulations.

The $B_s^0\pi^\pm$ background with a real B_s^0 meson is modeled using a Monte Carlo (MC) simulation. The background with a false B_s^0 meson is modeled using the sideband events obtained from data. The two background components are found to have similar shapes and are mixed in the right proportion to obtain the combined background. It was modeled by a function of the parameter $m_0 = m(B_s^0\pi^\pm) - \Delta, \Delta = 5.5$ GeV$/c^2$, of the form

$$F_{\rm bgr}(m) = (c_1 + c_2 \cdot m_0^2 + c_3 \cdot m_0^3 + c_4 \cdot m_0^4) \cdot e^{c_5+c_6\cdot m_0+c_7\cdot m_0^2}, \quad (1)$$

This empirical function gives a good description of the combined backgrounds.

The $B_s^0\pi^\pm$ invariant mass spectrum is shown in Fig. 1 a) with the cone cut and b) without the cone cut. An enhancement is seen near 5.57 GeV$/c^2$. To extract the signal parameters, the distributions are fitted with a function

$$F(m) = f_{\rm sig} \cdot F_{\rm sig}(m, M_X, \Gamma_X) + f_{\rm bgr} \cdot F_{\rm bgr}(m), \quad (2)$$

where $F_{\rm sig}(m, M_X, \Gamma_X)$ is a signal function, M_X and Γ_X are the mass and natural width of the resonance and $f_{\rm sig}, f_{\rm bgr}$ are normalization factors. The shape parameters in the background term $F_{\rm bgr}$ are fixed to the values obtained from fitting the combined background distribution. Signal function is modeled by a relativistic Breit-Wigner function convolved with a Gaussian detector resolution function (with $\sigma_{\rm res} = 3.8$ MeV$/c^2$ taken from simulations) and the mass-dependent efficiency of the cone cut. The fit shown in Fig. 1a yields the mass and width of $M_X = 5567.8 \pm 2.9$ MeV$/c^2, \Gamma_X = 21.9 \pm 6.4$ MeV$/c^2$, and the number of signal events of $N = 133 \pm 31$. The local statistical significance of the signal, defined as $\sqrt{-2\ln(\mathcal{L}_0/\mathcal{L}_{\rm max})}$, where $\mathcal{L}_0, \mathcal{L}_{\rm max}$ are likelihood values at the signal yield fixed to zero and at the best-fit signal yield obtained from a binned maximum likelihood fit, is 6.6σ. With the look-elsewhere effect [3] taken into account, the global statistical significance is 6.1σ.

We obtain the systematic uncertainties for the measured values of the $X(5568)$ state mass, width, and the number of signal events. The dominant uncertainties are due to the background and signal shapes. When the systematic uncertainties are included, the

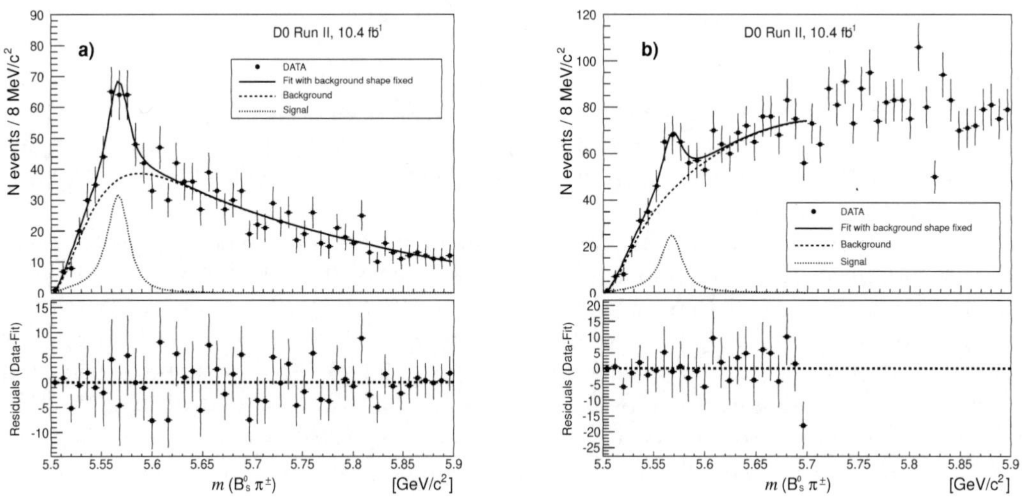

Figure 1. The $m(B_s^0\pi^\pm)$ distribution with the background distribution and the fitting function superimposed a) after applying $\Delta R < 0.3$ cone cut and b) without the cone cut.

significance of the observed signal, including the look-elsewhere effect, is reduced to 5.1σ. For the analysis without the ΔR cut (Fig. 1b) we obtain a significance including the systematic uncertainty and look-elsewhere effect of 3.9σ. Note, that the fit range in this case is reduced to $m(B_s^0\pi^\pm) = 5.7$ GeV/c^2 due to the possible background processes not present in our MC calculations, which lead to a tendency for data to exceed background at higher masses. The mass and natural width of the $X(5568)$ state with obtained systematic uncertainties are $M_X = 5567.8 \pm 2.9(\text{stat.})^{+0.9}_{-1.9}(\text{syst.})$ MeV/c^2 and $\Gamma_X = 21.9 \pm 6.4(\text{stat.})^{+5.0}_{-2.5}(\text{syst.})$ MeV/c^2.

The ratio of the yield of the new state $X(5568)$ to the yield of the B_s^0 meson is measured to be $[8.6 \pm 1.9(\text{stat.}) \pm 1.4(\text{syst.})]\%$.

A possible interpretation of the observed structure is a four-quark state made up of a diquark-antidiquark pair. With the $B_s^0\pi^\pm$ produced in an S wave, its quantum numbers would be $J^P = 0^+$. On the other hand, the state can decay trough the chain $B_s^*\pi^\pm, B_s^* \to B_s^0\gamma$, where the low-energy photon is not detected. In this case, the quantum number of this state would be $J^P = 1^+$. The mass of the new state would be shifted by addition of the nominal mass difference $m(B_s^*) - m(B_s^0)$, while its width would remain unchanged. If this structure is interpreted as a tetraquark state, it would be the first state with four different valence quark flavors b, s, u, d.

3. Confirmation of the $X(5568)$ with semileptonic decays of B_s meson

Subsequent analysis by LHCb [4] and CMS [5] have not confirmed the existence of $X(5568)$ in pp interactions at $\sqrt{s} = 7$ and 8 TeV/c^2. Thus, the additional studies of $B_s^0\pi^\pm$ final state are very important.

We present here a search for $X(5568)$ in the decay to $B_s^0\pi^\pm$ using semileptonic B_s^0 decays, $B_s^0 \to \mu^\mp D_s^\pm + X$ (charge conjugate states are assumed) from the full Run II integrated luminosity of 10.4 fb^{-1} in $p\bar{p}$ collisions. Here X includes the unseen neutrino and possibly other hadrons from the B_s^0 decay. The background in semileptonic channel are mostly independent of those in the hadronic channel. The presence of the neutrino in the final state leads to the wider mass resolution for the signal. The character of possible reflections of other resonant structures is quite different in the semileptonic and hadronic channels. Thus, a study of $X(5568)$ in the semileptonic decay channel may provide an independent confirmation of its existensc.

The muons are required to have $3 < p_T < 25$ GeV/c. The $D_s^- \to \phi\pi^-, \phi \to K^+K^-$ is reconstructed as follows. The two particles from ϕ decay are assumed to be kaons and are required to have $p_T > 1.0$ GeV/c, opposite charge and mass $1.012 < m(K^+K^-) < 1.03$ GeV/c^2. The charge of the third particle, assumed to be a pion, has to be opposite to that of the muon. The three tracks are combined to form a common D_s^- vertex. The D_s^- and B_s^0 decay vertices should be well separated from the primary vertex. The transverse momentum of the $\mu^+D_s^-$ system is required to satisfy the condition $p_T > 10$ GeV/c. To minimize the effect of the missing neutrino in the final state the effective mass is limited to 4.5 GeV/$c^2 < m(\mu^+D_s^-) < m(B_s^0)$. A track representing the pion in the $B_s^0\pi^\pm$ combination is required to have transverse momentum $0.5 < p_T(\pi) < 25$ GeV/c. To improve the resolution of the invariant mass of the $B_s^0\pi^\pm$ system we define the invariant mass as $m(B_s^0\pi) = m(\mu^\mp D_s^\pm\pi) - m(\mu^\mp D_s^\pm) + m(B_s^0)$, where $m(B_s^0) = 5.3667$ GeV/c^2. We study events as a function of mass in the range $5.506 < m(B_s^0\pi^\pm) < 5.906$ GeV/c^2. The ΔR cone cut was not used in this analysis.

The background parametrization was taken from the Monte-Carlo (MC) background sample generated using the PYTHIA [6] inclusive jet production model. To correct for the effects of the trigger selection and the reconstruction on the data, we weight each MC event so that the transverse momentum of the reconstructed muon and the μD_s system agree with thouse in the data. The $m(B_s^0\pi^\pm)$ invariant mass distribution of the MC background is modeled by:

$$F_{\text{bgr}}(m) = (c_1 \cdot m_0 + c_2 \cdot m_0^2 + c_3 \cdot m_0^3 + c_4 \cdot m_0^4) \cdot e^{c_5 \cdot m_0 + c_6 \cdot m_0^2}, \quad (3)$$

where $m = m(B_s^0\pi^\pm)$, $m_0 = m - m_{\text{thr}}$, and $m_{\text{thr}} = 5.5063$ GeV$/c^2$ is the $B_s^0\pi^\pm$ mass threshold. Several alternative parametrizations were also used to check the fit stability and for background shape systematics estimation.

The signal is modeled by a relativistic Breit-Wigner function convolved with a Gaussian detector resolution function which also takes into account the impact of the unseen neutrino. The fit function has the form similar to shown in Eq. 2. In the fit shown in Fig. 2 (left plot) the normalization parameters f_{sig}, f_{bgr} and the Breit-Wigner parameters M_X, Γ_X are allowed to vary. The fit yields the mass and natural width of $M_X = 5566.7^{+3.6}_{-3.4}(\text{stat.})^{+1.0}_{-1.0}(\text{syst.})$ MeV$/c^2$, $\Gamma_X = 6.0^{+9.5}_{-6.0}(\text{stat.})^{+1.9}_{-4.6}(\text{syst.})$ MeV$/c^2$, and the number of signal events of $N = 139^{+51}_{-63}(\text{stat.})^{+11}_{-32}(\text{syst.})$. These numbers include the systematic uncertainties due to background shape description, background reweighting, B_s^0 mass scale in MC and data, detector resolution and the missing neutrino effect, signal modeling. The dominant systematic uncertainty is due to the background shape description. The local statistical significance of the signal is 4.5σ. If the systematic uncertainties are taken into account, the statistical significance of the signal is 3.2σ.

The ratio of the number of B_s^0 which come from the decay of $X(5568)$ to all B_s^0 ($X(5568)$ production ratio) was measured to be $[7.3^{+2.8}_{-2.4}(\text{stat.})^{+0.6}_{-1.7}(\text{syst.})]\%$ in the semileptonic channel.

The mass, natural width and $X(5568)$ production ratio are in the agreement between the semileptonic and hadronic channels within uncertainties [2]. If we assume that we observe the same object in both channels and the semileptonic and hadronic measurements are independent, we can calculate the combined significance for the two channels using the formula to calculate the combined p-value: $p_{\text{comb}} = p_{\text{had}} \cdot p_{\text{sl}} \cdot [1 - \ln(p_{\text{had}} \cdot p_{\text{sl}})]$, where p_{had} is the p-value for the hadronic channel, and p_{sl} is the p-value for the semileptonic channel. The obtained combined p-value corresponds to a significance of 5.7σ for the combination of the hadronic channel with the ΔR cut and the semileptonic channel (4.7σ without ΔR cut in the hadronic channel). More details about this analysis can be found at [7].

4. Search for exotic baryons decaying to $J/\psi\Lambda$

Recently the LHCb collaboration has reported an observation of two $J/\psi p$ resonances named P_c at masses 4380 MeV$/c^2$ and 4450 MeV$/c^2$, consistent with pentaquark states, in $\Lambda_b^0 \to J/\psi K^- p$ decays (charge conjugate states are assumed) [8]. Both states are within 400 MeV$/c^2$ of the $J/\psi p$ threshold. Numerous states with the quark contents including a $c\bar{c}$ pair and three light quarks are expected to exist within 500 MeV$/c^2$ of the appropriate threshold. This section presents a search for inclusive production of pentaquarks with hidden charm in the final state $J/\psi\Lambda$, with $J/\psi \to \mu^+\mu^-$ and $\Lambda \to p\pi^-$. Such states could be produced promptly, directly in

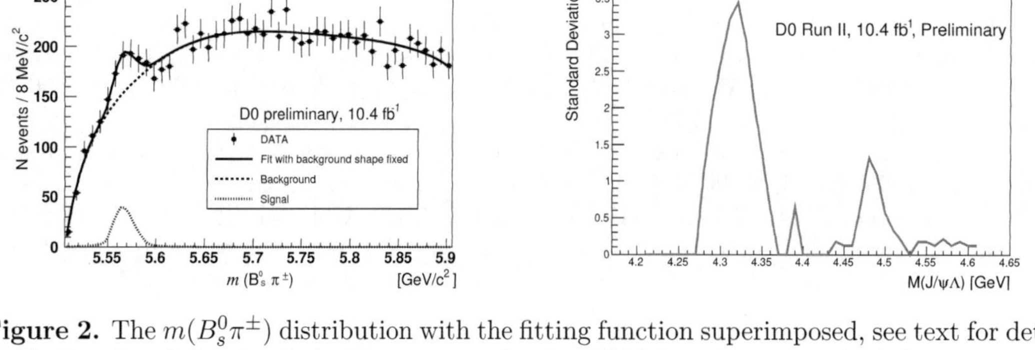

Figure 2. The $m(B_s^0\pi^\pm)$ distribution with the fitting function superimposed, see text for details (left). Results of the signal significance scan for the $J/\psi\Lambda$ invariant mass (right).

$p\bar{p}$ collisions by QCD processes or non-promptly, in weak decays of b-baryons. In the following we focus on the non-prompt production.

Events used in this analysis are selected from a Run II data sample of 10.4 fb^{-1}. To reconstruct a J/ψ candidate, two muons with the opposite charge, transverse momentum $p_T(\mu) > 1.0$ GeV/c, total transverse momentum $p_T(\mu\mu) > 4.0$ GeV/c, originated from a common vertex, and have an invariant mass in the range $2.92 < m(\mu\mu) < 3.25$ GeV/c^2, are selected. The reconstruction and selection of $\Lambda \to p\pi^-$ candidates is similar to that in the D0 measurement of the Λ_b^0 lifetime [9]. To select $J/\psi\Lambda$ candidates, we define the Λ signal window as $1.110 - 1.122$ GeV/c^2, the lower limit on the Λ transverse momentum is set to 0.7 GeV/c.

We define the search mass region as the range from the $J/\psi\Lambda$ threshold to 4.7 GeV/c^2 and perform binned maximum likelihood fits to the distributions of the $J/\psi\Lambda$ invariant mass with the function, defined in Eq. 2. The signal is modeled by a Gaussian function with mass M_X and width (standard deviation) Γ_X. The background is modeled with the following function:

$$F_{\mathrm{bgr}}(m) \propto m \cdot (m^2/m^2_{\mathrm{thr}} - 1)^{c_1} \cdot e^{-c_2 m} \cdot (1 - e^{-(m - m_{\mathrm{thr}})/c_3}), \quad (4)$$

where m stands for $m(J/\psi\Lambda)$ and m_{thr} is its threshold value.

Using the signal and background models described above, we perform mass fits of the sum of signal and background or background only to the data, with the signal mass set at fixed values in 10 MeV/c^2 steps. The signal width and the background shape parameters c_1, c_2, c_3 are varied. In each mass point the local statistical significance is determined. The result of this scan is showed on Fig. 2 (right plot). The highest local significance of 3.45σ occurs at $m(J/\psi\Lambda) = 4.32$ GeV/c^2. If looks-elsewhere effect is taken into account, the corresponding global significance is estimated to be 2.8σ. We thus have no evidence for new baryons decaying to $J/\psi\Lambda$. More details about this analysis can be found at [10].

References

[1] Choi S K *et al.* 2003 *Phys. Rev. Lett.* **91** 262001
[2] Abazov V M *et al.* 2016 *Phys. Rev. Lett.* **117** 022003
[3] Gross E and Vitells O 2010 *Eur. Phys. J.* C **70** 525
[4] Aaij R *et al.* 2016 *Phys. Rev. Lett.* **117** 152003
[5] CMS Collaboration 2016 CMS-PAS-BPH-16-002 (Geneva: CERN)
[6] Sjöstrand T, Mrenna S and Skands P 2006 *JHEP* **05** 026
[7] *https://www-d0.fnal.gov/Run2Physics/WWW/results/prelim/B/B68/*
[8] Aaij R *et al.* 2015 *Phys. Rev. Lett.* **115** 072001
[9] Abazov V M *et al.* 2012 *Phys. Rev.* D **85** 112003
[10] *https://www-d0.fnal.gov/Run2Physics/WWW/results/prelim/B/B69/*

Study of $K^+ \to \pi^0 e^+ \nu_e \gamma$ decay with OKA setup.

A.Yu. Polyarush
(on behalf of OKA collaboration)
Institute for Nuclear Research of the Russian Academy of Sciences, Moscow, Russia
E-mail: polyarush@inr.ac

Abstract. Results of study of the $K^+ \to \pi^0 e^+ \nu \gamma$ decay at OKA setup are presented. 13118 events of this decay have been observed. The branching ratio with cuts $E^*_\gamma > 10$ MeV, $0.6 < cos\Theta^*_{e\gamma} < 0.9$ is calculated $R = \frac{Br(K^+ \to \pi^0 e^+ \nu_e \gamma)}{Br(K^+ \to \pi^0 e^+ \nu_e)} = (0.59 \pm 0.02(stat.) \pm 0.03(syst.)) \times 10^{-2}$. For the asymmetry A_ξ we get $A_\xi = -0.019 \pm 0.020(stat.) \pm 0.027(syst.)$

1. Introduction

The decay $K^+ \to \pi^0 e^+ \nu \gamma$ provides fertile testing ground for the Chiral Perturbation Theory (ChPT) [1,2]. $K^+ \to \pi^0 e^+ \nu \gamma$ decay amplitudes are calculated at order ChPT $O(p^4)$ in [1], and branching ratios are evaluated in [3]. Recently, next-to-leading $O(p^6)$ corrections were calculated for the corresponding neutral kaon decay [4].

The matrix element for $K^+ \to \pi^0 e^+ \nu \gamma$ has general structure

$$T = \frac{G_F}{\sqrt{2}} e V_{us} \varepsilon^\mu(q) \Big\{ (V_{\mu\nu} - A_{\mu\nu}) \overline{u}(p_\nu) \gamma^\nu (1-\gamma_5) v(p_l)$$
$$+ \frac{F_\nu}{2 p_l q} \overline{u}(p_\nu) \gamma^\nu (1-\gamma_5)(m_l - \hat{p_l} - \hat{q}) \gamma_\mu v(p_l) \Big\} \equiv \epsilon^\mu A_\mu. \quad (1)$$

First term of the matrix element describes Bremsstrahlung of kaon and direct emission. The lepton Bremsstrahlung is presented by the second part of Eq.(1). This decay is especially interesting as it is sensitive to T–odd contributions. According to CPT theorem, observation of T violation is equivalent to observation of CP–violating effects. Important experimental observable used in CP violation searches is the T–odd correlation, for $K^- \to \pi^0 e^- \overline{\nu} \gamma$ decay it is defined as

$$\xi = \frac{1}{M_K^3} p_\gamma \cdot [p_\pi \times p_e]. \quad (2)$$

First suggestion to investigate T–odd triple-product correlations was done in [5]. To establish the presence of a nonzero triple-product correlations, one constructs a T–odd asymmetry of the form

$$A_\xi = \frac{N_+ - N_-}{N_+ + N_-}, \quad (3)$$

where N_+ and N_- are numbers of events with $\xi > 0$ and $\xi < 0$.

2. OKA setup

OKA collaboration operate at IHEP Protvino U-70 Proton Synchrotron of NRC "Kurchatov Institute"-IHEP, Protvino. OKA detector (see figure 1) is located in positive RF-separated beam with 12.5% of K-meson. The detailed description of the OKA detector is given in our previous publications [6, 7].

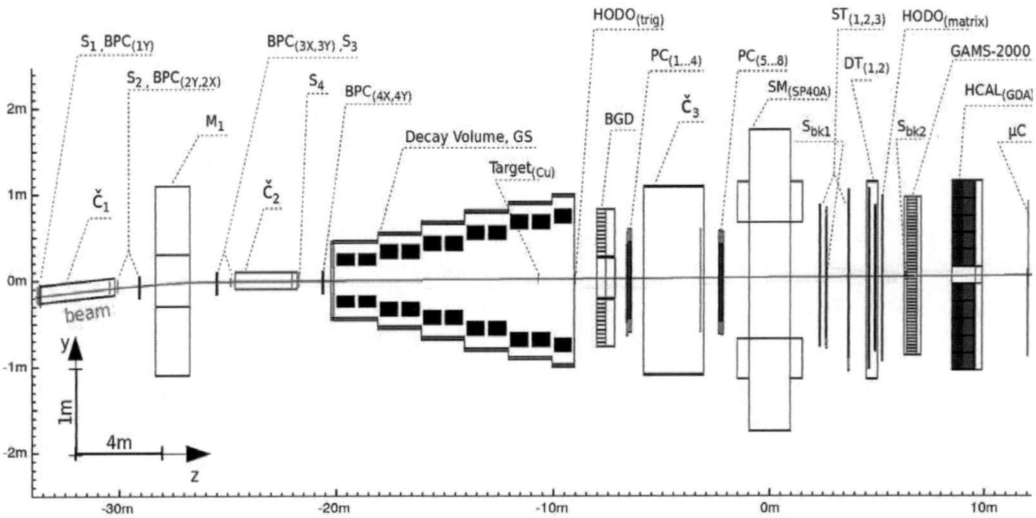

Figure 1. Layout of the OKA detector

3. $K^+ \to \pi^0 e^+ \nu \gamma$ events selection

A study of the $K^+ \to \pi^0 e^+ \nu \gamma$ decay is done with the data set accumulated in November 2012 run with a 17.7 GeV/c beam momentum. The Monte-Carlo simulation based on Geant3 package [8] includes a realistic description of the experimental setup. To select $K^+ \to \pi^0 e^+ \nu \gamma$ decay channel a set of requirements is applied:

(1) One positive charged track detected in tracking system and 4 showers detected in electromagnetic calorimeters GAMS- 2000 and EGS.
(2) One shower must be associated with charged track.
(3) Charged track identified as positron. The electron identification is done using the ratio of the energy of the shower to the momentum of the associated charged track. The particles with 0.8< $E/p < 1.2$ are accepted as electrons.
(4) Vertex situated within the decay volume.
(5) The effective mass $M_{\gamma\gamma}$ for one $\gamma\gamma$ – pair is $0.12 < M_{\gamma\gamma} < 0.15$ GeV.

Absence of signals in veto system above noise threshold is required.

4. Background suppression

The main background decay channels for the decay $K^+ \to \pi^0 e^+ \nu_e \gamma$ are:

(1) $K^+ \to \pi^0 e^+ \nu$ with extra photon. The main source of additional photons is an positron interactions in the substance of the detector.

(2) $K^+ \to \pi^+\pi^0\pi^0$ where one of the π^0 photons is not detected and π^+ decays to $e\nu$ or mistakenly identified as an positron.

(3) $K^+ \to \pi^+\pi^0$ with "fake photon" and π^+ decayed or mistakenly identified as positron. Fake photon clusters can come from the interactions π^+ in the material of the detector, external bremsstrahlung upstream of the magnet, accidentals. All these sources are included in our MC calculations.

(4) $K^+ \to \pi^+\pi^0\gamma$ when π^+ decays or is mistakenly identified as positron.

(5) $K^+ \to \pi^0\pi^0e^+\nu$ when one γ is lost.

To suppress the background channels, we used a set of kinematic cuts:

Cut 1: $E_{miss} > 0.5$ GeV. Requirement on the missing energy in the decay reduces mainly background (4).

Cut 2: $|\Delta y| = |y_\gamma - y_{e+}| > 4$ cm, where y_γ, y_{e+} are the y-coordinates of the γ and e^+ intersection of the front plane of the calorimeter GAMS.

Cut 3: $| \ x, y \ | < 100$ cm, where x, y are the coordinates of the reconstructed neutrino intersection of the front plane of the calorimeter GAMS.

Cut 4: $M_K > 0.045$ GeV, where M_K - K meson mass restored from the kinematics of the event.

In order to suppress all the main backgrounds, we use a cut on the missing mass squared $M^2(\pi^0 e\gamma) = (P_K - P_{\pi^0} - P_e - P_\gamma)^2$. For the signal events this variable corresponds to the square of the neutrino mass and must be zero within measurement accuracy.

Cut 5 $-0.005 < M^2(\pi^0e^+\gamma) < 0.005$.

The dominant background to $K_{e3\gamma}$ arises from K_{e3} with extra photon. The background (1) is suppressed by cut 2 and requirement on the angle between electron and photon in the laboratory frame $\Theta_{e\gamma}$. The distribution of the K_{e3}-background events has very sharp peak at zero angle. This peak is significantly narrower than that for signal events. This happens, in particular, because the emission of the photons by the electron from K_{e3} decay occurs in the setup material downstream the decay vertex, but angle is still calculated as if emission comes from the vertex.

Cut 6 $0.004 < \Theta_{e\gamma} < 0.040$.

Right part of this cut is introduced for suppression of backgrounds (2,3,4,5). After applying all the cuts, 13118 events were selected, with a background of 1628 events. Background normalization was done by comparison numbers of events for K_{e3} decay in MC and real data samples. Event reduction statistics is summarized in table 1.

Table 1. Event reduction statistics for the real data, the background MC and signal MC.

Cut	Real data	$K^+ \to \pi^0e^+\nu$	$K^+ \to \pi^+\pi^0\pi^0$	$K^+ \to \pi^+\pi^0$	Signal
$E_{miss} > 0.5$GeV	515612	395930	19869	7138	62670
$\Delta y > 4$ cm	65093	10396	13975	1373	29190
$\mid x_\nu, y_\nu \mid < 100$cm	44363	7385	8726	1027	21950
$M_K > 0.045$GeV	23446	3801	2901	333	17030
$\mid M^2(\pi^0e^+\gamma) \mid < 0.005$	20739	3520	398	298	16120
$0.004 < \Theta_{e\gamma} < 0.040$	13118	1239	225	138	12040

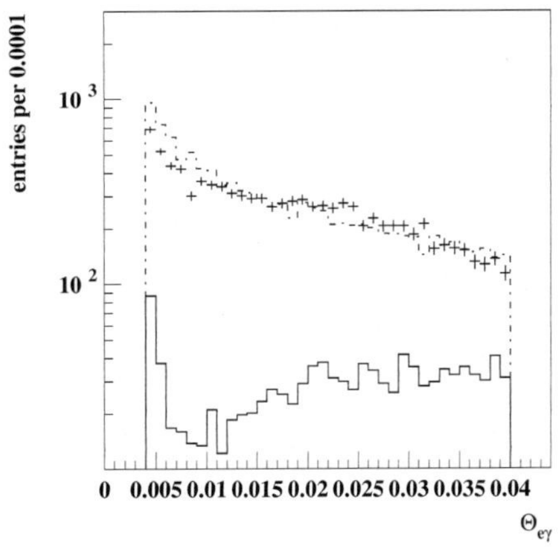

Figure 2. Distribution over $\Theta_{e\gamma}$ - the angle between electron and photon in lab. system. Real data (points with errors), MC background (solid line histogram) and signal plus MC background (dashed line histogram).

5. Results

Distribution over $\Theta_{e\gamma}$ - the angle between electron and photon in laboratory system (see figure 2). Reasonable agreement of the data with MC is seen. When generating the signal MC, a generator based on $O(p^2)$ [9] is used.

To obtain the branching ratio of the $K_{\pi^0 e^+ \nu_e \gamma}$ relative to the K_{e3} (R), the background and efficiency corrected number of $K_{e3\gamma}$ events is compared to that of 2812875 K_{e3} events found with the similar selection criteria. Further, the branching ratio with cuts $E^*_\gamma > 10$ MeV, $0.6 < cos\Theta^*_{e\gamma} < 0.9$, chosen for comparability with the previous experiments is calculated.

$$R = \frac{Br(K^+ \to \pi^0 e^+ \nu_e \gamma)}{Br(K^+ \to \pi^0 e^+ \nu_e)} = (0.59 \pm 0.02(stat.) \pm 0.03(syst.)) \times 10^{-2} \quad (4)$$

Systematic errors are estimated by variation of the cuts of table 1 and using two different ways of backgrounds normalization. A comparison with the results of previous experiments is shown in table 2. Statistics more than doubles compared to the previous measurement.

Table 2. Br($K^+ \to \pi^0 e^+ \nu_e \gamma$)/Br($K^+ \to \pi^0 e^+ \nu_e$) for $E^*_\gamma > 10$ MeV,$0.6 < cos\Theta^*_{e\gamma} < 0.9$ in comparison with previous data.

$R_{exp} \times 10^2$	N_{ev}	experiment
0.59 ± 0.02	3435	this experiment
0.48 ± 0.02	1423	ISTRA+ [10]
0.46 ± 0.08	82	XEBC [11]
0.56 ± 0.04	192	ISTRA [12]
0.76 ± 0.28	13	HLBC [13]

For the asymmetry A_ξ(for the same cuts as in table 2) we get

$$A_\xi = -0.019 \pm 0.020(stat.) \pm 0.027(syst.) \quad (5)$$

IOP Conf. Series: Journal of Physics: Conf. Series **934** (2017) 012027 doi:10.1088/1742-6596/934/1/012027

References

[1] Bijnens J, Echer G and Gasser J 1993 *Nucl.Phys.* B **396** 81
[2] Pitch A 1995 *Rep. Prog. Phys.* **58** 563
[3] Maiani L, Pancheri G and Paver N 1988 *The Second DAFNE Physics Handbook* (Rome: SIS-Ufficio Pubblicazioni)
[4] Gasser J *et al.* 2005 *Eur. Phys. J.* C **40** 205
[5] Gevas J, Iliopolus J and Kaplan J 1966 *Phys. Lett.* **20** 432
[6] Sadovsky A S *et al.* 2017 *arXiv*:1709.01473
[7] Yushchenko O P *et al.* 2017 *arXiv*:1708.09587
[8] Brun R *et al.* 1984 GEANT 3: user's guide *Preprint* CERN-DD/EE/84-1
[9] Braguta V V, Likhoded A A and Chalov A E 2002 *Phys. Rev.* D **65** 054038
[10] Ajinenko I V *et al.* 2007 *Yad. Fiz* **70** 2125
[11] Barmin V V *et al.* 1991 *SJNP* **53** 606
[12] Bolotov V N *et al.* 1986 *JETP Lett.* **42** 68
[13] Romano F *et al.* 1971 *Phys Lett.* B **36** 525

IOP Conf. Series: Journal of Physics: Conf. Series **934** (2017) 012028 doi:10.1088/1742-6596/934/1/012028

Usage of machine learning for the separation of electroweak and strong $Z\gamma$ production at the LHC experiments

A M Petukhov and E Yu Soldatov

National Research Nuclear University MEPhI (Moscow Engineering Physics Institute), Kashirskoe highway 31, Moscow, 115409, Russia

E-mail: aleksandr.petukhov@cern.ch, EYSoldatov@mephi.ru

Abstract. Separation of electroweak component from strong component of associated $Z\gamma$ production on hadron colliders is a very challenging task due to identical final states of such processes. The only difference is the origin of two leading jets in these two processes. Rectangular cuts on jet kinematic variables from ATLAS/CMS 8 TeV $Z\gamma$ experimental analyses were improved using machine learning techniques. New selection variables were also tested. The expected significance of separation for LHC experiments conditions at the second data-taking period (Run2) and 120 fb^{-1} amount of data reaches more than 5σ. Future experimental observation of electroweak $Z\gamma$ production can also lead to the observation physics beyond Standard Model.

1. Introduction

After completion of the Standard Model (SM) in 2012 by the experimental observation of the Higgs boson in ATLAS and CMS experiments at the LHC [1, 2], the main focus of experimental high energy physics moved to the search of the manifestations of the physics beyond the Standard Model (BSM).

Direct search of the BSM physics basically implies the observation of the brand new resonances, which is constraned by the collision energy limitations on the modern accelerators. Indirect search implies the precision study of the rare SM processes and observation of any deviations, caused by the new interraction or the new resonance in the inaccessible energy region.

In this work we study such kind of rare process - the electroweak Z-boson production with association of the photon. The electroweak (EWK) $Z\gamma$ production consists of processes with fourth-order electroweak coupling. It contains vector boson scattering (VBS) as well as other non-VBS processes, e.g. initial- or final-state vector boson radiation. The VBS has high interference with other EWK-production processes and therefore can not be studied separately. Vector boson scattering is a $VV \to VV$ process with $V = Z/W/\gamma$, that allows to probe the vector boson self-couplings, namely triple and quartic gauge-boson couplings (TGCs and QGCs). The latter is of particular interest, because new phenomena could generate additional contributions to QGCs with respect to the Standard Model predictions [3, 4]. In the experiment, such process (see figure 1 (a)) can be deduced from events with Z, γ and two or more jets in the final state.

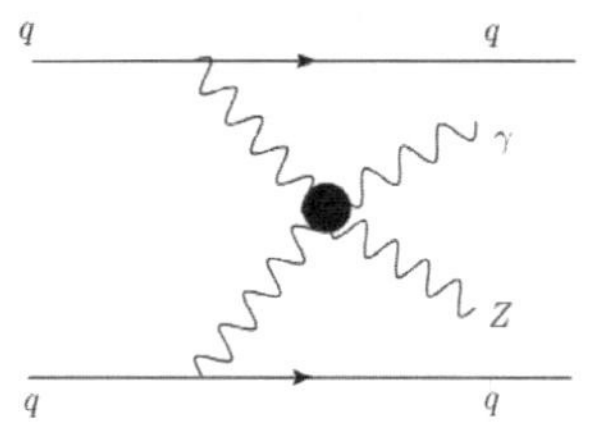

(a)

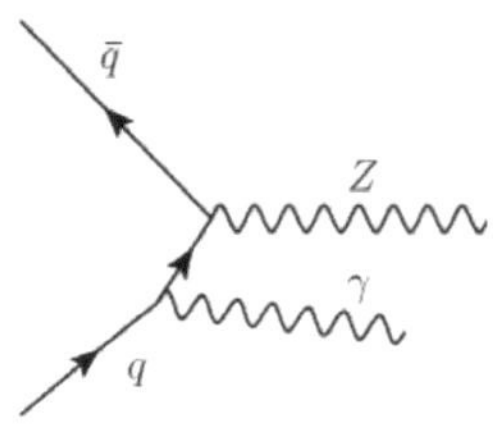

(b)

Figure 1. Feynman diagrams for VBS (a) and QCD (b) production of $Z\gamma$.

Same final state could be produced by QCD mediated processes — called QCD-production in this paper — with second-order electroweak and second-order strong couplings. These processes include gluon mediated scattering as well as initial- and final-state radiated gluons. The most important diagram of this process is given in figure 1 (b). The problem is that QCD process cross section is approximately 2 order of magnitude higher than EWK, which makes an EWK component extraction really complicated.

The main difference between EWK and QCD is the kinematic parameters of the jets. In the VBS signal process, the two jets are produced from valence quarks and recoil against each other due to the emission of bosons. Jets are then well separated and bosons emitted in the centre of jet rapidity gap. This is not the case for the QCD process where the bosons are radiated from fermions lines.

The ATLAS and CMS experiments recently studied the electroweak production of $Z\gamma$ final state [5, 6], but were not able to observe the VBS process with significance larger than 3 standard deviations.

In this study, simulated data of proton-proton collisions at the Run2 LHC center-of-mass energy $\sqrt{s} = 13$ TeV was used to estimate the significance of EWK- and QCD-production processes separation in the $Z \to \nu\bar{\nu}$ channel. Results achieved with rectangular cuts were compared with results for Boosted Decision Tree (BDT) algorithms.

2. Simulated samples

The Monte Carlo event samples were generated with MadGraph5_aMC@NLO MC generator [7] interfaced to PYTHIA6 and PYTHIA8 [8] for parton showering and Delphes [9] for the detector simulation.

The EWK- and QCD-production samples were generated separately, at the Leading-order and the Next-to-leading-order accuracy respectively. The interference between these processes was previously found to be less than 10% [5], decreasing with the di-jet invariant mass, and was not taken into account.

The generated data was normalized to the integral luminosity of 36.1 fb^{-1} corresponding to the amount of data collected by the ATLAS experiment in 2015-2016. The expected luminosity of whole ATLAS/CMS Run2 equal to 120 fb^{-1} was also used for to estimate the significance.

3. Significance estimates

3.1. Event preselection

In this study, only events with one photon, at least two jets and high missing transverse momentum were considered. The latter is the signature of the Z-boson decaying to the neutrino-antineutrino pair.

Basing on the [5, 6] and due to the fact, that single photon trigger threshold increased for both ATLAS and CMS in Run2, the initial experimental selections to suppress non-$Z\gamma$ backgrounds are:

- Electron and muon veto in the event;
- $p_T(\gamma) > 150$ GeV;
- $p_T^{miss} > 150$ GeV;
- $\Delta\varphi(\gamma, p_T^{miss}) > 2.1$.

3.2. Rectangular cuts

Basing on the ATLAS 8 TeV $Z\gamma$ analysis [5], to estimate the significance achieved with simple rectangular cuts the following kinematic variables were used:

- m_{jj}, di-jet invariant mass;
- ΔY_{jj}, absolute rapidity difference between two jets;
- $\zeta_\gamma = |\frac{y(\gamma)-(y(j_1)+y(j_2))/2}{y(j_1)-y(j_2)}|$, γ-centrality;
- $p_T^{balance} = \frac{|\vec{p}_T^{miss}+\vec{p}_T^{\gamma}+\vec{p}_T^{j1}+\vec{p}_T^{j2}|}{|\vec{p}_T^{miss}|+|\vec{p}_T^{\gamma}|+|\vec{p}_T^{j1}|+|\vec{p}_T^{j2}|}$.

Distributions of the EWK and QCD production processes for these variables are shown in figure 2.

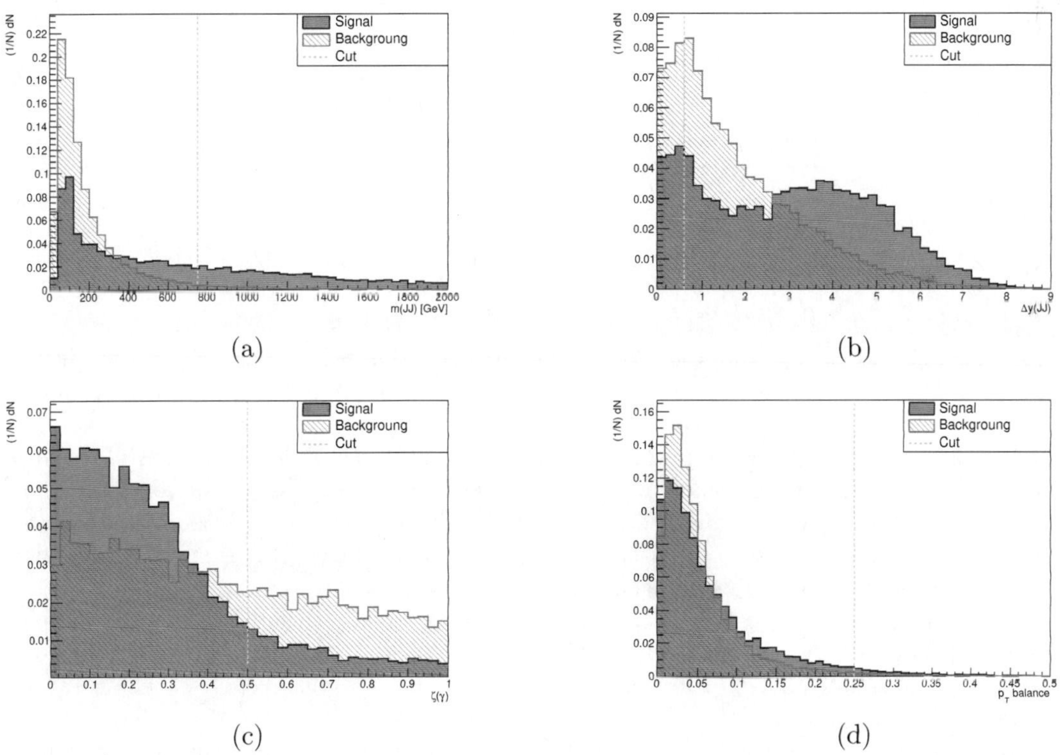

Figure 2. Distributions of the EWK (Signal) and QCD (Background) $Z\gamma$ production components versus m_{jj} (a), ΔY_{jj} (b), ζ_γ (c) and $p_T^{balance}$ (d). The vertical green dashed line shows the value for the best rectangular cut applied.

The resulting significance is given in table 1. The main source of significance uncertainty comes from systematic uncertainty of the cross section calculation caused by the PDF uncertainty and the choice of the renormalization/factorization scales.

Table 1. Statistical significance ($S/\sqrt{S+B}$) for different separation algorithms, with normalization to 36.1 fb^{-1} of the integrated luminosity. S is the number of EWK process events, B is the number of the QCD process events.

Algorithm	$S/\sqrt{S+B}$
Rectangular cuts	2.65 ± 0.04
4 variables, AdaBoost	2.99 ± 0.04
4 variables, Gradient Boost	2.89 ± 0.04
Additional variables, AdaBoost	3.08 ± 0.04
Additional variables, Gradient Boost	2.90 ± 0.04

3.3. Boosted desicion trees

To increase the significance of signal separation the ROOT-based machine learning kit TMVA [10] was used. Namely, the Boosted Decision Trees classifier with two boost algorithms — AdaBoost and Gradient Boost. It was chosen because it allows for a better signal separation than simple linear classifiers while not requiring laborious tuning like Artificial Neural Network algorithms.

Two sets of input variables were used — the same as in the section 3.2 and with the following variables added:

- $\eta_{j_1}\eta_{j_2}$, where η is the pseudorapidity of the jet;
- ΔR_{jj}, where $\Delta R = \sqrt{(\Delta\varphi)^2 + (\Delta\eta)^2}$;
- $\Delta R_{j\gamma}$;
- Δp_T^{jj};
- $\sin|\frac{\Delta\varphi(JJ)}{2}|$.

These variables were chosen among the rest within the framework of the conducted optimization study as the most powerful ones in discrimination of the processes under consideration.

The γ-centrality upper cut was found to increase the significance of the separation. The cause of it is that the jet rapidity difference in the denominator makes it possible for some events to have ζ_γ few orders of magnitude larger than the bulk of the measurements. This makes it harder for the BDT algorithm to make a proper cut on ζ_γ and in result it leads to less effective classifier. The cut providing the maximum significance was found to be $\zeta_\gamma < 1.5$.

The figure 3 shows the classifier response distributions for signal and background samples (EWK and QCD-production respectively) for the extended set of variables. The figure 4 shows dependence of the significance on the value of the cut on the classifier response. The resulting significances are given in table 1.

The AdaBDT provides separation with higher significance, although its response seems to be less stable than the one of the Gradient Boost.

Significance was also estimated for L =120 fb^{-1} - the expected integrated luminosity of the whole ATLAS/CMS Run2. The full results are given in table 2.

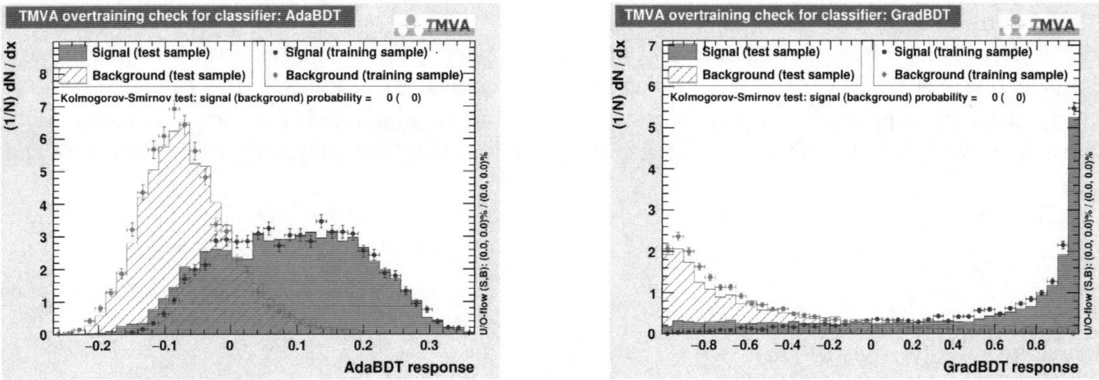

(a) (b)

Figure 3. The classifier response distributions for signal and background (EWK and QCD-production respectively) samples for the extended set of variables: (a) AdaBoost (b) Gradient Boost.

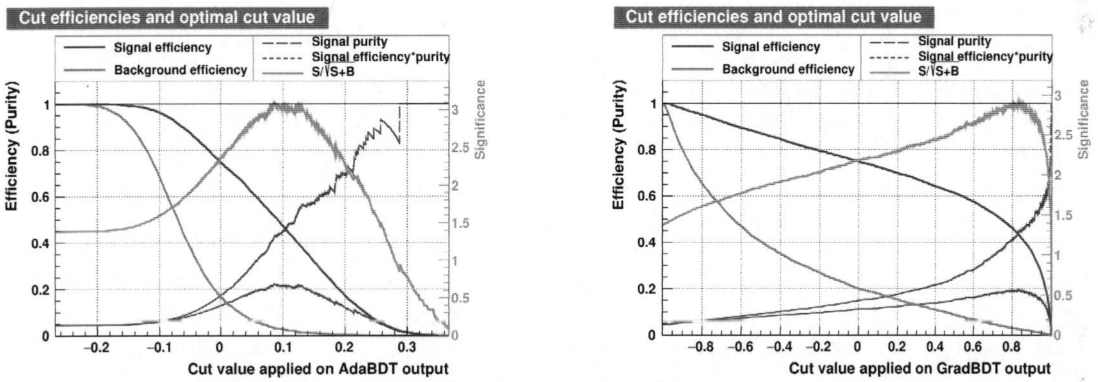

(a) (b)

Figure 4. The dependence of significance on the value of the cut on the classifier response for the extended set of variables: (a) AdaBoost (b) Gradient Boost.

Table 2. Statistical significance ($S/\sqrt{S+B}$) for different separation algorithms, with normalization to 120 fb^{-1} of the integrated luminosity. S is the number of EWK process events, B is the number of the QCD process events.

Algorithm	$S/\sqrt{S+B}$
Rectangular cuts	4.84 ± 0.08
4 variables, AdaBoost	5.45 ± 0.08
4 variables, Gradient Boost	5.27 ± 0.08
Additional variables, AdaBoost	5.62 ± 0.08
Additional variables, Gradient Boost	5.28 ± 0.08

4. Conclusions

This work demonstrates the real possibility to distinguish the EWK component of $Z\gamma$ production from the dominant QCD one. A set of discriminative variables was chosen after the optimization study and tests. Separation was done using simple rectangular cuts approach and using BDT algorithms AdaBoost and Gradient Boost. BDT application gives increase in significance of the order of 12-13%.

Maximal reached significance for 36.1 fb^{-1} is $3.08 \pm 0.04\ \sigma$ and for 120 fb^{-1} is $5.62 \pm 0.08\ \sigma$. Thus classifier developed in this study can lead to the experimental observation of the electroweak $Z\gamma$ production.

5. Acknowledgements

We gratefully acknowledge the financial support from Russian Science Foundation grant No.17-72-10021.

References

[1] Aad G *et al.* 2012 *Science* **338** 1576-82
[2] Chatrchyan S *et al.* 2012 *Science* **338** 1569-75
[3] Eboli O J P *et al.* 2004 *Phys. Rev.* D **69** 095005
[4] Eboli O J P *et al.* 2006 *Phys. Rev.* D **74** 073005
[5] Aad G *et al.* 2017 *JHEP* **07** 107
[6] Chatrchyan S *et al.* 2017 *Phys. Lett.* B **770** 380-402
[7] Alwall J *et al.* 2014 *JHEP* **07** 079
[8] Sjöstrand T *et al.* 2008 *Comp. Phys. Comm.* **178** 852-67
[9] de Favereau J *et al.* 2014 *JHEP* **02** 057
[10] Hoecker A *et al.* 2007 *arXiv*:physics/0703039

Top quark measurements at ATLAS

S Grancagnolo
Humboldt University, Newtonstr 15, 12489 Berlin, Germany
E-mail: sergio.grancagnolo@cern.ch

Abstract. The top quark is the heaviest known fundamental particle. As it is the only quark that decays before it hadronizes, it allows us to probe the properties of bare quarks at the Large Hadron Collider. Highlights of a few recent precision measurements by the ATLAS Collaboration of the top quark using 13 TeV and 8 TeV collision data will be presented: top-quark pair and single top production cross sections including differential distributions will be presented alongside measurements of top-quark properties, including results using boosted top quarks, probe our understanding of top-quark production in the TeV regime. Measurements of the top-quark mass and searches for rare top quark decays are also presented.

1. Introduction
The large number of top quarks available in the data collected at various energies with the ATLAS detector [1], allows one to study properties of this particle and to measure its production and differential cross sections with high precision. This document summarises recent results in top-quark physics, organized as follows. For each analysis, motivations and overview of specific strategies are given: mass measurement in Chapter 2, properties in Chapter 3, $t\bar{t}$ cross section in Chapter 4, and single top production in Chapter 5. Conclusions are taken in Chapter 6.

2. Top-quark mass
The mass of top quark is an important input to the global fit of electroweak parameters, cross-checking the consistency and constraining possible extensions of the Standard Model (SM). Its value has also cosmological implications, affecting the stability of the Higgs potential. The analysis presented here [2] considers the decay $t\bar{t} \rightarrow W^+W^-b\bar{b} \rightarrow \ell\nu q\bar{q}'b\bar{b}$. Events are selected in a 20.2 fb^{-1} dataset collected at the collision energy of 8 TeV, by requiring exactly one isolated high-momentum lepton and at least four jets with transverse momentum $p_{\text{T}} > 25$ GeV and $|\eta| < 2.5$,[1] exactly two of which contain hadrons with b quarks (b-tagging). The lepton (ℓ) can be a μ with $p_{\text{T}} > 25$ GeV or an e with transverse energy $E_{\text{T}} > 25$ GeV, also including those from a tau-lepton decay. To take into account the presence of the neutrino, minimum $E_{\text{T}}^{\text{miss}}$ and m_{T}^{W} are required[2] depending on the lepton channel.

[1] ATLAS uses a right-handed coordinate system with its origin at the nominal interaction point (IP) in the center of the detector and the z-axis along the beam pipe. The x-axis points from the IP to the center of the LHC ring, and the y-axis points upward. Cylindrical coordinates (r,ϕ) are used in the transverse plane, ϕ being the azimuthal angle around the z-axis. The pseudorapitidy is defined in terms of the polar angle θ as $\eta = -\ln\tan(\theta/2)$.

[2] The missing transverse momentum $E_{\text{T}}^{\text{miss}}$ is the absolute value of the vector $\overrightarrow{E_{\text{T}}^{\text{miss}}}$ given by the negative vectorial sum of the transverse component of all calibrated energy deposits in the calorimeters and momentum reconstructed

The two channels (e and μ) are finally merged and the preselection is further refined using a boosted decision tree (BDT) with 13 input variables. The most sensitive one is the opening angle ΔR between the two light jets associated to the W boson decay. The full reconstruction of the event is performed using a kinematic fit, assuming that the best assignment of jets to partons maximises a likelihood where a common mass parameter $\mathrm{m}_{\mathrm{top}}^{\mathrm{reco}}$ is associated to the leptonically and hadronically decaying top quarks. To take in account dominant effects due to jet energy scale and differences in jets originating from light or b quarks, templates were generated by scaling the energy of the jets. The corresponding scale factors are determined together with $\mathrm{m}_{\mathrm{top}}$ in a three-dimensional final fit, resulting in $\mathrm{m}_{\mathrm{top}} = 172.08 \pm 0.39(\mathrm{stat}) \pm 0.82(\mathrm{syst})$ GeV. This is combined with previous ATLAS measurements at 7 TeV, including the orthogonal $t\bar{t} \rightarrow$ dilepton channel, taking in account all correlations. Using the best linear unbiased estimate (BLUE), the combined result is $\mathrm{m}_{\mathrm{top}} = 172.51 \pm 0.27(\mathrm{stat}) \pm 0.42(\mathrm{syst})$ GeV with a χ^2 probability of 78% (Figure 1) [2].

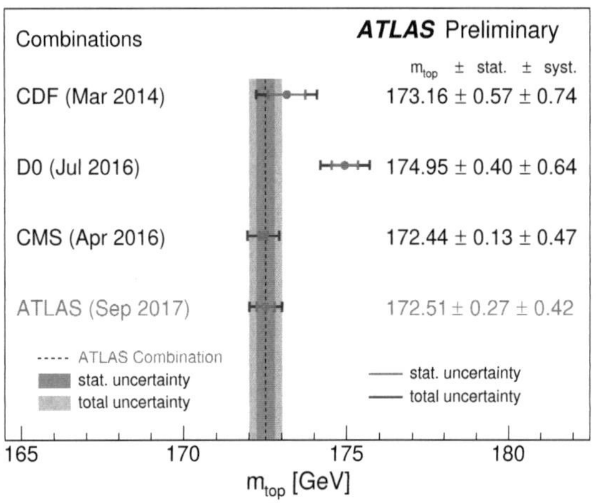

Figure 1. Comparison of combined $\mathrm{m}_{\mathrm{top}}$ results [2].

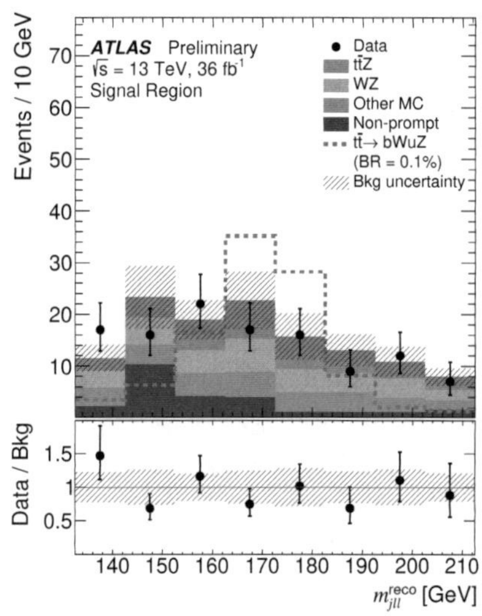

Figure 2. Reconstructed top-quark mass in the signal region compared with a FCNC signal normalised to $BR(t \rightarrow uZ) = 0.1\%$ [3].

3. Top-quark properties

3.1. Top-quark decays through flavour-changing neutral current

Decays such as $t \rightarrow qZ$ are forbidden in the SM at tree level, and they are strongly suppressed at loop level by the GIM mechanism. Several extensions beyond the SM (BSM) predict a branching ratio (BR) order of magnitude larger for this Flavour-changing neutral current (FCNC) decay. This search considers the chain $t\bar{t} \rightarrow bW(\rightarrow \ell\nu)qZ(\rightarrow \ell\ell)$, in a 36 fb^{-1} dataset collected at 13 TeV. With a similar object reconstruction as in Chapter 2, events are selected by requiring exactly three isolated charged leptons with $p_{\mathrm{T}} > 15$ GeV, two of them with the same flavour and opposite charge and an invariant mass within 15 GeV of the Z boson mass, at least two jets with $p_{\mathrm{T}} > 25$ GeV, exactly one of them b-tagged, and $E_{\mathrm{T}}^{\mathrm{miss}} > 40$ GeV. Applying

in tracking detectors. The W-boson transverse mass m_{T}^{W} is defined as $\sqrt{2p_{\mathrm{T}}^{\ell}E_{\mathrm{T}}^{\mathrm{miss}}(1-\cos\phi(\ell,\overrightarrow{E_{\mathrm{T}}^{\mathrm{miss}}}))}$.

energy-momentum conservation, a χ^2 is calculated from the reconstructed masses of qZ, bW, and $\ell\nu$ systems. After minimisation, the assignments between reconstructed top quarks and simulated particles is around 80% for FCNC. Five additional control regions are defined and used in a simultaneous fit to better estimate background contributions and constrain systematic uncertainties. No evidence of signal events is found (Figure 2). Observed (expected) limits BR($t \to uZ$)$< 1.7 \times 10^{-4}$ (2.4×10^{-4}) and BR($t \to cZ$)$< 2.3 \times 10^{-4}$ (3.2×10^{-4}) are set at 95% confidence level, constraining as well corresponding effective field theory operators [3].

3.2. Charge asymmetry

The rapidity distribution of top quarks, being produced from both valence and sea quarks in the proton, is broader than that of top antiquarks, that in average have a larger momentum than sea quarks only. Asymmetries in the angular distributions in the $q\bar{q} \to t\bar{t}$ process are introduced by higher-order effects in QCD, and they can become large at high values of the invariant mass of the $t\bar{t}$ system in the presence of BSM contributions. The sensitive observable at LHC is the charge asymmetry: $A_C = \frac{N^{\Delta|y|>0} - N^{\Delta|y|<0}}{N^{\Delta|y|>0} + N^{\Delta|y|<0}}$, with N indicating the number of events with a positive or negative rapidity difference $\Delta|y| = |y_t| - |y_{\bar{t}}|$. ATLAS and CMS combined their measurements with 5 fb^{-1} at 7 TeV and 20 fb^{-1} at 8 TeV. The final state considered is $t\bar{t} \to \ell$+jets as in Chapter 2. Using the BLUE method, three separate combinations are obtained. The combination of inclusive results at 7 TeV is: $A_C^{\mathrm{LHC7}} = 0.005 \pm 0.007(\mathrm{stat}) \pm 0.006(\mathrm{syst})$. The combination of inclusive results at 8 TeV is: $A_C^{\mathrm{LHC8}} = 0.0055 \pm 0.0023(\mathrm{stat}) \pm 0.0025(\mathrm{syst})$. The differential charge asymmetry as a function of the invariant mass of the $t\bar{t}$ system at 8 TeV is shown in Figure 3 [4].

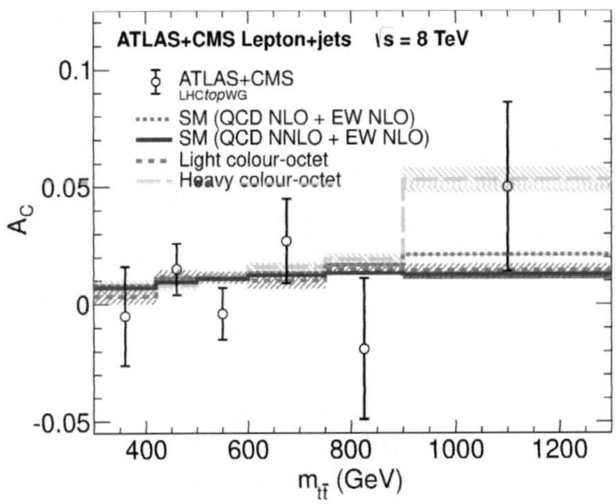

Figure 3. Combined ATLAS+CMS charge asymmetry in six $t\bar{t}$ invariant-mass bins, compared with SM and BSM predictions [4].

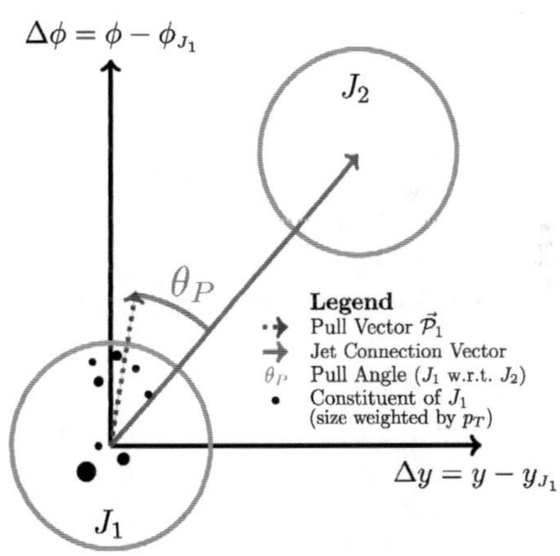

Figure 4. Jet pull observables for a di-jet system. The pull angle θ_P is between the pull vector and the vector connecting two jets [5].

3.3. Colour flow

In nature, only colour-neutral hadrons can be observed, and direct measument of the quantum chromodynamics (QCD) interaction of their constituents is not possible. Quark and gluons are measured as jets, but colour connection influences jet shape, structure and event topology. The distribution of particles within a jet is related with the colour carried by corresponding partons. An observable describing this information is the pull vector, a p_T-weighted radial moment of the jet, computed from its constituents. For a dijet system, the pull angle between the pull vector and the vector connecting the two jets is considered (Figure 4). In the $t\bar{t} \to \ell$+jets event topology, the jets originating from the hadronic W decay are colour connected, while the two b-tagged jets are not.

These observables are studied using the same dataset as in Chapter 3.1, and similar object reconstruction and event selection as in Chapter 2. A corresponding one is also defined at particle level, using the Monte Carlo (MC) truth record. After bin-by-bin background subtraction, and removing effects of the detector smearing as obtained from simulation, the signal distributions are then unfolded to a particle-level spectrum, using an Iterative Bayesian (IB) method. A general good agreement is found with SM predictions, even if observables sensitive to colour flow remain poorly modelled [5].

3.4. Decay width

The decay width of the top quark is the largest of all fermions, and any deviation from SM expectation might hint to non-SM decays. The next-to-leading-order and next-to-next-to-leading-order calculation predicts $\Gamma_{\text{NLO}} = 1.33$ GeV, and $\Gamma_{\text{NNLO}} = 1.322$ GeV respectively [6]. The same 8 TeV dataset and a similar object reconstruction and event selection of the $t\bar{t} \to \ell$+jets channel is performed as in Chapter 2. A likelihood-based assignment of reconstructed jets to partons is used. The events are categorised based on the flavour of the selected lepton, if they have exactly one or at least two b-tagged jets, and if all four associated jets have $|\eta| < 1$ or at least one has $|\eta| \geq 1$. Two observables are used in the final template fit in this eight regions. The reconstructed invariant mass $m_{\ell b}$ of the b-jet and the lepton from $t \to bW(\to \ell\nu)$, and the distance $\Delta R_{\text{min}}(j_b, j_l)$ between the b-jet and the closest light jet from the hadronic top decay. The signal templates are generated by reweighting events with alternative width hypotheses. The measured top decay width is $\Gamma_t = 1.76 \pm 0.33(\text{stat})^{+0.79}_{-0.68}(\text{syst})$ GeV [7].

4. Top-quark pair production cross section

BSM effects can appear in $t\bar{t}$ differential distributions, while being undetected by inclusive cross section measurements. For large Lorentz boosts of the Lorentz boost of the top quarks, their decay products are collimated into a large-radius jet. This study uses 14.7 fb^{-1} of data at 13 TeV looking for an all-hadronic boosted topology where both W bosons decay into a pair of quarks. These jets must satisfy $p_T^{t,1} > 500$ GeV, $p_T^{t,2} > 350$ GeV, $|\eta| < 2.0$, and a top-tagging algorithm, each being also associated with a b-tagged jet having $p_T > 25$ GeV. No lepton with $p_T > 25$ GeV is allowed in the events selected. Background is estimated from data using nine regions defined by the number of top-tagged and b-tagged jets. After its subtraction, an IB unfolding technique is used to correct for detector acceptance and gives particle-level distributions of normalised cross sections for various observables, like the production angle ($\chi^{t\bar{t}} = \exp(2|y^*|)$, where $y^* = 1/2(y^{t,1} - y^{t,2})$) sensitive to $t\bar{t}$ production modelling, and the rapidity boost ($y_{\text{B}}^{t\bar{t}} = 1/2(y^{t,1} + y^{t,2})$) sensitive to parton distribution functions. The total cross section in the fiducial phase space is $\sigma^{\text{fid}}_{t\bar{t}(\text{boost})} = 374 \pm 13(\text{stat})^{+111}_{-92}(\text{syst})$ fb, in good agreement with SM predictions [8].

5. Single Top-quark production

5.1. Production in association with a W boson

The Wtb-vertex coupling is parameterised by the V_{tb} Cabibbo-Kobakashi-Maskawa matrix element and f_{V}^{L} form factor. One of the three leading-order electroweak single top quark channels in the SM is the tW (see Figure 5). The differential cross section is measured with the same dataset as in Chapter 3.3. The final state is required to have one b-tagged jet, $E_{\text{T}}^{\text{miss}}$, and exactly two opposite sign leptons, with an invariant mass outside Z-mass window if they have

IOP Conf. Series: Journal of Physics: Conf. Series **934** (2017) 012029 doi:10.1088/1742-6596/934/1/012029

the same flavour. A BDT is used to separate from $t\bar{t}$ background, optimised to reduce systematic uncertainties, with the input variable $p_{\mathrm{T}}(\ell_1\ell_2E_{\mathrm{T}}^{\mathrm{miss}}b)$ giving largest separating power. Unfolding using the IB technique gives particle-level distributions of normalised cross sections for various observables, like $m_{\mathrm{T}}(\ell\ell\nu\nu b)$ and $E(\ell\ell b)$. Compared to MC predictions with different generators, parton shower, hadronisation and underlying-event models, results are found in good agreement with SM predictions [9].

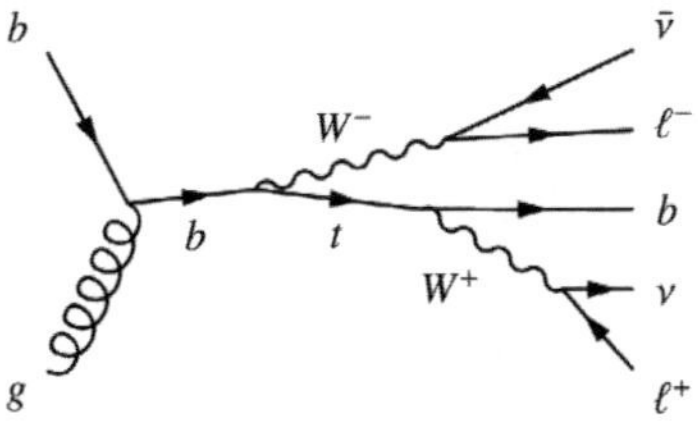

Figure 5. Single top production Feynman diagram in tW channel and leptonic decays for both W bosons [9].

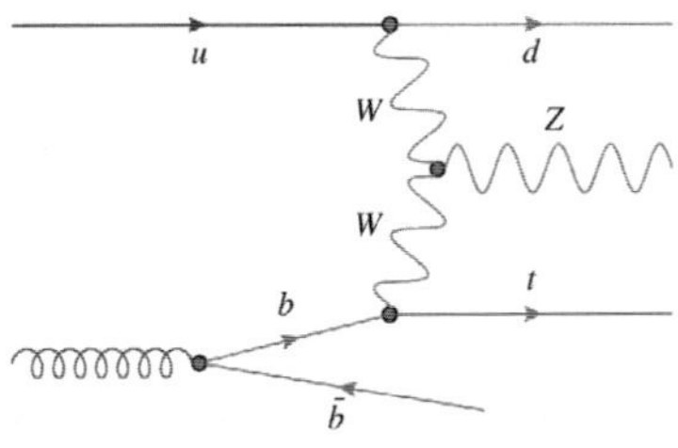

Figure 6. Feynman diagram for the tZq process probing the WWZ vertex [10].

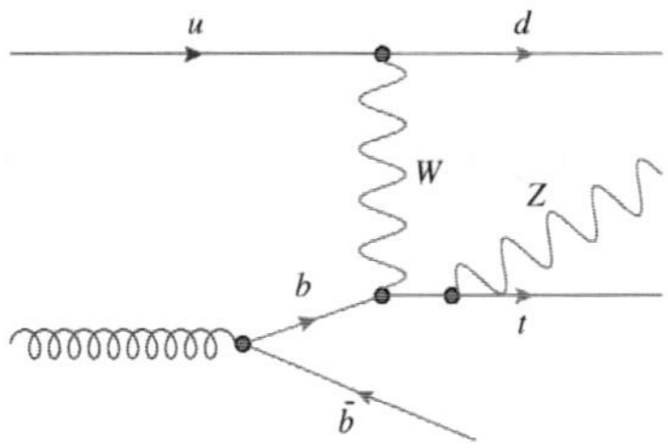

Figure 7. Feynman diagram for the tZq process probing the tZ vertex [10].

5.2. Production in association with a Z boson

The rare SM single-top production process $pp \to tZq$, probes WWZ and tZ couplings (Figures 6, 7). It is measured using the same dataset as in Chapter 3.3. The final state contains exactly three charged leptons with $p_{\mathrm{T}} > 28$, 25, and 15 GeV respectively, two jets with $p_{\mathrm{T}} >$ 30 GeV, only one of them b-tagged. A same-flavour, opposite-sign lepton pair with invariant mass within 10 GeV of the Z boson is required. The W boson is reconstructed from the remaining lepton and $E_{\mathrm{T}}^{\mathrm{miss}}$, with $m_{\mathrm{T}}^{W} > 20$ GeV. A multivariate analysis using a three-layer neural network (NN) combines several variables into one discriminant. The output, O_{NN}, separate signal from background events having values closer to 1 and 0 respectively. Best separation power is given by the variable $|\eta|$ of the untagged jet. A binned likelihood fit is performed to extract the tZq signal strength, maximised on the NN output, to normalise signal and background events in each bin. The measured cross section is $\sigma_{tZ(3\ell)} = 600 \pm 170(\mathrm{stat}) \pm 140(\mathrm{syst})$ fb, assuming $\mathrm{m_{top}} = 172.5$ GeV, with a significance of 4.2σ (5.4σ expected) [10].

6. Conclusions

All results presented are in agreement with the SM predictions. The good performances of LHC and the high-efficiency data-taking of the experiments indicates that additional large statistics will be available in the future, and even more precision can be achieved on rare decays and property measurements sensitive to BSM effects.

References

[1] ATLAS Collaboration 2008 *JINST* **3** S08003
[2] ATLAS Collaboration 2017 ATLAS-CONF-2017-071 (Geneva: CERN)
[3] ATLAS Collaboration 2017 ATLAS-CONF-2017-070 (Geneva: CERN)
[4] ATLAS and CMS Collaborations 2017 *Preprint* 1709.05327
[5] ATLAS Collaboration 2015 *Phys. Lett.* B **750** 475
[6] Gao J *et al.* 2013 *Phys. Rev. Lett.* **110** 042001
[7] ATLAS Collaboration 2017 *Preprint* 1709.04207
[8] ATLAS Collaboration 2016 ATLAS-CONF-2016-100 (Geneva: CERN)
[9] ATLAS Collaboration 2017 ATL-COM-PHYS-2017-515 (Geneva: CERN)
[10] ATLAS Collaboration 2017 *Preprint* 1710.03659

Study of kinematic observables sentitive to the Higgs boson production channel in $pp \to Hjj$ process

N Belyaev[1], **R Konoplich**[2,4] **and K Prokofiev**[3]

[1] National Research Nuclear University MEPhI (Moscow Engineering Physics Institute), Kashirskoe highway 31, Moscow, 115409, Russia
[2] Department of Physics, New York University, 4 Washington Place, New York, NY 10003, USA
[3] Department of Physics and Institute for Advanced Study, the Hong Kong University of Science and Technology

E-mail: NLBelyayev@mephi.ru, RK60@nyu.edu, Kirill.Prokofiev@cern.ch

Abstract. After the Higgs boson discovery at the LHC, a lot of additional measurements should be performed to understand in details the properties of the observed particle. These measurements include cross sections measurements, couplings measurements, studies of the interaction vertex structures etc. One of the most perspective subjects to study is the kinematics of the production jets, associated with the Higgs boson. It is demonstrated, that the kinematic correlations of such jets can be used to distinguish different production channels of Higgs boson: gluon-gluon fusion (ggF) and vector boson fusion (VBF). Such separation plays an important role because possible beyond the Standard Model contributions in ggF and VBF channels lead to different effects, which should be taken into account in searches for BSM physics.

1. Introduction

After the Higgs boson was discovered by the ATLAS and CMS collaborations at the LHC [1,2], the particle content of Standard Model (SM) became complete. Searches for physics beyond the Standard Model (BSM) are currently ongoing. Several impressive discoveries based on the LHC data were done in last years, namely the pentaquark observation [3], the observation of CP violation in rare baryon decays [4] and measurement of the Higgs boson mass [5].

Separation of gluon-fusion (ggF) and vector boson fucion (VBF) Higgs boson production channels was the main subject of study in many works [6–11]. However, due to the progress in neural networks and boosted decision tree (BDT) classifiers [12], there are plenty of new possibilities to improve the ggF-VBF separation even more. The separation of these channels is important because many BSM theories predict different effects for ggF and VBF vertices [13–16]. The separate investigation of both ggF and VBF production mechanisms will thus help to reveal the properties of discovered Higgs boson.

[4] Also at Physics Department, Manhattan College, 4513 Manhattan College Parkway, Riverdale, New York, NY 10471, USA

IOP Conf. Series: Journal of Physics: Conf. Series **934** (2017) 012030 doi:10.1088/1742-6596/934/1/012030

2. Higgs model

One of the efficient ways to study possible BSM effects is to use the Effective field theory approach (EFT). This approach is based on the effective Lagrangian of particular model. Effective Lagrangian usually includes high dimensional operators which contribute to different vertices. In this study, ggF and VBF Higgs boson production channels are considered in the framework of the widely used Higgs Characterisation model [8] for the aMC@NLO Monte Carlo generator [17]. The tree-level HVV Lagrangian of this model can be written as following:

$$\begin{aligned}
\mathcal{L}_0^V = \Big\{ & c_\alpha \kappa_{\rm SM} \big[\tfrac{1}{2} g_{HZZ}\, Z_\mu Z^\mu + g_{HWW}\, W_\mu^+ W^{-\mu}\big] \\
& - \frac{1}{4}\big[c_\alpha \kappa_{H\gamma\gamma} g_{H\gamma\gamma}\, A_{\mu\nu} A^{\mu\nu} + s_\alpha \kappa_{A\gamma\gamma} g_{A\gamma\gamma}\, A_{\mu\nu} \widetilde{A}^{\mu\nu}\big] \\
& - \frac{1}{2}\big[c_\alpha \kappa_{HZ\gamma} g_{HZ\gamma}\, Z_{\mu\nu} A^{\mu\nu} + s_\alpha \kappa_{AZ\gamma} g_{AZ\gamma}\, Z_{\mu\nu} \widetilde{A}^{\mu\nu}\big] \\
& - \frac{1}{4}\big[c_\alpha \kappa_{Hgg} g_{Hgg}\, G_{\mu\nu}^a G^{a,\mu\nu} + s_\alpha \kappa_{Agg} g_{Agg}\, G_{\mu\nu}^a \widetilde{G}^{a,\mu\nu}\big] \\
& - \frac{1}{4}\frac{1}{\Lambda}\big[c_\alpha \kappa_{HZZ}\, Z_{\mu\nu} Z^{\mu\nu} + s_\alpha \kappa_{AZZ}\, Z_{\mu\nu} \widetilde{Z}^{\mu\nu}\big] \\
& - \frac{1}{2}\frac{1}{\Lambda}\big[c_\alpha \kappa_{HWW}\, W_{\mu\nu}^+ W^{-\mu\nu} + s_\alpha \kappa_{AWW}\, W_{\mu\nu}^+ \widetilde{W}^{-\mu\nu}\big] \\
& - \frac{1}{\Lambda} c_\alpha \big[\kappa_{H\partial\gamma}\, Z_\nu \partial_\mu A^{\mu\nu} + \kappa_{H\partial Z}\, Z_\nu \partial_\mu Z^{\mu\nu} \\
& + \big(\kappa_{H\partial W}\, W_\nu^+ \partial_\mu W^{-\mu\nu} + h.c.\big)\big] \Big\} X_0 \,,
\end{aligned} \tag{1}$$

where Λ is the new physics energy scale and the field strength tensors are defined as follows:

$$\begin{aligned}
V_{\mu\nu} &= \partial_\mu V_\nu - \partial_\nu V_\mu \quad (V = A, Z, W^\pm)\,, \\
G_{\mu\nu}^a &= \partial_\mu G_\nu^a - \partial_\nu G_\mu^a + g_s f^{abc} G_\mu^b G_\nu^c \,.
\end{aligned}$$

The dual tensor $\widetilde{V}_{\mu\nu}$ is defined as:

$$\widetilde{V}_{\mu\nu} = \frac{1}{2}\epsilon_{\mu\nu\rho\sigma} V^{\rho\sigma}.$$

Lagrangian (1) contains CP-even and CP-odd Higgs boson states. Case $c_\alpha = 1$ corresponds to a pure Standard Model CP-even Higgs boson state while $c_\alpha = 0$ corresponds to a pure BSM CP-odd Higgs boson state. Higgs boson states with different CP-parities can be mixed by setting the c_α parameter to some value inside the interval $(0, 1)$

3. Higgs boson production

Kinematic distributions were studied at Monte Carlo level with the Next-to-leading order with quantum chromodynamics corrections (NLO QCD). Two initial data samples in the .lhe file format were generated with the aMC@NLO Monte Carlo generator for $\sqrt{s} = 13$ TeV proton-proton collisions. The first data set contains 10^5 SM ggF events with the following couplings configuration: $c_\alpha = 1$; $k_{Hgg} = 1$; $k_{SM} = 1$ and $\Lambda = 10^3$ GeV. The second data set contains 10^5 SM VBF events with the following couplings configuration: $c_\alpha = 1$; $k_{SM} = 1$; $k_{H\gamma\gamma} = 1$; $k_{HZ\gamma} = 1$; $k_{HZZ} = 1$; $k_{HWW} = 1$ and $\Lambda = 10^3$ GeV. Feynman diagrams of Leading-order ggF and VBF processes are shown at figure 1. After the .lhe files were generated, parton showering was applied by PYTHIA8 generator [18] with .stdhep files as the output.

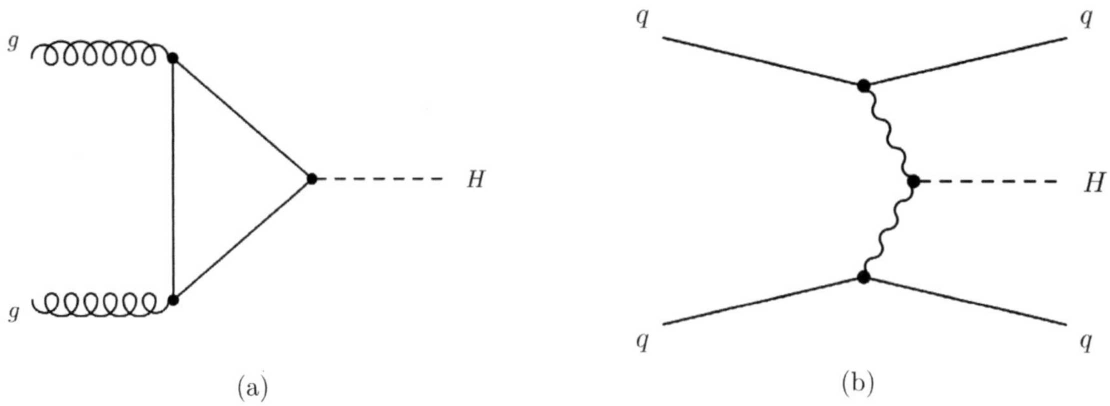

Figure 1. Feynman diagrams of the Leading-order ggF channel (a) and VBF channel (b).

At the last step, .stdhep files were processed by DELPHES3 simulation package [19], which took into account simple detector effects of averaged Large hadron collider (LHC) detector. Data files produced by DELPHES3 simulation package were converted into .root format and then processed with the scientific framework ROOT [20].

4. Jet distributions

For this study, only reconstructed jets with transverse momentum $p_T^{jet} > 30$ GeV and pseudorapidity $\eta_j < 5$ were taken into account. Jet correlations were studied for the following kinematic variables:

- The invariant mass of the final state jets: m_{jj};
- The transverse momentum of X_0: $p_{T,H}$;
- The number of jets in the event N_{jets};
- The absolute value of difference between pseudorapidities of two jets: $|\Delta\eta_{jj}| = |\eta_{j_1} - \eta_{j_2}|$;
- The sine of the absolute value of half the difference between phi angle of two jets: $\mathrm{Sin}\left(\left|\frac{\Delta\phi_{jj}}{2}\right|\right) = \mathrm{Sin}\left(\left|\frac{\phi_{j_1}-\phi_{j_2}}{2}\right|\right)$;
- The angular separation between two jets: $\Delta R_{jj} = \sqrt{\Delta\eta_{jj}^2 + \Delta\phi_{jj}^2}$;
- The product of pseudorapidities of two jets: $\eta_{j_1} \cdot \eta_{j_2}$;
- The η-centrality of $H + 2j$ system: η-centrality $= \left|\frac{\eta_H - \overline{\eta}_{jj}}{\Delta\eta_{jj}}\right| = \left|\frac{\eta_H - \frac{\eta_{j_1}+\eta_{j_2}}{2}}{\Delta\eta_{jj}}\right|$;
- The $\vec{p}_T$-centrality of $H + 2j$ system: $\vec{p}_T$-centrality $= \frac{|\vec{p}_{T,H} - \overline{\vec{p}}_{T,jj}|}{|\Delta\vec{p}_{T,jj}|} = \frac{\left|\vec{p}_{T,H} - \frac{\vec{p}_{T,j_1}+\vec{p}_{T,j_2}}{2}\right|}{|\Delta\vec{p}_{T,jj}|}$;
- The balance of transverse momentum of $H + 2j$ system: p_T-balance $= \frac{|\vec{p}_{T,H}+\vec{p}_{T,j_1}+\vec{p}_{T,j_2}|}{|\vec{p}_{T,H}|+|\vec{p}_{T,j_1}|+|\vec{p}_{T,j_2}|}$;
- The normalized sum of jet momenta: $\xi_{jets} = \frac{p_{T,j_1}+p_{T,j_2}}{E_{j_1}+E_{j_2}}$.
- The thrust variable: $T = max\left(\sum_{N_{jets}} \frac{|\hat{T}\cdot\vec{p}_{i,jet}|}{|\vec{p}_i|}\right)$;

The thrust axis of a jet, $\hat{T}$, is defined as the direction which maximizes the sum of the longitudinal momenta of the energy clusters. Calculated distributions of the above-defined variables are shown at figure 2 and 3. The observables which are based on the transverse momentum of the

IOP Conf. Series: Journal of Physics: Conf. Series **934** (2017) 012030 doi:10.1088/1742-6596/934/1/012030

Higgs boson and outcoming jets have better separation power compared with the observables which are based on the angular variables.

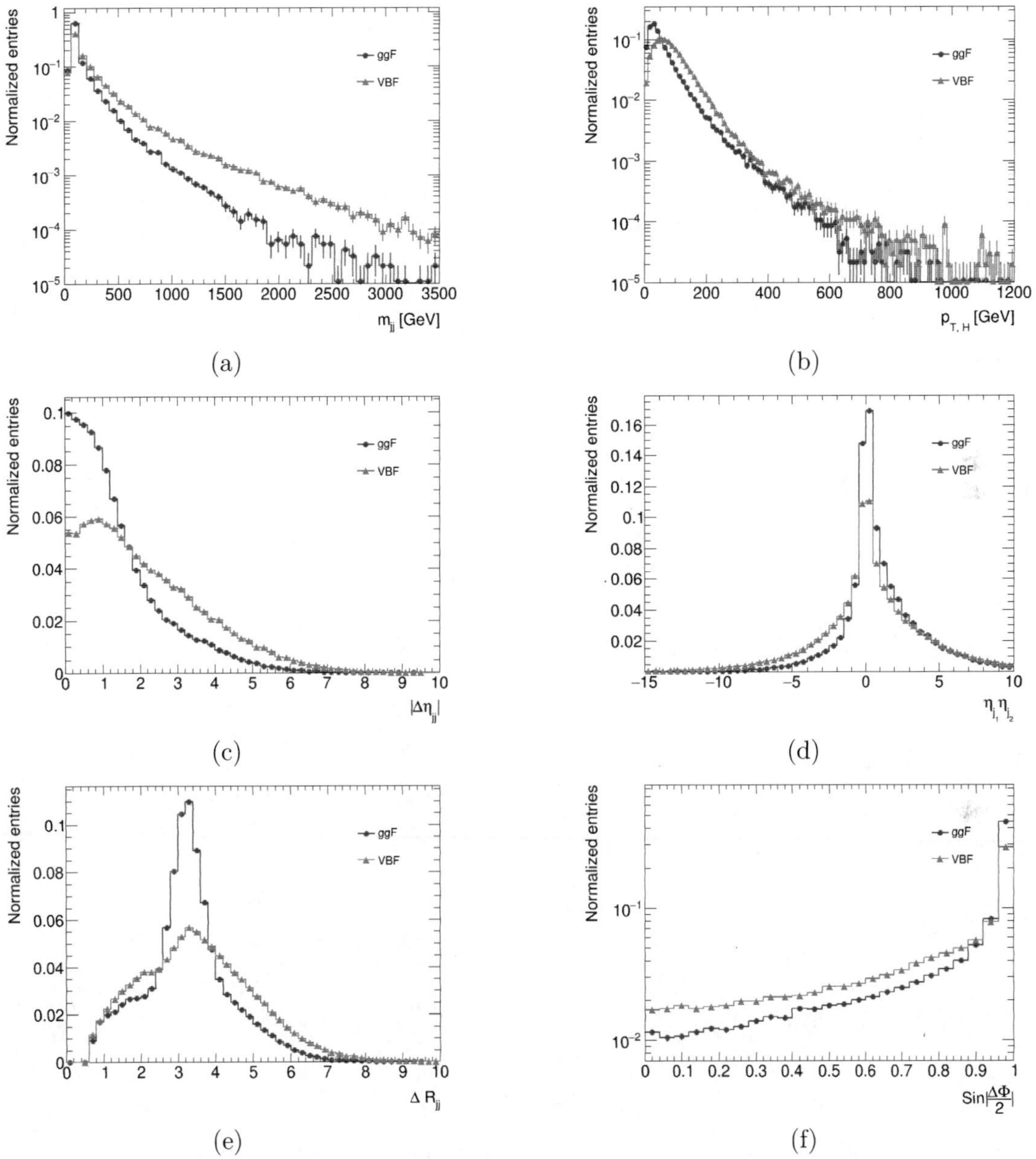

Figure 2. Jet distributions over different kinematic variables: m_{jj} (a), $p_{T,H}$ (b), $|\Delta\eta_{jj}|$ (c), $\eta_{j_1}\eta_{j_2}$ (d), ΔR_{jj} (e) and $\mathrm{Sin}(|\frac{\Delta\Phi}{2}|)$ (f). Blue lines correspond to ggF channel, while the red ones correspond to VBF.

Variables m_{jj} and $p_{T,H}$ are expected to have different shapes due to kinematics distinctions of parton fusion (ggF) and scattering (VBF) processes, but the effect is moderate. Variables $|\Delta\eta_{jj}|$ and $|\Delta R_{jj}|$ are more useful for the ggF-VBF separation due to greater difference in shapes. Variables $\mathrm{Sin}\left(\left|\frac{\Delta\phi_{jj}}{2}\right|\right)$ and $\eta_{j_1} \cdot \eta_{j_2}$ have limited sensitivity to the Higgs boson production channel and their impact on the separation performance should be tested with the BDT discriminators.

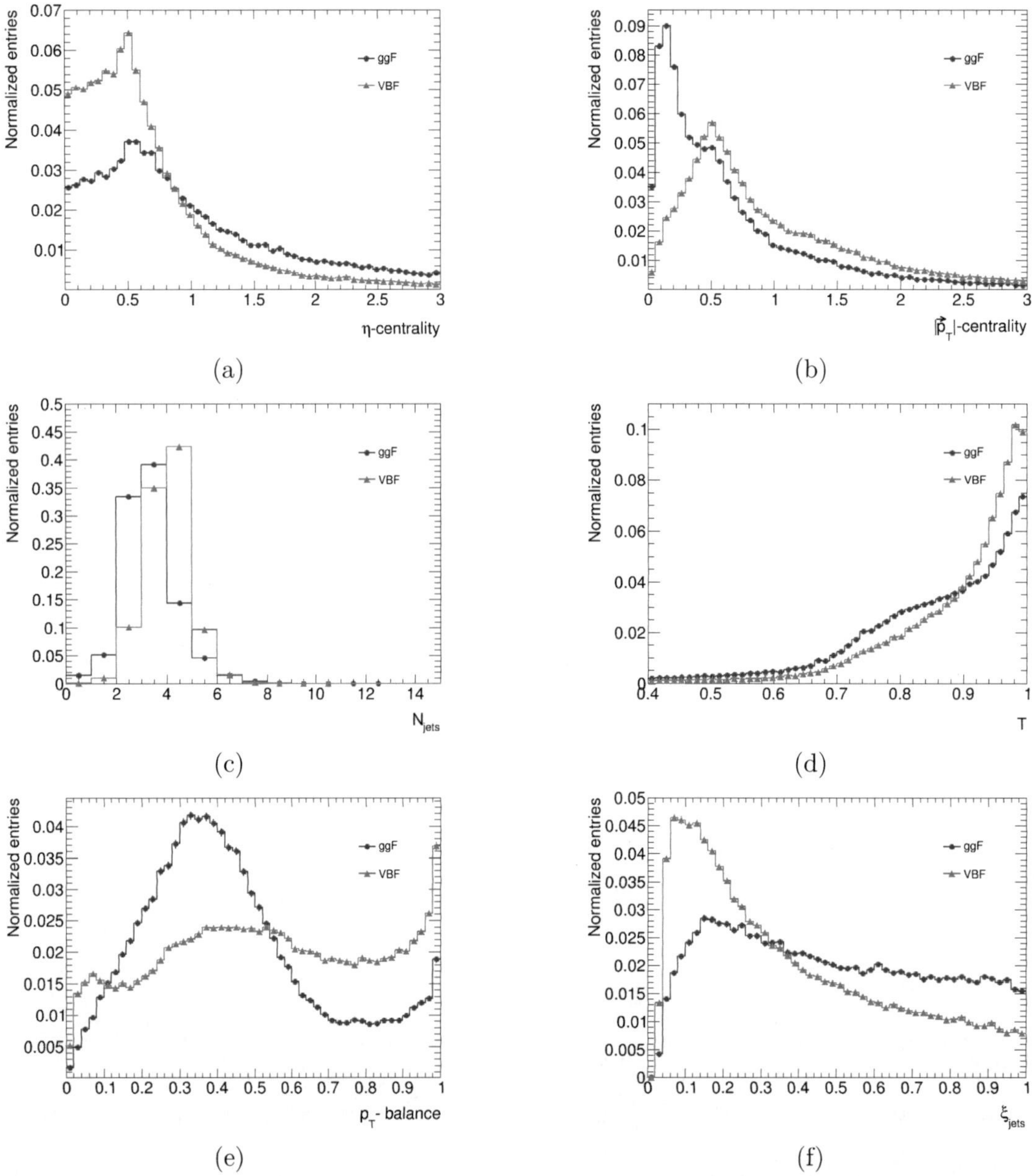

Figure 3. Jet distributions over different kinematic variables: η-centrality (a), $\vec{p}_T$-centrality (b), N_{jets} (c), T (d), p_T-balance (e) and ξ_{jets} (f). Blue lines correspond to ggF channel, while the red ones corresponds to VBF.

Variables η-centrality, $\vec{p}_T$-centrality, N_{jets}, ξ_{jets} and p_T-balance demonstrated in figure 3 have much more sensitivity than the ones shown in figure 2. Observables $\vec{p}_T$-centrality and p_T-balance have qualitatively different shapes which makes them powerful discriminators. Tight cuts can be also set at η-centrality, ξ_{jets} and N_{jets} observables, while thrust variable T have limited sensitivity. All the reviewed variables can be directly implemented into BDT classification algorithm, which will definately improve the ggF-VBF separation.

5. Conclusion

In this paper, the Effective field theory approach was used in order to probe the variables sensitive to the Higgs boson production channel. Kinematic distributions were calculated for both ggF and VBF channels in Standard Model case. The obtained results demonstrate high potential for precise ggF-VBF separation based on the BDT algorithms. Reviewed kinematic variables can be also used to distinguish possible CP violation effects in the Higgs sector. The most important variables in terms of ggF-VBF separation are η-centrality, $\vec{p}_T$-centrality, p_T-balance and ξ_{jets}.

6. Acknowledgements

The work of N. Belyaev on the selection of discriminating variables was performed within the financial support from Russian Science Foundation - grant No.17-72-10021. The work of R. Konoplich is partially supported by the US National Science Foundation under Grant No.PHY-1402964. The work of K. Prokofiev is partially supported by a grant from the Research Grant Council of the Hong Kong Special Administrative Region, China (Project Nos. CUHK4/CRF/13G).

References

[1] Aad G *et al.* 2012 *Phys. Lett.* B **716** 1–29
[2] Chatrchyan S *et al.* 2012 *Phys. Lett.* B **716** 30–61
[3] Aaij R *et al.* 2015 *Phys. Rev. Lett.* **115** 072001
[4] Aaij R *et al.* 2017 *Nature Physics* **13** 391
[5] Aaboud M *et al.* 2017 *Preprint* ATLAS-CONF-2017-046
[6] Kruse A *et al.* 2015 *Phys. Rev.* D **91** 053009
[7] Aaboud M *et al.* 2016 *JHEP* **11** 112
[8] Maltoni F *et al.* 2014 *Eur. Phys. J.* C **74** 3065
[9] Klamke G and Zeppenfeld D 2007 *JHEP* **04** 052
[10] Aaboud M *et al.* 2016 *Preprint* ATL-PHYS-PUB-2016-008
[11] Khachatryan V *et al.* 2015 *Phys. Rev.* D **92** 032008
[12] Bramer M 2007 *Principles of Data Mining* (London: Springer)
[13] Hagiwara K, Li Q and Mawatari K 2009 *JHEP* **07** 101
[14] Hankele V *et al.* 2006 *Phys. Rev.* D **74** 095001
[15] Agashe K, Contino R and Pomarol A 2005 *Nucl. Phys.* B **719** 165–187
[16] Dolan M J *et al.* 2014 *Phys. Rev.* D **90** 073008
[17] Alwall J *et al.* 2011 *JHEP* **06** 128
[18] Ask S *et al.* 2014 *Comp. Phys. Comm.* **191** 159–177
[19] Delaere C *et al.* 2014 *JHEP* **02** 057
[20] Brun R *et al.* 2009 *Comp. Phys. Comm.* **180** 2499–2512

IOP Conf. Series: Journal of Physics: Conf. Series **934** (2017) 012031 doi:10.1088/1742-6596/934/1/012031

Recent results from the NA48 experiment at CERN

Sergey Shkarovskiy on behalf of the NA48/2 Collaboration
Joint Institute for Nuclear Research, Joliot-Curie, 6 Dubna, Moscow region, Russia, 141980
E-mail: Sergey.Shkarovskiy@cern.ch

Abstract. The NA48/2 experiment presents a final result of the charged kaon semileptonic decays form factors measurement based on 4.28 million $K^{\pm}_{e3}$ and 2.91 million $K^{\pm}_{\mu 3}$ selected decays collected in 2004. The result is competetive with other measurements in $K^{\pm}_{\mu 3}$ mode and has a smallest uncertainty for $K^{\pm}_{e3}$, that leads to the most precise combined $K^{\pm}_{l3}$ result and allows to reduce the form factor uncertainty of $|V_{US}|$. The NA48/2 experiment at CERN collected a very large sample of charged kaon decays into multiple final states. From this data sample we have reconstructed about 1663 events of the very rare decay $K^{\pm} \to \mu^{\pm}\nu e^{+}e^{-}$ over almost negligible background in the region with $m(e^{+}e^{-})$ above 140 MeV, which is of great interest in Chiral Perturbation Theory. We present the m_{ee} spectrum and a model-independent measurement of the decay rate for this region.

1. Introduction

The main purpose of the NA48/2 experiment at the CERN SPS was a search for the direct CP violation in $K^{\pm}$ decay to three pions [1]. The experiment used simultaneous K^{+} and K^{-} beams with momenta of 60 GeV/c propagating through the detector along the same beam line.

The main components of the NA48/2 detector were a magnetic spectrometer, composed by four drift chambers and a dipole magnet deflecting the charged particles in the horizontal plane, and a liquid krypton electromagnetic calorimeter (LKr) with an energy resolution of about 1% for 20 GeV photons and electrons. For the selection of muons, a muon veto system (MUV) was essential to distinguish muons from pions.

The data used for the form factor (FF) analysis were collected in 2004 during a dedicated run with a special trigger setup which required one or more tracks in the magnetic spectrometer and an energy deposit of at least 10 GeV/c in the electromagnetic calorimeter.

Radiative decays of K mesons can be used to test theories describing low energy Quantum Chromodynamics (QCD). Chiral Perturbation Theory (ChPT) is one of the successful frameworks for such calculations. This theory is an an effective field theory that describes QCD below the scale Λ of $O(1GeV)$. The pseudo-scalar meson masses are below the cutoff scale and an expansion can be performed in terms of (p/Λ), called the chiral expansion. The Standard Model can be tested using radiative decays in next-to-leading order in the chiral expansion without any further assumptions.

The phase space is dominated by Inner Bremsstrahlung from the final state muon, which is a pure QED process and its contribution to the branching ratio can be calculated exactly. The intriguing part of the phase space is the invariant mass of the $e^{+}e^{-}$ pair in the region $M_{ee} > 140MeV/c^{2}$, where the ChPT form factors have an important contribution to the branching ratio. We use a MC generator based on [2] with all form factors computed up to $O(p^{4})$ in ChPT. Radiative corrections are included in the MC generation using the PHOTOS package. However, above $M_{ee} = 140MeV/c^{2}$ the phase space is greatly reduced, therefore a large number of decayed kaons is needed in order to measure precisely the very small branching ratio.

2. K_{l3} form factors

Semileptonic kaon decays $K^{\pm} \rightarrow \pi^0 l^{\pm} \nu$ (K_{l3}) offer the most precise determination of the CKM matrix element $|V_{US}|$ [3], that require both a branching ratio and a FFs experimental measurement. K_{l3} precision FFs measurement results based on the NA48/2 data analysis are presented in this paper.

The K_{l3} decay width in the absence of electromagnetic effects can be represented by the Dalitz plot density depending on the lepton and pion energies in kaon rest frame E_l and E_π respectively [4]:

$$\frac{d^2\Gamma_0(K_{l3})}{dE_l dE_\pi} = N(Af_+^2(t) + Bf_+(t)f_-(t) + Cf_-^2(t)), \quad (1)$$

where $t = (P_K - P_\pi)^2 = m_K^2 + m_\pi^2 - 2m_K E_\pi$, N is a normalization constant and $f_-(t) = (f_+(t) - f_0(t))(m_K^2 - m_{\pi+}^2)/t$. Here $f_+(t)$ and $f_0(t)$ are the so called vector and scalar K_{l3} FFs, respectively. m_K is a mass of K^+, m_π is a mass of π^0 and m_{π^+} is a mass of π^+. Definitions of the implemented parameterizatios are shown

	$f_+(t)$	$f_0(t)$
Quadratic	$1 + \lambda'_+ t/m_\pi^2 + \frac{1}{2}\lambda''_+(t/m_\pi^2)^2$	$1 + \lambda'_0 t/m_\pi^2$
Pole	$\frac{M_V^2}{(M_V^2 - t)}$	$\frac{M_S^2}{(M_S^2 - t)}$
Dispersive	$\exp\left(\frac{(\Lambda_+ + H(t))t}{m_\pi^2}\right)$	$\exp\left(\frac{(ln[C] - G(t))t}{(m_K^2 - m_\pi^2)}\right)$

Table 1. Definitions of the FF parameterizations used in the analysis

in the table 1: the Quadratic [5] parameterization (fit parameters $\lambda'_+, \lambda''_+, \lambda'_0$), the Pole [6] (fit parameters M_V, M_S) and the Dispersive [7] one (fit parameters $\Lambda_+, ln[C]$).

2.1. *The K_{l3} events reconstruction and selection*

The data selection requires one track in the magnetic spectrometer and a time coincidence with at least two clusters in the LKr from the π^0 decay. The track had to be inside the geometrical acceptance of the detector, and needed a good reconstructed decay vertex, proper timing and a momentum $p > 5GeV/c$ in case of electrons. For muons the momentum needed to be greater than $10GeV/c$ to ensure proper efficiency of the MUV system. To identify a track as the electron we require $2.0 > E/p > 0.9$, where E is the energy deposited in the LKr and p is the momentum measured in the spectrometer, and no signal in the MUV system. To identify the track as a muon we require an associated hit in the MUV system and $E/p < 0.9$.

Longitudinal K_{l3} decay position Z_n (neutral vertex Z coordinate) is defined as a longitudinal position of π^0 decay, reconstructed from LKr data assuming PDG [8] value for π^0 mass. The transverse neutral vertex coordinates (X_n, Y_n) are calculated as the impact point position of the reconstructed charged track on the Z_n plane.

For the kaon momentum (P_K) measurement we direct Z axis along the beam average position in space, measured from $3\pi^{\pm}$ data. In the assumptions of $m(\nu) = 0$ and kaon flight along the beam axis (that means availability of the measured $P_t(\nu) = -P_t$), two solutions of quadratic equation for P_K exist: $P_K = P_{1,2} = (\phi P_Z \pm \sqrt{d})/(E^2 - P_Z^2)$, where $\phi = 0.5(M_K^2 + E^2 - P_t^2 - P_Z^2)$, $d = \phi^2 P_Z^2 - (E^2 - cxP_Z^2)(M_K^2 E^2 - \phi^2))$ and E, P_t, P_Z are the total energy and total momentum of all the registered particles π^0, l.

The background contribution has been estimated using the NA48/2 Monte Carlo. For K_{e3} the background from $K^{\pm} \rightarrow \pi^{\pm}\pi^0$ (2π) significantly contributes to the signal. A cut in the $P_t(\nu) >= 0.03\ GeV/c$ of the event reduced this background to less than 0.027%. For $K_{\mu3}$ selection, essential background may come from 2π decays with a subsequent $\pi^{\pm} \rightarrow \mu^{\pm}\bar{\nu}$. The cuts in $m(\pi^{\pm}\pi^0)$ and $m(\mu\bar{\nu})$

IOP Conf. Series: Journal of Physics: Conf. Series **934** (2017) 012031 doi:10.1088/1742-6596/934/1/012031

reduces the contamination to 0.0264%. For both $K_{\mu 3}$ and K_{e3} samples the $K^{\pm} \rightarrow \pi^{\pm}\pi^{0}\pi^{0}$ decays can contribute to the background. Its estimated contribution is about 0.183% for $K_{\mu 3}$ and 0.0286% for K_{e3}.

The total statistics of selected data is 4.28×10^{6} events for K_{e3}, and 2.91×10^{6} events for $K_{\mu 3}$ selection.

2.2. Preliminary results on the K_{l3} form factors

To extract the FF a events-weighting fit is performed in $5 \times 5 MeV$ cells in the Dalitz plot of $E^{*}_{\pi^0}$ vs E^{*}_{l} energies, computed in the kaon rest frame. The K_{e3} and $K_{\mu 3}$ Dalitz plots were fitted simultaneously with a common set of the fit parameters MINUIT [9] package called from the ROOT [10] interface minimizes χ^2 by means of parameters variation, and in such a way the resulting fit parameter values, their errors (table 2) and correlation coefficients (table 3) are found. The fit quality is defined by the following values

Quadratic	λ'_+	λ''_+	λ_0
$K_{\mu 3}$	$23.32 \pm 3.08_{stat} \pm 3.50_{syst}$	$2.14 \pm 1.06_{stat} \pm 0.96_{syst}$	$14.33 \pm 1.11_{stat} \pm 1.25_{syst}$
K_{e3}	$23.52 \pm 0.78_{stat} \pm 1.29_{syst}$	$1.60 \pm 0.30_{stat} \pm 0.39_{syst}$	
K_{l3}	$23.35 \pm 0.75_{stat} \pm 1.23_{syst}$	$1.73 \pm 0.29_{stat} \pm 0.41_{syst}$	$14.90 \pm 0.55_{stat} \pm 0.80_{syst}$
Pole	m_V	m_S	
$K_{\mu 3}$	$879.1 \pm 8.1_{stat} \pm 13.5_{syst}$	$1196.4 \pm 18.1_{stat} \pm 28.8_{syst}$	
K_{e3}	$896.8 \pm 3.4_{stat} \pm 7.6_{syst}$		
K_{l3}	$894.3 \pm 3.2_{stat} \pm 5.4_{syst}$	$1185.5 \pm 16.6_{stat} \pm 35.5_{syst}$	
Dispersive	Λ_+	$ln[C]$	
$K_{\mu 3}$	$23.55 \pm 0.50_{stat} \pm 0.97_{syst}$	$186.68 \pm 5.12_{stat} \pm 9.23_{syst}$	
K_{e3}	$22.54 \pm 0.20_{stat} \pm 0.62_{syst}$		
K_{l3}	$22.67 \pm 0.18_{stat} \pm 0.55_{syst}$	$186.12 \pm 4.91_{stat} \pm 11.09_{syst}$	

Table 2. Fit results for the Quadratic ($\times10^3$), Pole (MeV/c^2) and Dispersive ($\times10^3$) Parameterisation

of χ^2/ndf : 1004.6/1073 for the Quadratic Parameterization, 1001.1/1074 for the Pole and 998.3/1074 for the Dispersive one.

	λ''_+	λ_0		m_S	$ln[C]$
λ'_+	-0.954	-0.076	m_V	-0.278	
λ''_+		0.035	Λ_+		-0.035

Table 3. Full uncertainty correlation coefficients for K_{l3} Quadratic, Pole and Dispersive Parametrizations

The NA48/2 is the first experiment measuring the FF using both K^+ and K^-. In $K_{\mu 3}$ the result is dominated by the statistical error, for K_{e3} by the systematic. The NA48/2 K_{e3} and $K_{\mu 3}$ in agreement within each other and our combined results are competitive with the current world average.

3. The $K^{\pm} \rightarrow \mu^{\pm}\nu_{\mu}e^{+}e^{-}$ analysis

The analysis of the mode $K^{\pm} \rightarrow \mu^{\pm}\nu_{\mu}e^{+}e^{-}$ ($K_{\mu vee}$) is based on the reconstruction of a three-track vertex. Two of those tracks have to be identified as electrons using the LKr calorimeter and to have an invariant mass $M_{ee} > 140 MeV/c^2$. The third track has to be compatible with a muon, acting as a minimum ionising particle (MIP) in the LKr and leaving a signal in the MUV detector.

IOP Conf. Series: Journal of Physics: Conf. Series **934** (2017) 012031 doi:10.1088/1742-6596/934/1/012031

3.1. Event selection

The $K_{\mu\nu ee}$ selection is characterized by the presence of three charged tracks and a missing momentum carried away by the undetected neutrino.The missing mass for such a mode is equivalent to a neutrino mass squared and should be zero for signal events. The phase space region of $M_{ee} > 140MeV/c^2$ is experimentally clean, because decays with e^+e^- pairs coming from $K^\pm \to \pi^\pm\pi^0$ or $K^\pm \to \pi^0\mu^\pm\nu_\mu$ decay with a following $\pi^0 \to e^+e^-\gamma$ decay ($m_{\pi^0} = 135MeV/c^2$) are fully suppressed.

The three tracks have to form a good vertex inside the 98m long decay volume of the experiment and to have a total charge of $| Q |= \pm 1$. Each of the tracks has to pass through the geometrical acceptance of the DCH, HOD, LKr and MUV detector and to have momenta in the range between 3 and $50GeV/c$. The showers in the LKr produced by the charged tracks have to be isolated from each other by at least 20cm to avoid overlapping showers. The total momentum of the three tracks has to be $p_{3track} < 66GeV/c$ in order to be consistent with the beam kaons. A cut is applied on the invariant mass of the muon-neutrino system of $M_{\mu\nu} > 170MeV/c^2$, to suppress events coming from $K^\pm \to \pi^\pm e^+e^-$ followed by $\pi^\pm \to \mu^\pm\nu_\mu$.

The particle identification is based on the LKr calorimeter and the MUV system. The ratio E/p of the energy deposited in the LKr divided by the total momentum distinguishes between muons, pions and electrons. The muon is required to leave only a small fraction of its energy in the LKr ($E/p < 0.2$) and to have a positive identification in the MUV detector. Electrons should leave all their energy in the LKr with an energy deposition shower shape different from the shape of hadronic showers produced by pions. To distinguish between genuine electrons and misidentified pions we require $0.95 < E/p < 1.05$ and use a linear discriminant variable, which makes use of the different shape and starting point of the created showers. This discriminant provides almost complete suppression of pions coming from $K_{3\pi}$ or $K^\pm \to \pi^+\pi^-e^\pm\nu_e$ decays.

The mode $K^\pm \to \pi^\pm\pi^+\pi^-$ ($K_{3\pi}$) is chosen as normalization for the $K_{\mu\nu ee}$ branching ratio measurement, because the topologies of both decays are similar.

The NA48/2 detector recorded more than 10^{11} decayed kaons in the fiducial volume of the experiment, which makes it suitable for rare decay rate measurements as $K_{\mu\nu ee}$. The acceptance for this decay mode is M_{ee} dependent and of the order of 12-15%

3.2. Background

After the full selection only small fraction of background processes survives. The remaining background is composed of decays with multiple pions in the final state misidentified as either muons or electrons. It has three components: $K^\pm \to \pi^\pm\pi^+\pi^-$, $K^\pm \to \pi^+\pi^-e^\pm\nu_e$ and $K^\pm \to \pi^\pm\pi^0\pi^0$, followed by two Dalitz decays of the two π^0s ($e^+e^-\gamma$). The two electrons have to come from two different π^0 decays so the M_{ee} can be higher than $140MeV/c^2$. The background contamination coming from those modes can be studied directly in data using Wrong Sign (WS) selection. The WS selection is the same as the signal selection, but requiring two same-sign electrons/positrons and an opposite sign muon ($\mu^+e^-e^-$ or $\mu^-e^+e^+$). Those Lepton Flavor Violating modes do not exist in the SM, therefore the only surviving events come from processes constituting the background for the $K_{\mu\nu ee}$ selection. A scaling factor is applied to the WS selected data in order to take into account that the modes with three pions in the final state have two possibilities to enter the signal selection and only one to enter the WS selection.

3.3. Results

After the event selection, 1663 signal candidates are observed with an estimated background contamination of $54 \pm 10(stat) \pm 5(syst)$ events. The total number of kaons decayed in the fiducial volume of the experiment is $(1.56 \pm 0.01) \times 10^{11}$ as obtained from the number of normalisation events corrected for acceptance, trigger efficiency and branching ratio. The branching ratio is computed for each of the $z = (M_{ee}/M_K)^2$ bins (15 bins in total) and shown on figure 1. The results obtained in each bin are then summed to get the total branching ratio. This minimizes the effect of the z dependence of the signal acceptance. The total model-independent branching ratio in the phase space $M_{ee} \geq 140MeV/c^2$ is $Br(K^\pm \to \mu^\pm\nu_\mu e^+e^-) = (7.84 \pm 0.21(stat) \pm 0.08(syst) \pm 0.06(ext)) \times 10^{-8}$. Radiative corrections, not available in [2] have been included in the acceptance calculation using the PHOTOS package. The systematic uncertainty is 1.2% and is dominated by the effect of radiative corrections on the signal acceptance and the background contamination. The external error is due to the uncertainty on the normalization channel branching ratio $Br(K\pm \to \pi^{\pm}\pi^{+}\pi^{-})$ [8].

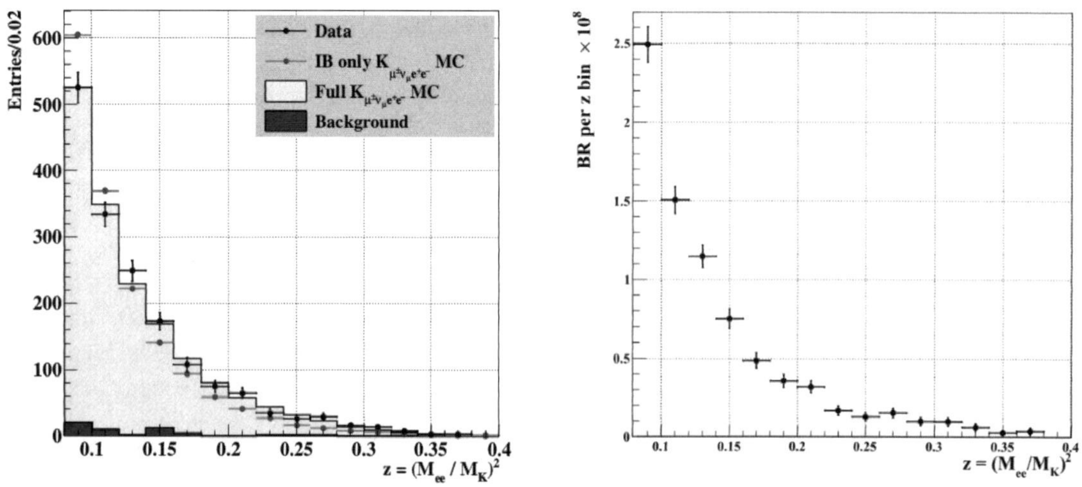

Figure 1. On the left: z distribution after the final selection. Data is presented with points with errorbars. On the right: Branching ratio of the process $K^{\pm} \to \mu^{\pm}\nu_{\mu}e^{+}e^{-}$ calculated for each bin of the z distribution. The final result is the sum of the individual contributions in each bin.

Conclusion

K_{l3} form factors measurement is performed by NA48/2 experiment on the basis of 2004 run data. Result is competetive with the other ones in $K^{\pm}_{\mu3}$, and a smallest error in $K^{\pm}_{e3}$ has been reached, that gives us also the combined result with the smallest error.

The branching ratio of the $K^{\pm} \to \mu^{\pm}\nu_{\mu}e^{+}e^{-}$ decay mode has been measured with 3% background contamination in the phase space region $M_{ee} > 140MeV/c^2$. The achieved uncertainty is improved by a factor 1.5 with respect to the previous most precise measurement [11] and is statistically dominated. It is the first measurement of this mode where radiative corrections are included. The observed spectrum is in agreement with the predictions by ChPT [2]. The result is also compatible with previous measurements [11],[12].

References

[1] Batley J R *et al.* (NA48/2) 2007 *Eur. Phys. J.* **C52** 875–891 (*Preprint* 0707.0697)
[2] Bijnens J Ecker G and Gasser J 1993 *Nucl. Phys* **B396** 81–118 (*Preprint* 9209261)
[3] Antonelli M *et al.* (FlaviaNet Working Group on Kaon Decays) 2010 *Eur.Phys.J.* **C69** 399–424 (*Preprint* 1005.2323)
[4] Chounet L M, Gaillard J M and Gaillard M 1972 *Phys.Rept.* **4** 199–324
[5] Olive K *et al.* (Particle Data Group) 2014 *Chin.Phys.* **C38** 090001
[6] Lichard P 1997 *Phys.Rev.* **D55** 5385–5407 (*Preprint* hep-ph/9702345)
[7] Bernard V, Oertel M, Passemar E and Stern J 2009 *Phys.Rev.* **D80** 034034 (*Preprint* 0903.1654)
[8] Patrignani C *et al.* (Particle Data Group) 2016 *Chin. Phys.* **C40** 100001
[9] James F and Roos M 1975 *Comput. Phys. Commun.* **10** 343–367
[10] Brun R and Rademakers F 1997 *Nucl. Instrum. Meth.* **A389** 81–86
[11] Poblaguev A A *et al.* (E865) 2002 *Phys. Rev. Lett* **89** 4 (*Preprint* 0204006)
[12] Diamant-Berger A M *et al.* (E865) 1976 *Phys. Lett* **62B** 485–490

Scattering of kinks in a non-polynomial model

D Bazeia,[1] E Belendryasova[2] and V A Gani[2,3]

[1]Departamento de Física, Universidade Federal da Paraíba, 58051-900 João Pessoa, Paraíba, Brazil

[2]National Research Nuclear University MEPhI (Moscow Engineering Physics Institute), 115409 Moscow, Russia

[3]National Research Center Kurchatov Institute, Institute for Theoretical and Experimental Physics, 117218 Moscow, Russia

E-mail: vagani@mephi.ru

Abstract. We study a model described by a single real scalar field in the two-dimensional space-time. The model is specified by a potential which is non-polynomial and supports analytical kink-like solutions that are similar to the standard kink-like solutions that appear in the φ^4 model when it develops spontaneous symmetry breaking. We investigate the kink-antikink scattering problem in the non-polynomial model numerically and highlight some specific features, which are not present in the standard case.

1. Introduction

Topological defects are extremely important for theoretical physics. They are used for describing models in cosmology high energy physics, quantum and classical field theory, condensed matter and so on. In this context, the $(1+1)$-dimensional models are of special interest, because the dynamics of many three- and two-dimensional systems can be reduced to the one-dimensional models [1]–[23].

Consider a field-theoretical model in $(1+1)$-dimensional space-time with a real scalar field $\varphi(x,t)$ with the Lagrangian density

$$\mathcal{L} = \frac{1}{2}\left(\frac{\partial\varphi}{\partial t}\right)^2 - \frac{1}{2}\left(\frac{\partial\varphi}{\partial x}\right)^2 - U(\varphi), \tag{1}$$

where the potential $U(\varphi)$ is a non-negative function with two or more degenerate minima (vacua of the model): $\varphi_1^{(0)}$, $\varphi_2^{(0)}$, $\dots$, $U(\varphi_1^{(0)}) = U(\varphi_2^{(0)}) = ... = 0$. The equation of motion for the field φ is

$$\frac{\partial^2\varphi}{\partial t^2} - \frac{\partial^2\varphi}{\partial x^2} + \frac{dU}{d\varphi} = 0. \tag{2}$$

The energy functional corresponding to the Lagrangian (1) reads

$$E[\varphi] = \int_{-\infty}^{\infty}\left[\frac{1}{2}\left(\frac{\partial\varphi}{\partial t}\right)^2 + \frac{1}{2}\left(\frac{\partial\varphi}{\partial x}\right)^2 + U(\varphi)\right] dx. \tag{3}$$

In this model, kinks and antikinks are static solutions which minimize energy and interpolate between neighboring vacua.

IOP Conf. Series: Journal of Physics: Conf. Series **934** (2017) 012032 doi:10.1088/1742-6596/934/1/012032

2. The sinh-deformed φ^4 model

This paper deals with the model which we call the sinh-deformed φ^4 model. The potential $U(\varphi)$ of this model can be obtained from the potential $V(\varphi) = \frac{1}{2}(1-\varphi^2)^2$ of the standard φ^4 model by deformation procedure which was elaborated and applied in [24]–[28]:

$$U(\varphi) = \frac{V(\varphi \to f(\varphi))}{(df/d\varphi)^2}, \tag{4}$$

where "$f(\varphi)$" — the deforming function and "$\varphi \to f(\varphi)$" means that φ is replaced by $f(\varphi)$. So,

$$U(\varphi) = \frac{1}{2}\,\mathrm{sech}^2\varphi\,\left(1 - \sinh^2\varphi\right)^2. \tag{5}$$

This potential has two minima with $\varphi_\pm = \pm\mathrm{arsinh}\,1$, $U(\varphi_\pm) = 0$, figure 1(a). The kink of the new model, $\varphi_{\mathrm{k}}^{(\mathrm{new})}(x)$, is related with kink of the φ^4 model, $\varphi_{\mathrm{k}}^{(\mathrm{old})}(x)$, by the inverse deforming function f^{-1},

$$\varphi_{\mathrm{k}}^{(\mathrm{new})}(x) = f^{-1}(\varphi_{\mathrm{k}}^{(\mathrm{old})}(x)). \tag{6}$$

Using $\varphi_{\mathrm{k}}^{(\mathrm{old})}(x) = \tanh x$, we obtain kink and antikink of the sinh-deformed φ^4 model:

$$\varphi_{\mathrm{k}}(x) = \mathrm{arsinh}(\tanh x), \quad \varphi_{\bar{\mathrm{k}}}(x) = -\mathrm{arsinh}(\tanh x), \tag{7}$$

see figure 1(b).

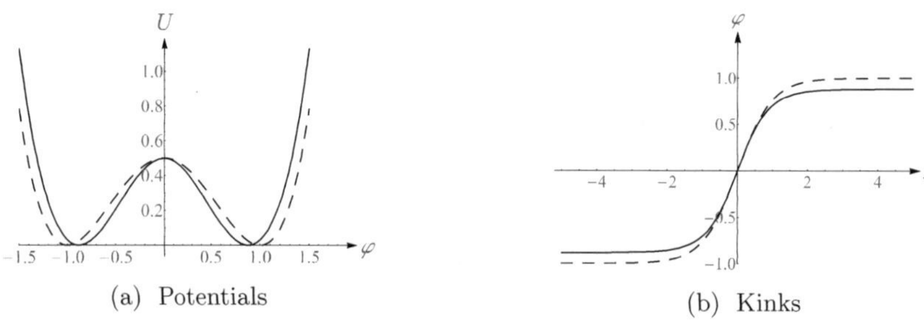

Figure 1. Potentials and kinks of the φ^4 (dashed curves) and the sinh-deformed φ^4 (solid curves) models.

3. Kink-antikink scattering

We studied the kink-antikink scattering in the sinh-deformed φ^4 model using the initial configuration

$$\varphi(x,t) = \mathrm{arsinh}\left[\tanh\left(\frac{x + x_0 - v_{\mathrm{in}}t}{\sqrt{1-v_{\mathrm{in}}^2}}\right)\right] - \mathrm{arsinh}\left[\tanh\left(\frac{x - x_0 + v_{\mathrm{in}}t}{\sqrt{1-v_{\mathrm{in}}^2}}\right)\right] - \mathrm{arsinh}\,1, \tag{8}$$

where x_0 is half-distance between two kinks, and they start moving towards each other with the initial velocities v_{in}. We used $2x_0 = 10$ and the following parameters of the numerical scheme: $\delta t = 0.008$, and $\delta x = 0.01$

We found the critical value of the initial velocity $v_{\mathrm{cr}} \approx 0.4639$, which separates two different regimes of the kinks scattering. At $v_{\mathrm{in}} > v_{\mathrm{cr}}$ the kinks escape to infinities after one collision.

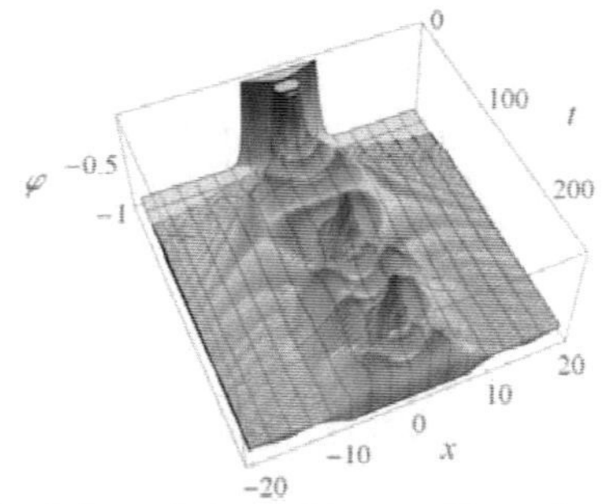

(a) A bound state of two oscillons at $v_{\text{in}} = 0.44190$

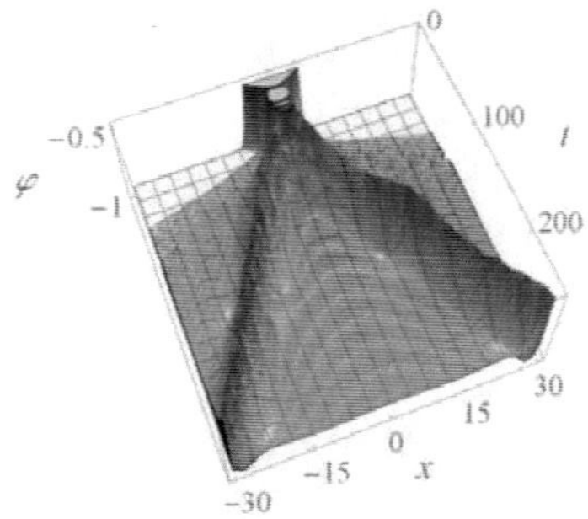

(b) An escape of two oscillons at $v_{\text{in}} = 0.44195$

Figure 2. Formation of two oscillons in the kink-antikink collision.

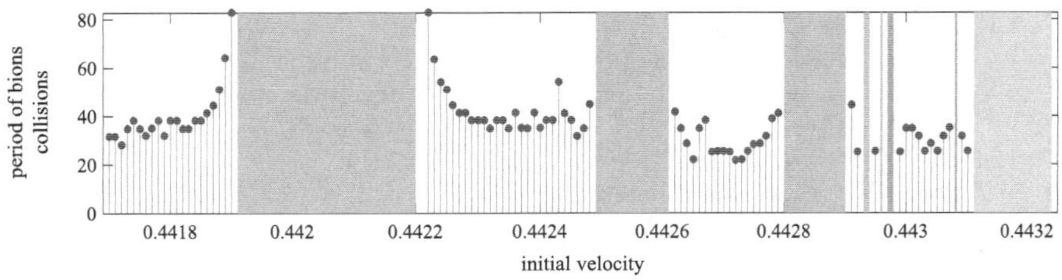

Figure 3. The period of oscillations of the oscillons as a function of the initial velocity of the colliding kinks. The blue-shaded areas denote the escape of oscillons. The green-shaded area denotes the 2-bounce escape window for kinks, the pink-shaded areas denote the 4-bounce escape windows for kinks.

At the initial velocities $v_{\text{in}} < v_{\text{cr}}$ there is the kinks' capture and formation of their bound state, and many escape windows. The picture looks similar to the φ^4 case.

Besides that, in the range $v_{\text{in}} < v_{\text{cr}}$ we have found new interesting result, which has not been observed for the φ^4 kinks. In many cases the final configuration looked like two oscillons, which can form a bound state, figure 2(a), or escape to infinity, figure 2(b). In a bound state the oscillons' oscillate near each other with the period which depends on the initial velocity of the colliding kinks, see figure 3.

4. Conclusion

We studied the kink-antikink scattering in the sinh-deformed φ^4 model.

There is the critical initial velocity $v_{\text{cr}} \approx 0.4639$, which separates two regimes of the kinks scattering. At $v_{\text{in}} > v_{\text{cr}}$ the kinks escape to infinities after one collision. At the initial velocities $v_{\text{in}} < v_{\text{cr}}$ we observed formation of kinks' bound state and resonance phenomena. This pattern is similar to that found in the φ^4 model. Moreover we found new interesting phenomenon — formation of two oscillons which can form a bound state or escape to infinities. It would be very interesting to investigate the nature of appearance of oscillons. We believe that it could become a subject of further study.

5. Acknowledgments

This work was performed using resources of the NRNU MEPhI high-performance computing center. The research was supported by the Brazilian agency CNPq under contracts 455931/2014-3 and 306614/2014-6, and by the MEPhI Academic Excellence Project under contract No. 02.a03.21.0005, 27.08.2013.

References

[1] Gani V A, Ksenzov V G and Kudryavtsev A E 2010 *Phys. Atom. Nucl.* **73** 1889 [*Yad. Fiz.* **73** 1940]
[2] Gani V A, Ksenzov V G and Kudryavtsev A E 2011 *Phys. Atom. Nucl.* **74** 771 [*Yad. Fiz.* **74** 797]
[3] Bazeia D and Mohammadi A 2017 *Eur. Phys. J.* **C 77** 203
[4] Bazeia D, Marques M A and Menezes R 2016 *Eur. Phys. J.* **C 76** 241
[5] Lensky V A, Gani V A and Kudryavtsev A E 2001 *Sov. Phys. JETP* **93** 677 *Zh. Eksp. Teor. Fiz.* **120** 778
[6] Bazeia D, Freire W, Losano L and Ribeiro R F 2002 *Mod. Phys. Lett.* **A 17** 1945
[7] Gani V A, Konyukhova N B, Kurochkin S V and Lensky V A 2004 *Comput. Math. Math. Phys.* **44** 1968 [*Zh. Vychisl. Mat. Mat. Fiz.* **44** 2069] [(*Preprint* arXiv: 0710.2975)]
[8] Bazeia D, Lobão Jr A S, Losano L and Menezes R 2014 *Eur. Phys. J.* **C 74** 2755
[9] Gani V A, Lizunova M A and Radomskiy R V 2016 *JHEP* **04** 043
[10] Gani V A, Lizunova M A and Radomskiy R V 2016 *J. Phys.: Conf. Ser.* **675** 012020
[11] Gani V A, Kudryavtsev A E and Lizunova M A 2014 *Phys. Rev.* **D 89** 125009
[12] Moradi Marjaneh A, Gani V A, Saadatmand D, Dmitriev S V and Javidan K 2017 *JHEP* **07** 028
[13] Gani V A, Lensky V and Lizunova M A 2015 *JHEP* **08** 147
[14] Gani V A, Lensky V, Lizunova M A and Mrozovskaya E V 2016 *J. Phys.: Conf. Ser.* **675** 012019
[15] Bazeia D, González León M A, Losano L and Mateos Guilarte J 2006 *Phys. Rev.* **D 73** (2006) 105008
[16] Radomskiy R V, Mrozovskaya E V, Gani V A and Christov I C 2017 *J. Phys.: Conf. Ser.* **798** 012087
[17] Belendryasova E and Gani V A *Scattering of the φ^8 kinks with power-law asymptotics*, *Preprint* `arXiv:1708.00403`.
[18] Bazeia D, Belendryasova E and Gani V A *Scattering of kinks of the sinh-deformed φ^4 model*, *Preprint* `arXiv:1710.04993`.
[19] Gani V A, Moradi Marjaneh A, Askari A, Belendryasova E and Saadatmand D *Scattering of the double sine-Gordon kinks*, *Preprint* `arXiv:1711.01918`.
[20] Gani V A and Kudryavtsev A E 1999 *Phys. Rev.* **E 60** 3305
[21] Gani V A et al 1999 *Phys. Atom. Nucl.* **62** 895 [*Yad. Fiz.* **62** 956] [(*Preprint* `hep-ph/9712526`)]
[22] Belova T I, Gani V A and Kudyavtsev A E 2001 *Phys. Atom. Nucl.* **64** 140 [*Yad. Fiz.* **64** 143] [(*Preprint* `hep-ph/0003308`)]
[23] Gani V A and Kudryavtsev A E 2001 *Phys. Atom. Nucl.* **64** 2043 [*Yad. Fiz.* **64** 2130]
[24] Bazeia D, Losano L and Malbouisson J M C 2002 *Phys. Rev.* **D 66** 101701
[25] Almeida C A, Bazeia D, Losano L and Malbouisson J M C 2004 *Phys. Rev.* **D 69** 067702
[26] Bazeia D and Losano L 2006 *Phys. Rev.* **D 73** 025016
[27] Bazeia D, González León M A, Losano L, and Mateos Guilarte J 2006 *Phys. Rev.* **D 73** 105008
[28] Bazeia D, Lima E E M and Losano L 2017 *Int. J. Mod. Phys.* **A** 32 1750163

IOP Conf. Series: Journal of Physics: Conf. Series **934** (2017) 012033 doi:10.1088/1742-6596/934/1/012033

Mass spectra of bottomonia using relativistic corrections to the potential

Ajay Kumar Rai and Raghav Chaturvedi

Department of Applied Physics , Sardar Vallabhbhai National Institute of Technology, Surat, Gujarat-395 007, India

E-mail: raiajayk@gmail.com, raghavr.chaturvedi@gmail.com

Abstract. We have calculated the masses of different states of bottomonia in the framework of potential non-relativistic QCD, non-perturbatively. The potential consists of two terms, a static term incorporating Coulombic plus confinement part along with a constant and a relativistic correction term classified in powers of the inverse of heavy quark mass $O(1/m)$. Spin-spin interaction has been considered to calculate the masses of the singlet and the triplet states of bottomonia. The Schrödinger equation is solved numerically to obtain the masses from 1S to 6S states, the calculated masses for $\eta_b(1S)$ is 9.390 GeV and for $\Upsilon(1S)$ is 9.459 GeV and is found to be in good accordance with the experimental results.

1. Introduction

Bottomonia is a bound state of bottom quark and its antiquark, it was first discovered at Fermilab in the bound state of γ in 1977. Bottomonia system is known to have non-relativistic nature it can be treated as two body system of heavy quark and anti-quark in quantum mechanics. Many experimental groups such as BABAR, Belle, CLEO-III, ATLAS, CMS and HERA-B are producing and expected to produce more precise data [1]. It is a powerful tool for understanding the quark anti-quark interaction in particular QCD. Many theoretical methods have successfully explained the mass spectra and decay of heavy quarkonia namely, LQCD, QCD sum rules, Perturbative QCD, Lattice NRQCD, field theories, Phenomenological potential models [2–7].

It is difficult to obtain the quarkonia potential starting from the basic principle of quantum chromodynamics(QCD) so it is necessary to account for non-perturbative and non-relativistic effects connected with complicated structure of QCD vacuum. The light quarks in a meson move at relativistic speeds. However, the speed of the charm and bottom quark is sufficiently smaller, so that relativistic effects affect these states much less, also the binding energy is small compared to the rest energy of the constituents. All this led to a theoretical uncertainty in the potential at large and intermediate distances which led to the beginning of potential models. Among many theoretical attempts to explain the hadron properties based on its quark structure very few models were successful in predicting the hadronic properties starting from mass spectra to decay widths. We start with the non-relativistic Hamiltonian consider a potential and solve Schrödinger equation numerically to obtain the masses. Most common of all the potential is Coulomb plus confining, it has been supported by static potential computed by using lattice QCD simulations, which offers powerful tool for the nonperturbative study of QCD.

IOP Conf. Series: Journal of Physics: Conf. Series **934** (2017) 012033 doi:10.1088/1742-6596/934/1/012033

As the mass of bottom quark is not infinitely heavy, in the present study in addition to the Coulomb plus confining potential a relativistic correction has been considered, which depends on quark spin. This relativistic correction to the mass has been derived [2, 3] in the framework of pNRQCD which is an effective field theory, and are usually classified in powers of the inverse of heavy quark mass or velocity. In heavy quarkonia system, there exists a hierarchy $m >> mv >> mv^2$, with $m >> \Lambda_{QCD}$, Λ_{QCD} is a QCD scale parameter assumed to be few hundred MeV. To have better control of such hierarchy is to employ effective field theory for high precision calculation. Two such EFT's have been used by many authors in past namely nonrelativistic QCD(NRQCD) [4, 5] and potential NRQCD (pNRQCD) [6]. NRQCD has been derived by integrating the energy scale above m in QCD, and pNRQCD by by integrating further the energy scale above mv in NRQCD. The higher energy contribution is incorporated in the effective couplings called the matching coefficients.

2. Theoretical framework

Considering it as non-relativistic system in order to calculate the mass of bottomonia we consider it as a molecule. We use the following Hamiltonia [7]

$$H = M + \frac{P^2}{2m} + V_{pNRQCD}(r) + V_{SD}(r) \tag{1}$$

Here, M is the total mass of the system and m is the reduced mass of the system. The interaction potential $V_{pNRQCD}(r)$ encompasses three terms a coulombic term V_v(r)(vector), a confinement term V_s(scalar) and relativistic correction in the framework of pNRQCD; $V_p(r)$ [8]. In order to calculate the S wave mass spectra we only consider spin-spin interaction in spin dependent term $V_{SD}(r)$ [9–11]

$$V_{pNRQCD}(r) = V_v(r) + V_s(r) + V_p(r) = -\frac{4\alpha_s}{3r} + \frac{Ar}{n^{0.16}} + \frac{1}{m}V^{(1)}(r) \tag{2}$$

$$V^{(1)}(r) = -\frac{9\alpha_c^2}{8r^2} + a\log r + C(const.) \tag{3}$$

In order to obtain the spin singlet-triplet hyperfine splitting of S wave states of bottomonia, we consider spin dependent part of the usual one gluon exchange potential given by,

$$V_{SD}(r) = V_{SS}(r)\left[S(S+1) - \frac{3}{2}\right] \tag{4}$$

Here, we first compute the mass spectra of bottomonia states to determine quark mass and confinement strength after fitting the spin averaged ground state masses with experimental data of respective mesons. The values of the potential parameters have been calculated and used in the calculation of mass spectra by solving the Schrödinger equation considering the eigne value problem and solve by RK4 method numerically. Also the normalised wave function for respective states can be calculated numerically.

3. Result and Discussion

In this paper we have calculated the masses of S wave states of Bottomonia by solving the Schrödinger equation numerically considering Coulomb plus confining potential along with relativistic correction. This relativistic correction to the mass has been derived in the framework

Table 1. S wave spectra of $b\bar{b}$ meson (in GeV)

state	Our	PDG [1]	[12]	[13]	[14]	[15]
1^1S_0	9.399	9.398	9.398	9.402	9.390	9.455
1^3S_1	9.470	9.460	9.460	9.465	9.460	9.502
2^1S_0	9.986	9.999	9.990	9.976	9.990	9.990
2^3S_1	10.033	10.023	10.023	10.003	10.015	10.015
3^1S_0	10.315	–	10.329	10.336	10.326	10.330
3^3S_1	10.352	10.355	10.355	10.354	10.343	10.349
4^1S_0	10.523	–	10.573	10.523	10.584	–
4^3S_1	10.615	10.579	10.586	10.635	10.597	10.607
5^1S_0	10.816	–	10.851	10.869	10.800	–
5^3S_1	10.845	10.876	10.869	10.878	10.811	10.818
6^1S_0	11.024	–	11.061	11.097	10.997	–
6^3S_1	11.051	11.019	11.088	11.102	10.988	10.995

of pNRQCD which is an effective field theory, and are usually classified in powers of the inverse of heavy quark mass or velocity.

It can be seen that the mass spectra calculated is in close resemblance with the latest PDG results and that with the results of other available theoretical approaches, thus it can be concluded that this potential gives satisfying results for S wave spectra. In order to bring the potential to full justice the entire mass spectra i.e P, D and F states and also various decays have to be calculated.

4. References

[1] Patrignani C *et. al.* 2016 *Chin. Phys.* C **40** 100001
[2] Pineda A and Vairo A 2001 *Phys. Rev.* D **63** 054007
[3] Pineda A and Vairo A 2001 *Phys. Rev.* D **64** 039902
[4] Casewell W and Lepage G 1986 *Phys. Lett.* B **167** 437
[5] Bodwin G T, Brateen E and Leoage G P 1995 *Phys. Rev.* D **51** 1125
[6] Brambilla N, Pineda A, Sotto J and Vario A 2005 *Mod. Phys.* **77** 1423
[7] Gupta S N and Johnson J N 1995 *Phys. Rev.* D **51** 168
[8] Koma Y and Koma M 2013 *Few Body Sys.* **54** 1027
[9] Rai A K , Parmar R H and Vinodkumar P C 2002 *J. Phys.* G **28** 2275
[10] Rai A K, Pandya J K and Vinodkumar P C 2005 *J. Phys.* G **31** 1453
[11] Rai A K, Patel B and Vinodkumar P C 2008 *Phys. Rev.* C **78** 055202
[12] Ebert D, Faustov R and Galkin V O 2011 *Eur. Phys. J* C **71** 1825
[13] Godfrey S and Moats K 2015 *Phys. Rev.* D **92** 054034
[14] Deng W J, Liu H, Gui L C and Zhong X H 2017 Phys. Rev. D **95** 074002
[15] Segovia J, Ortega P G, Entem D R and Fernandez F 2016 Phys. Rev. D **93** 074027

Diffractive scattering: problems in theory and practice

V A Petrov

A. A. Logunov Institute for High Energy Physics, NRC "Kurchatov Institute", Protvino,RF

E-mail: `Vladimir.Petrov@ihep.ru`

Abstract. We give a retrospective overview of the conceptual content of the strong interaction scattering at high energies with account of the latest experimental findings made at the LHC.

1. Introduction

All experiments which are being performed at the LHC can be roughgly divided into two classes: "Old Physics" and "New Physics". "Old physics" comprises any kind of measurements to complete the Standard Model (the brightest example is the Higgs discovery) while "New Physics", starting from the evidently vital and pressing problems like neutrino masses, but extends much far away with searches for supersymmetry, "stringy" effects, dark matter, lepto-quarks, compositeness ant even further: micro black holes and extra dimensions of space-time. "New physicists" are obsessed with a dream to find at least some sort of exotic. However, the years pass, more and more powerful machines come into operation and upgrade: ISR, Tevatron, $S\bar{p}pS$, LEP I, LEP II, at last LHC but no signatures of something fudamentally breaking the SM paradigm appear. Does it mean that what we really observe today can be referred as a boring, "Old Physics" stuff? It is true that the Standard Models leaves many questions about its architecture and composants asking for a new theoretical basis. Yet the very SM is actually a new, relatively recently born discipline, like electrodynamics just after the Maxwell equations were formulated. Similarly to the first period of the electrodynamics development we have to wait for a lot of new effects. This asks for a thorough and concentrated efforts and one should decide which part of financial means, intellectual and technical forces should be devoted to the the searches in "Old Physics" . At the moment there is an impression that too much of (always limited) resources is being spent for obtaining invariably negative results in the searches for, often quite ill-founded, fantasies of imminent "gurus". In this talk I would like to tell about a specific part of the Standard Model, namely the strong interactions at high energies. Their salient features are revealed best in diffractive scattering.

2. What is diffractive scattering?

At Fig. 1 the angular distribution of the light scattering off a polystyrene latex particle is shown. The energy of the incident photons is $\approx 2.5eV$. The size of the particles-scatterers is near 0.3 $\mu m = 3 \cdot 10^8 fm$.

Let us compare this pattern with that of the proton-proton scattering at Fig. 2. This time the energy of collision is much higher: $10^{11} \div 10^{12} eV$ while the effective scatterer size is of order

IOP Conf. Series: Journal of Physics: Conf. Series **934** (2017) 012034 doi:10.1088/1742-6596/934/1/012034

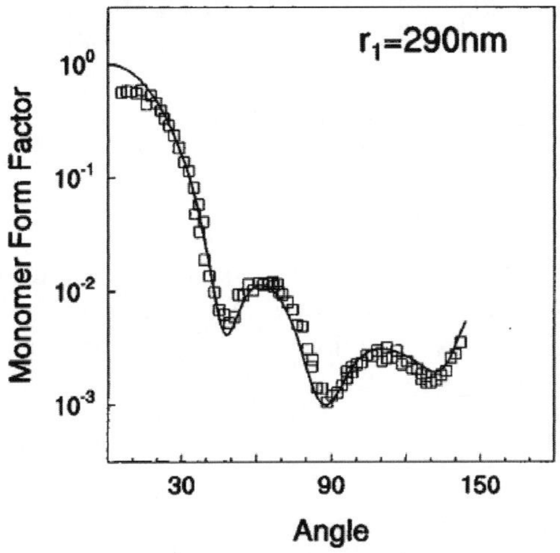

Figure 1. Diffractive pattern of the light scattering[1].

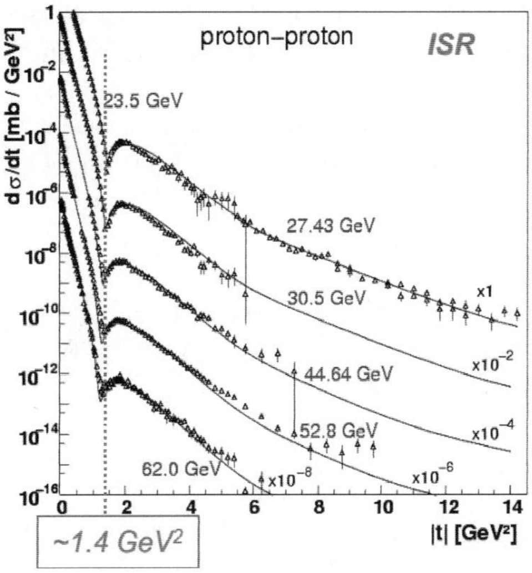

Figure 2. Diffractive pattern of the proton scattering [2] .

of 1 fm.

We see that at such huge difference in energy and scatterer sizes the diffraction angular distributions are very similar.

Why should we study the diffraction in the proton-proton collision at powerful accelerators if we can observe it at much less costly facilities? Moreover, why to study elastic scattering when "nothing happens" except the deviations of the colliding protons after their mutual "shock" akin to billiard balls we can play in a pub? There are several reasons.

First of all protons aren't hard balls. According to our present views the proton consists of the "valence" core comprised of three quarks (u,u,d)which swim in the "sea" of permanently changing quantum fluctuations consisting of arbitrary number of quark-antiquark pairs and gluons.The average size of the valence quark core is fixed while that of the "sea" is not Lorentz invariant and depends on the momentum of the proton. Moreover the latter expands with the energy growth and at some energy comes outside and continues to grow as pictured in Fig. 3.

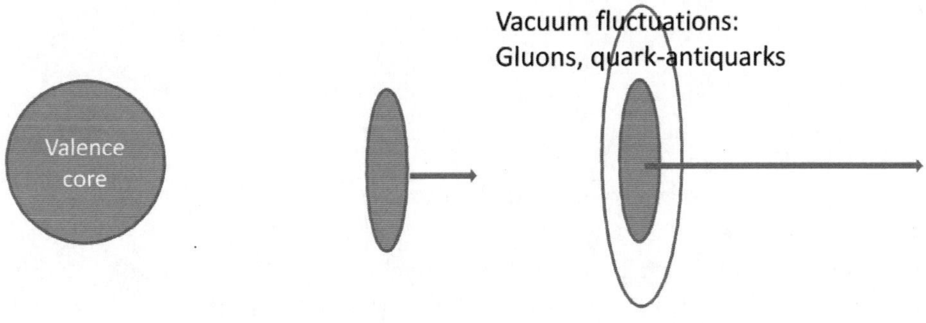

Figure 3. The shape of the moving proton

This fact makes the proton diffraction picture qualitatively different from that of the

IOP Conf. Series: Journal of Physics: Conf. Series **934** (2017) 012034 doi:10.1088/1742-6596/934/1/012034

diffraction picture of the light scattering off a polystyrene latex particle. With the energy growth the position of the diffractive minimum slowly moves to the left. Starting at the ISR at $-t = 1.4GeV^2$ it achieves the value near $0.5GeV^2$ at the LHC. Namely this feature reflects the complicated quantum structure of the proton and explains why we should study this seemingly simple process.

3. Interaction region

Actually the characteristic size of the effective scatterer — normally called "interaction region" —is defined not from the position of the minimum in the angular distribution of the scattered protons but from some derivative of the differential cross-sections $d\sigma/dt$, the "forward slope" $B(s)$ defined as

$$B(s) \equiv \frac{1}{d\sigma/dt}\frac{\partial[d\sigma/dt]}{\partial t}|_{t=0} = \frac{\partial \ln[d\sigma/dt]}{\partial t}|_{t=0}$$

Physical meaning of this quantity is given by the relation $2B(s) = \langle \mathbf{b}^2 \rangle$ where $\mathbf{b}$ is the impact parameter of the collision. So the experimentally observable quantity $B(s)$ defines the size of the interaction region, i.e. the region outside of which scattered particles move away freely. The growth of the interaction size mentioned above proceeds extremely slow. For instance, at the energies of the Protvino proton synchrotron U-70($0.01TeV$) $\langle \mathbf{b}^2 \rangle^{1/2} \approx 0.9fm$ while at the highest present energies of the LHC ($O(10TeV)$) $\langle \mathbf{b}^2 \rangle^{1/2}$ achieves only $1.3fm$. What can we learn from observation of the forward slope evolution at Fig. 4? The point indicated by

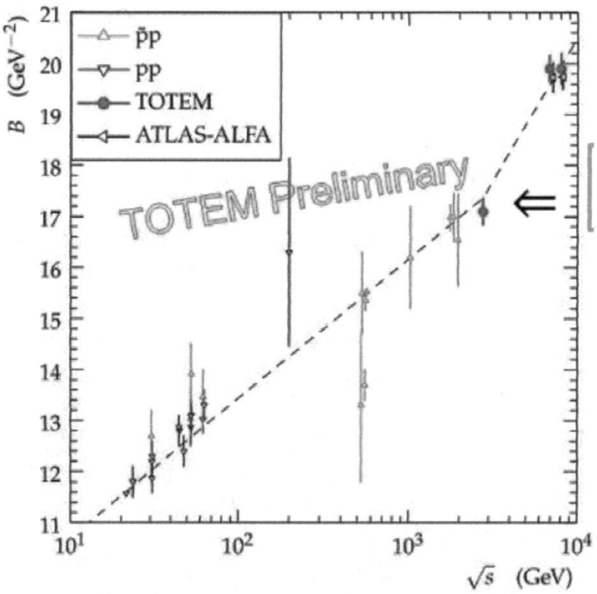

Figure 4. Energy evolution of the transverse interaction region [4] .

the arrow at Fig. 4 is obtained at $\sqrt{s} = 2.76TeV$ and shows $B = 17.2GeV^{-2}$. The energy evolution of $B(s)$ at lower energies looks as it points a linear log dependence on energy. But at higher energies ($7TeV$ and higher) the values of B jump to $B \geq 20GeV^{-2}$. So, something happens at energies near $2.5 - 3TeV$. At the moment we can only guess but one thing deserves our attention. If we calculate the 3D length corresponding to its 2D projection given by B we obtain exactly the pion Compton length $1/m_\pi$! Thus, if this results will be finally confirmed,

IOP Conf. Series: Journal of Physics: Conf. Series **934** (2017) 012034 doi:10.1088/1742-6596/934/1/012034

we have to conclude that some mysterious change of the interaction regime happens between 2.5 and 3.0 TeV.

The behaviour of the *longitudinal* interaction region is also very unusual.Contrary to naive expectations of compression of all longitudinal sizes in the direction of the colliding proton beams we actually deal with a fast ($\sim \sqrt{s}$) growing size achieving the values of order of $10^3 fm$ at the LHC. Only this example shows how interesting field is diffractive scattering at high energies.

4. What theory gives?

Since the beginning of 1970s the basic theory of strong interaction is QCD. It would be naive to expect that somebody would be able to derive the exact strong-interaction S matrix from the QCD Lagrangian. But some important ingredients could be the subject of the QCD studies. I mean the famous Regge trajectories which define the energy dependence of the observed cross-section. If $T(s,t)$ is the scattering amplitude then in a rough approximation its energy and angular dependence is given by $\sim s^{\alpha}(t)$. The main interest is related to a special trajectory $\alpha_P(t)$, the Pomeron, which has the highest "intercept", $\alpha_P(0)$ defining the behaviour of the total cross-sections. Unfortunately , QCD "works" well only at very high Euclidean momenta, $-t \gg mass^2$ in our case. It was shown by R. Kirschner and L. N. Lipatov [3] that in this limit

$$\alpha_P(t) \approx 1 + 4N_c ln2/\pi \cdot \frac{g_s^2(t)}{4\pi} + ...$$

where $\frac{g_s^2(t)}{4\pi} \sim 1/\ln(-t/\Lambda^2)$ according to the famous asymptotic freedom property of QCD. Some general considerations imply that Regge trajectories should be monotonic functions and this leads us to an important conclusion: $\alpha_P(0) > 1$. This means that the total cross-section indefinitely grow with energy. Unfortunately, this qualitatively important fact still cannot find its quantitative embodiment: the calculation of the numerical value of the Pomeron intercept $\alpha_P(0)$ remains on the agenda already 40 years! Thus the study of the diffractive scattering is tightly related with the most fundamental problem of QCD behaviour at large distances, which are, in turn, unavoidably related to the seminal confinement problem.

5. Conclusions

I would like to quote from the New Testament the question of Saint Peter to Jesus Christ "Quo vadis, Domine?" We also can ask the High Energy Physics: "Where are You going to?" And there are two ways to go: inside (the most of the SM and beyond the SM)or outside (strong interactions at high energies). Both directions are important , we should not underestimate each of them.

Acknowledgments

I am grateful to the organizers of this excellent conference for inviting me to give this talk and friendly atmosphere.

References

[1] Holtoff H, Borkovec M and Schurtenberger P 1997 *Phys. Rev.* E **56** 6945
[2] Nagy E *et al.* 1979 *Nucl.Phys.* B **150** 221
[3] Kirschner R and Lipatov L N 1986 *Sov. Phys. JETP* **63** 904
[4] Deile M *et al.* 2017 *EDS Blois 2017* (Prague) *https://indico.cern.ch/event/577066/contributions/2565986/attachments/1482376/2299352/eds2017.pdf* (to be published in the Proceedings)

IOP Conf. Series: Journal of Physics: Conf. Series **934** (2017) 012035 doi:10.1088/1742-6596/934/1/012035

Regge Trajectories of triply heavy baryons

Ajay Kumar Rai[1], **Zalak Shah**[2]

Department of Applied Physics, Sardar Vallabhbhai National Institute of Technology, Surat, Gujarat 395007, India

E-mail: [1]`raiajayk@gmail.com`, [2]`zalak.physics@gmail.com`

Abstract. Ω_{ccc}, Ω_{bbb}, Ω_{bcc} and Ω_{ccb} baryons are considerable theoretical interest in a baryonic analogue of heavy quarkonium because of the color-singlet bound state of three heavy quark (c,b) combination inside. Regge trajectories are concerned with the mass spectrum of the particles so that the present study exhibits the regge trajectories obtained from excited states of four experimentally unknown triply heavy Ω baryons. The trajectories are plotted in (n, M^2) and (J, M^2) planes which are helpful to determine the unknown quantum number and J^P values. The calculations have computed in Hypercentral Constituent Quark Model with hyper coulomb plus linear potential.

1. Introduction

The three heavy quarks (c,b) united and generate the system called *triply heavy baryons*. The well-known light baryon Ω contains the three strange quark combination. While these lights quarks are replaced by the heavy quarks the four new baryons are generated: Ω_{ccc}, Ω_{bbb}, Ω_{bcc} and Ω_{ccb} [1]. The baryons with three same heavy quarks are spin symmetrical states and thus, the ground states and radial excited states of Ω_{ccc} and Ω_{bbb} baryons are determined only with $J^P = \frac{3}{2}^+$. Whereas, the rest two baryons; mixed combination of c and b can have these states with $J^P = \frac{1}{2}^+, \frac{3}{2}^+$ [2]. Any experimental evidence of these states have not evicted yet. Although, many authors have studied theoretically the mass spectra of triply heavy baryons using different approaches [3–8, 10–12]. In this paper, we have used the masses of Ω_{ccc}, Ω_{bcc} and Ω_{ccb} baryons to construct the regge trajectories after the brief introduction and short discussion of our model.

The Hypercentral Constituent Quark Model(hCQM) is used to generate the mass spectra of triply heavy baryons in this paper. The model has successfully determined the ground states and excited states of singly, doubly, some of the triply heavy baryons in charm and bottom sector in our previous work [2, 13–17].

The Hamiltonian of the three-quark system is taken as

$$H = \frac{P_x^2}{2m} + V(x) \tag{1}$$

The hyper radius $x = \sqrt{\rho^2 + \lambda^2}$ and ρ and λ are the Jacobi co-ordinates [18–20]. $m = \frac{2m_\rho m_\lambda}{m_\rho + m_\lambda}$, is the reduced mass and x is the six dimensional radial hyper central coordinate of the three body system . The quark masses taken in calculations are $m_c = 1.275$ and $m_b = 4.67$ (in GeV).

The hypercentral potential $V(x)$ is the hyper color coulomb plus linear potential with first order correction [21]. It also contains the spin-dependant part.

$$V(x) = V^0(x) + \left(\frac{1}{m_\rho} + \frac{1}{m_\lambda}\right) V^{(1)}(x) + V_{SD}(x) \tag{2}$$

The first order correction $V^{(1)}(x)$ has already been used to obtain the mass spectra of heavy baryons, heavy-light mesons and exotic states [2, 14–16, 22–24]. The detail description of Eqn. (2) can be found in Ref. [14] and the refs. theirin.

2. Regge Trajectories

Tullio Regge had introduced the topic of Regge trajectories to the hadron physics [25]. The dynamical origin of regge trajectories are reviewed in different model of hadron spectroscopy broadly. The Regge trajectories are related by the mass and the spin of a hadrons [26].

The mass spectra are calculated for 1S-5S, 1P-5P, 1D-4D and 1F-2F states of triply heavy Ω_{ccc}, Ω_{bcc} and Ω_{ccb} baryons. These generated masses are used to plot the regge trajectories in two different planes figures 1-2. Figure 1 shows the graph using the masses of triply charm baryon, whereas figure 2 shows the graph using the masses of two charm-bottom triply heavy baryons.

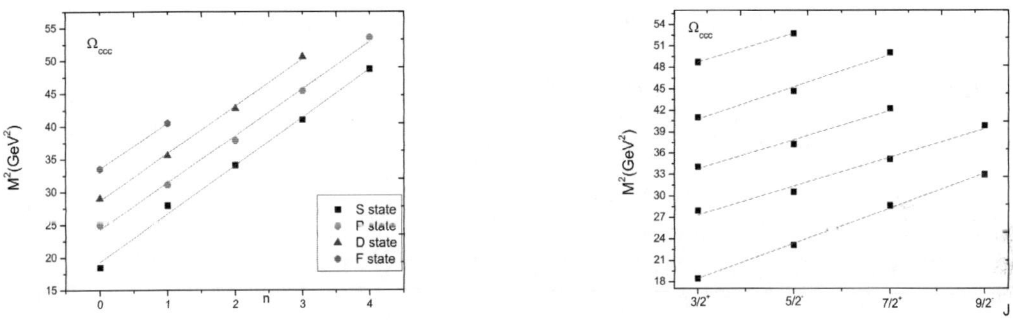

(a) (b)

Figure 1. Regge trajectories of Ω_{ccc} baryon in (n, M^2) (a) and (J, M^2) (b) planes [2].

Regge trajectories are related with the baryon masses(M), Isospin(I), principal quantum number(n) and total angular momentum(J). The relations are:

$$n = \beta M^2 + \beta_0 \qquad \& \qquad J = \alpha M^2 + \alpha_0 \tag{3}$$

where, β,α are slopes and β_0, α_0 are intercepts. With the help of slopes and intercepts one can identify the state with the J^P value.

The regge trajectories are plotted in (n, M^2) and (J, M^2) planes. In ($M^2 \to n$) graphs, the S, P, D and F states are plotted with J^P values $\frac{1}{2}^+$ $\frac{1}{2}^-$, $\frac{5}{2}^+$ and $\frac{7}{2}^-$. In ($M^2 \to J$) graphs, Ω_{ccc} is plotted for unnatural parities and Ω_{bcc} are plotted for natural parities. These graphs show the linearity between the states in same plane.

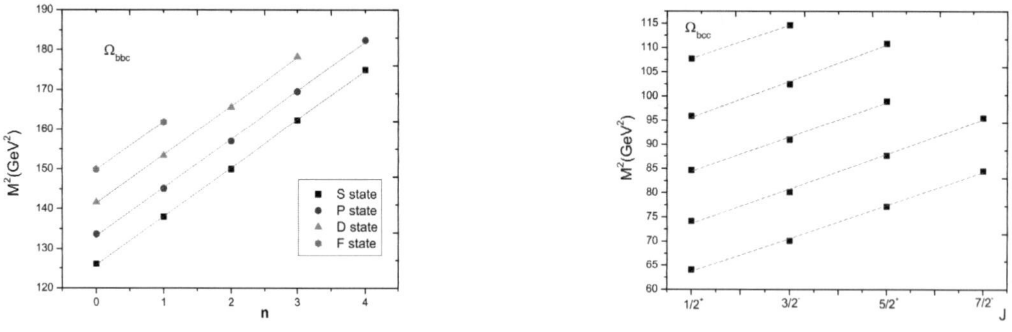

(a) (b)

Figure 2. Regge trajectory of Ω_{bbc} baryon in (n, M^2) plane (a) and for Ω_{bcc} baryon in (J, M^2) plane (b).

3. Conclusion

All three triply heavy baryons are experimentally unknown. Thus, the motive of these studies are identifying the states with particular J^P values. The trajectories are useful in that and the slope and intercepts of the graphs will also identify the principal quantum number of the state. These trajectories are still effective in determining spin and parity of discovered hadrons.

Acknowledgment

A. K. Rai acknowledges the financial support extended by DST, India, under SERB fast track scheme SR/FTP/PS- 152/2012.

References

[1] Patrignani C, *et al.* 2016 *Chin. Phys.* C **40** 100001
[2] Shah Z and Rai A K 2017 *Eur. Phys. J.* A **53** 195
[3] Wei K W, Chen B, Liu N and Guo X H 2015 *Phys. Rev.* D **92** 076008
[4] Thakkar K, Majethiya A, and Vinodkumar P C 2016 *Eur. Phys. J Plus* **131** 339
[5] Vijande J, Valcarce A and Garcilazo H 2015 *Phys. Rev.* D **91** 054011
[6] Brown Z S, Detmold W, Meinel S and Orginos K 2014 *Phys. Rev.* D **90** 094507
[7] Padmanath M, Edwards R G, Mathur N and Peardon M 2014 *Phys. Rev.* D **90** 074504
[8] Aliev T M, Azizi K and Savci M 2014 *J. Phys.* G **41** 065003
[9] Roberts W and Pervin M 2008 *Int. J. Mod. Phys.* A **23** 2817
[10] Bernotas A and Simonis V 2009 *Lith. J. Phys.* **49** 19
[11] Flynn J M, Hernandez E and Nieves J 2012 *Phys. Rev.* D **85** 014012
[12] Sun Z F, Liu Z W, Liu X and Zhu S L 2015 *Phys. Rev.* D **91** 094030
[13] Patel B, Rai A K and Vinodkumar P C 2008 *Pr. J. Phys.* **70** 797
[14] Shah Z, Thakkar K, Rai A K and Vinodkumar P C 2016 *Eur. Phys. J.* A **52** 313
[15] Thakkar K, Shah Z, Rai A K and Vinodkumar P C 2016 *Nucl. Phys.* A **965** 57
[16] Shah Z, Thakkar K and Rai A K 2016 *Eur. Phys. J.* C **76** 530
[17] Shah Z, Thakkar K, Rai A K and Vinodkumar P C 2016 *AIP. Conf. Proc.* **1728** 020096
[18] Bijker R, Iachello F and Leviatan A 2000 *Ann. Phys.* **284** 89
[19] Giannini M M and Santopinto E 2015 *Chin. J.Phys.* **53** 020301
[20] Bijker R, Iachello F and Santopinto E 1998 *J Phys.* A **31** 9041
[21] Koma Y, Koma M and Wittig H 2006 *Phys. Rev. Lett.* **97** 122003
[22] Devlani N, Kher V and Rai A K 2014 *Eur. Phys. J.* A **50** 154
[23] Kher V, Devlani N, and Rai A K 2017 *Chin. Phys.* C **41** 093101
[24] Rai A K and Rathaud D P 2015 *Eur. Phys. J.* C **75** 462
[25] Regge T 1959 *Nuovo Cim.* **14** 951

[26] Ebert D, Faustov R N and Galkin V O 2011 *Phys. Rev.* D **84** 014025

Radiative transitions and the mixing parameters of the D meson

Virendrasinh H. Kher* and Ajay Kumar Rai

Department of Applied Physics, Sardar Vallabhbhai National Institute of Technology, Surat, Gujarat, INDIA 395007
*Also at Applied Physics Department, Polytechnic, The M S University of Baroda, Vadodara 390002, Gujarat, INDIA.

E-mail: vhkher@gmail.com and raiajayk@gmail.com

Abstract. Spectroscopic parameters of heavy-light flavoured D meson are obtained within the framework of phenomenological quark-antiquark potential (Coulomb plus linear confinement) model using the Gaussian wave function. We incorporated $\mathcal{O}(1/m)$ corrections to the potential energy term and relativistic corrections to the kinetic energy term of the hamiltonian. We obtain the radiative (electric and magnetic) transitions and the mixing parameters of the DâĹŠ$\bar{D}$ oscillation. The results are compared with various experimental measurement as well as other theoretical predictions.

1. Introduction

The radiative transitions are important tools to determine the properties of heavy-light mesons. Using the radiative transitions one can probe the internal charge structure of hadrons, hence it is useful for determining the quantum numbers and hadronic structures of mesons. Particle anti-particle mixing has been of fundamental importance in testing the Standard Model and its proposed extensions [1–3].

2. Radiative transitions

The radiative transition amplitude is determined by the matrix element of the EM current between the initial quarkonium state i and the final state f, i.e., $\langle f \mid j^{\mu}_{em} \mid i\rangle$ [2,3].

2.1. E1 radiative transitions

The partial width for an E1 radiative transition between states is given by [4]

$$\Gamma_{(i\to f+\gamma)} = \frac{4\alpha}{3}\left(\frac{m_Q e_q - m_q e_{\overline{Q}}}{m_Q + m_q}\right)^2 \frac{\left(\frac{M_i^2-M_f^2}{M_f^2}\right)^3 E_f}{M_i} C_{fi}\delta_{SS'} \times \mid \langle f \mid r \mid i\rangle \mid^2 \tag{1}$$

The overlap integral is

$$\langle f\,|r|\,i\rangle = \int dr R_{n_i l_i}(r)\, R_{n_f l_f}(R) \tag{2}$$

Table 1. Electric dipole (E1) transitions widths of D meson.

Meson transition		This work		Other work (Γ in KeV)		
Initial	Final	E_γ(MeV)	Γ(KeV)	[6]	[7]	[8]
$D(1^3P_2)$	$D(1^3S_1)$	410	14.17	17.00	51	61.2
$D_1^{'}(1P)$	$D(1^3S_1)$	398	0.20	13.77	30.87	39.9
$D_1^{'}(1P)$	$D(1^1S_0)$	498	25.50	30.20	21.71	16.1
$D_1(1P)$	$D(1^3S_1)$	380	11.26	1.24	10.25	8.6
$D_1(1P)$	$D(1^1S_0)$	481	0.36	2.82	39.59	66
$D(1^3P_0)$	$D(1^3S_1)$	322	6.86	7.23	17	30
$D(2^3S_1)$	$D(1^3P_2)$	187	1.89	1.59		
$D(2^3S_1)$	$D_1^{'}(1P)$	200	0.02			
$D(2^3S_1)$	$D_1(1P)$	220	1.85			
$D(2^3S_1)$	$D(1^3P_0)$	281	1.29	1.52		
$D(2^1S_0)$	$D_1^{'}(1P)$	131	1.18	1.85		
$D(2^1S_0)$	$D_1(1P)$	152	0.03	2.69		
$D(1^3D_3)$	$D(1^3P_2)$	152	2.53			70.0
$D_2^{'}(1D)$	$D(1^3P_2)$	303	1.62			
$D_2(1D)$	$D(1^3P_2)$	277	5.06			
$D(1^3D_1)$	$D(1^3P_2)$	278	0.60			
$D(1^3D_1)$	$D_1^{'}(1P)$	291	7.00			
$D(1^3D_1)$	$D_1(1P)$	310	2.89			
$D(1^3D_1)$	$D(1^3P_0)$	369	20.13			53.8
$D_2^{'}(1D)$	$D_1^{'}(1P)$	316	16.13			
$D_2^{'}(1D)$	$D_1(1P)$	334	6.20			
$D_2(1D)$	$D_1^{'}(1P)$	290	4.27			
$D_2(1D)$	$D_1(1P)$	309	15.17			66.0

We employ Gaussian wave function for the calculation of the overlap integrals. The form of the wave function is given by

$$R_{nl}\left(r\right)=\mu^{\frac{3}{2}}\left(\frac{2\left(n-1\right)!}{\Gamma\left(n+l+1/2\right)}\right)^{\frac{1}{2}}\left(\mu r\right)^{l}\times e^{-\mu^{2}r^{2}/2}L_{n-1}^{l+1/2}\left(\mu^{2}r^{2}\right), \tag{3}$$

here μ is the variational parameter, n, l are the quantum numbers of the bound state and L are the Laguerre polynomials. The variational parameter μ was fitted using virial theorem by using a relativized Hamiltonian which includes $\mathcal{O}\left(1/m\right)$ corrections to the potential energy term and relativistic corrections to the kinetic energy term. For details see ref [5].

2.2. M1 radiative transitions

The rates for magnetic dipole transitions correspond to triplet-singlet between S-wave states of the same n quantum number as well as either triplet-singlet or singlet-triplet transitions between S-wave states of different n quantum numbers given by [9]

$$\Gamma_{M1}(i\to f+\gamma)=\frac{16\alpha}{3}\left(\frac{m_{\bar{q}}e_{Q}-m_{Q}e_{\bar{q}}}{4m_{\bar{q}}m_{Q}}\right)^{2}E_{\gamma}^{3}(2J_{f}+1)\left|\langle f|\,j_{0}(E_{\gamma}r/2)\,|i\rangle\right|^{2} \tag{4}$$

Here L = 0 for S-waves and $j_0(x)$ is the spherical Bessel function.

$$\langle f\left|r\right|i\rangle=\int drR_{n_{i}l_{i}}\left(r\right)j_{0}\left(E_{\gamma}r/2\right)R_{n_{f}l_{f}}\left(R\right) \tag{5}$$

Our calculated radiative magnetic dipole transitions widths for states of D meson are tabulated in Table 2 using the masses estimated by our model [5].

Table 2. Magnetic dipole (M1) transitions widths of D mesons.

Meson transition	This work		Others (Γ in KeV)		
Initial→Final	E_γ (MeV)	Γ(KeV)	[6]	[8]	[7]
$D(1^3S_1) \to D(1^1S_0)$	123	0.271	0.339	10.8	1.8
$D(2^3S_1) \to D(2^1S_0)$	72	0.055	0.007		
$D(3^3S_1) \to D(3^1S_0)$	53	0.021	0.001		
$D(2^3S_1) \to D(1^1S_0)$	659	6.371		100	
$D(2^1S_0) \to B(1^3S_1)$	508	8.594			
$D_1(1P) \to D(1^3P_0)$	102	2.340			
$D_1^{'}(1P) \to D(1^3P_0)$	14	0.020			
$D(1^3P_2) \to D_1(1P)$	36	0.314			
$D_1^{'}(1P) \to D(1^3P_2)$	88	1.524			
$D(1^3D_3) \to D_2(1D)$	15	0.040			
$D(1^3D_3) \to D(1^3D_1)$	43	0.936			
$D_2^{'}(1D) \to D(1^3D_1)$	43	0.539			
$D_2^{'}(1D) \to D_2(1D)$	28	0.150			

3. Mixing parameters

The neutral D meson, exhibit particle-antiparticles mixing, leading to oscillations between the light (L) and heavy (H) mass eigenstates [1]. With assuming CPT conservation throughout, in each system, the light (L) and heavy (H) mass eigenstates are,

$$\left|(D)_q^{L,H}\right\rangle = \frac{1}{\sqrt{1+\left|(q/p)_q\right|^2}}\left(\left|(D)_q\right\rangle \pm (q/p)_q\left|(\bar{D})_q\right\rangle\right) \tag{6}$$

have a mass difference $\Delta m_q = m_H - m_L > 0$, and a total decay width difference $\Delta\Gamma_q = \Gamma_L - \Gamma_H$. The time evolution of the neutral D meson doublet is described by the Schrodinger equation [1]

$$i\frac{d}{dt}\begin{pmatrix} D_q \\ D_q \end{pmatrix} = \left[\begin{pmatrix} M_{11}^q & M_{12}^{q\star} \\ M_{12}^q & M_{11}^q \end{pmatrix} - \frac{i}{2}\begin{pmatrix} \Gamma_{11}^q & \Gamma_{12}^{q\star} \\ \Gamma_{12}^q & \Gamma_{11}^q \end{pmatrix}\right]\begin{pmatrix} D_q \\ D_q \end{pmatrix} \tag{7}$$

where M_{12} and Γ_{12} are the off-diagnol elements of the mass and decay matrices.

The formula of predictions for the off-diagonal elements of the mass and the decay matrices are [1]

$$M_{12} = -\frac{G_F^2 m_W^2 \eta_D m_D B_D f_D^2}{12\pi^2} S_0\left(m_s^2/m_W^2\right)\left(V_{us}^\star V_{cs}\right)^2 \tag{8}$$

$$\Gamma_{12} = \frac{G_F^2 m_c^2 \eta_D' m_D B_D f_D^2}{8\pi}\left[\left(V_{us}^\star V_{cs}\right)^2\right] \tag{9}$$

The known function $S_0(x_t)$ can be approximated very well by $0.784 x_t^{0.76}$ [10], and V_{ij} are the elements of the CKM matrix. The parameters η_D and η_D' correspond to the gluonic corrections for respective meson.

In the absence of CP violation, the time -integrated mixing probability (χ_q) is given by

$$\chi_q = \frac{x_q^2 + y_q^2}{2\left(x_q^2+1\right)}, \qquad \text{where,} \quad x_q = \Delta m_q \tau_{D_q}, \qquad y_q = \frac{\Delta\Gamma_q \tau_{D_q}}{2} \tag{10}$$

IOP Conf. Series: Journal of Physics: Conf. Series **934** (2017) 012036 doi:10.1088/1742-6596/934/1/012036

The mass difference Δm is a measure of the frequency of the change from a $(D)^0$ into a $\bar{D}^0$ or vice versa. We have the time-integrated mixing rate for semi-leptonic decays as [1]

$$R_M \simeq \frac{1}{2}(x_q^2 + y_q^2). \tag{11}$$

For the estimation of the mixing parameters x_q, y_q, χ_q and R_M, we use $\eta_D = 0.86$, $\eta'_D = 0.21$ and the gluonic correction to the oscillation is given by Ref. [11, 12]. The bag parameter $B_{D_q} = 1.34$ is taken from the lattice result of [13], while the pseudoscalar mass (M_D, and the pseudoscalar decay constant (f_D) of the charmed mesons are taken from our previous study in the article [5]. The calculated mixing parameters x_q, y_q, χ_q and R_M are tabulated in table 3.

Table 3. Mixing parameters of the D meson.

This work	$\Delta m_q \times 10^{-15}$	$x_q \times 10^{-3}$	$y_q \times 10^{-3}$	$\chi_q \times 10^{-5}$	$R_M \times 10^{-3}$
f_P	2.8	2.96	17.2	0.15	0.152
f_{Pcor}	1.27	1.322	7.67	3.03	0.03027
Ref. [14]		1.6±2.3 ± 1.2	5.7±2.0±1.3 ± 0.7		
Ref. [15]		8.0±2.9	3.3±2.4		0.0864±0.0311
Ref. [16]					0.13±0.22±0.20
Ref. [17]					0.04−0.6+0.7
Ref. [18]					0.02±0.47±0.14
Ref. [19]					1.6±2.9±2.9

4. Conclusion

From comparison of our estimated radiative (E1 and M1 dipole) transitions widths with other theoretical estimations, we conclude that the various models have different predictions of E1 and M1 dipole transitions may be due to different parameters and treatments are used in the relativistic corrections in the model. We have compared our results of the mixing parameters to PDG [1] as well as other theoretical results and they are in agreement with our results.

References

[1] Patrignani C and Group P D 2016 *Chin. Phys.* C **40** 100001
[2] Ding G J, Zhu J J and Yan M L 2008 *Phys. Rev.* D **77** 014033
[3] Lu Q F *et al.* 2016 *Phys. Rev.* D **94** 074012
[4] Radford S F, Repko W W and Saelim M J 2009 *Phys. Rev.* D **80** 034012
[5] Kher V, Devlani N and Rai A K 2017 *Chin. Phys.* C **41** 073101
[6] Devlani N and Rai A K 2013 *Int. J. Theor. Phys.* **52** 2196–208
[7] Close F E and Swanson E S 2005 *Phys. Rev.* D **72** 094004
[8] Godfrey S and Moats K 2016 *Phys. Rev.* D **93** 034035
[9] Brambilla N *et al.* 2011 *Eur. Phys. J.* C **71** 1534
[10] Inami T and Lim C 1981 *Prog. Theor. Phys.* **65** 297
[11] Shah M, Patel B and Vinodkumar P 2016 *Phys. Rev.* D **93** 094028
[12] Buras A J, Jamin M and Weisz P H 1990 *Nucl. Phys.* B **347** 491–536
[13] Buras A J 2003 *Phys. Lett.* B **566** 115–9
[14] del Amo Sanchez P *et al.* 2010 *Phys. Rev. Lett.* **105** 081803
[15] Zhang L M *et al.* 2007 *Phys. Rev. Lett.* **99** 131803
[16] Bitenc U *et al.* 2008 *Phys. Rev.* D **77** 112003
[17] Aubert B *et al.* 2007 *Phys. Rev.* D **76** 014018
[18] Bitenc U *et al.* 2005 *Phys. Rev.* D **72** 071101
[19] Cawlfield C *et al.* 2005 *Phys. Rev.* D **71** 077101

IOP Conf. Series: Journal of Physics: Conf. Series **934** (2017) 012037 doi:10.1088/1742-6596/934/1/012037

The black hole at the Galactic Center: observations and models in a nutshell

Alexander Zakharov [1,2,3,4,5]

[1]National Astronomical Observatories of Chinese Academy of Sciences, 20A Datun Road, 100012 Beijing, China
[2]Institute of Theoretical and Experimental Physics, B. Cheremushkinskaya 25, 117218 Moscow, Russia
[3]Bogoliubov Laboratory of Theoretical Physics, Joint Institute for Nuclear Research, 141980 Dubna, Russia
[4]National Research Nuclear University MEPhI (Moscow Engineering Physics Institute), 115409, Moscow, Russia
[5]North Carolina Central University, Durham, NC 27707, USA

E-mail: zakharov@itep.ru

Abstract.
The Galactic Center (Sgr A^*) is a peculiar place in our Galaxy (Milky Way). Our Solar system is located at a distance around 8 kpc from the Galactic Center (GC). There were a number of different including exotic ones such as boson stars, fermion balls, neutrino balls, a cluster of neutron stars. Some of these models are significantly constrained with consequent observations and now supermassive black hole with mass around 4×10^6 $M_\odot$ is the preferable model for GC. Moreover, one can test alternative theories of gravity with observations of bright stars near the Galactic Center and and observations of bright structures near the black hole at the Galactic Center to reconstruct shadow structure around the black hole with current and future observational VLBI facilities such as the Event Horizon Telescope. In particular, we got a graviton mass constraint which is comparable and consistent with constraints obtained recently by the LIGO-Virgo collaboration.

1. Introduction

It is clear that the center of our Galaxy (Milky Way) is a very attractive target and astronomers observe it in different spectral bands from radio up to X-ray excluding optical and UV bands due to a presence of gas and dust along a line of sight [1]. Earlier, for the Galactic Center people proposed many theoretical models such as a dense cluster of stars, fermion balls, neutrino balls, boson stars etc, however, the preferable model is the supermassive black hole [2]. Clearly, that predictions for observations critically depend on choices of theoretical models, however, often it is very hard to distinguish theoretical models in observations as it was shown for shadow formation in the framework of boson stars and conventional black hole models [3]. To test gravitational potential one could analyze motions of test bodies. If astronomers use photons as test particles it would be possible to reconstruct a shadow shape in mm and sub-mm bands with current and future VLBI network such as the Event Horizon Telescope [4]. To interpret observational data one needs general relativity approach in the strong gravitational field limit [5, 6]. Another opportunity is to use bright stars or clouds of hot gas as tracers of gravitational potential. In this case astronomers objects with large telescopes in IR band. we discuss such an option in the next section.

2. Observations of bright stars at the Galactic Center

There are two groups of astronomers who successfully observe bright stars near the Galactic Center in IR band. One group led by A. Ghez (UCLA, USA) use the twin ten meter Keck telescope in Hawaii. Results of their observations and interpretations may be found in [8, 9, 10, 16]. In [12] the astronomers discussed opportunities to constrain alternative theories of gravity with data on trajectories of bright stars. Another group of astronomers use 4 eight meter telescopes at the Paranal Mountain. R. Genzel (MPE & Berkeley University) is the leader of the group, results of observations of the group may found in [13, 14, 15]. In paper [14] it was claimed that gas cloud G2 may be used as a tracer to test metric near the Galactic Center, however, consequent observations showed probably the G2 object may be a star but not a gas cloud. Several years ago ESO and MPE (Garching) started to built the Very Large Telescope Interferometer (GRAVITY) and astronomers expected to reach angular resolution at a level around 10 micro-arcseconds for bright sources with GRAVITY facilities. First results were published recently [18]. In the future Thirty Meter Telescope (TMT) and European Extremely Large Telescope (E-ELT) will be built and it will give an opportunity to improve current orbit reconstructions for bright stars, therefore, the Galactic Center model will be significantly clarified.

3. Constraints on black hole parameters and gravity theories from trajectories of bright stars at the Galactic Center

3.1. Constraints on black hole parameters and extended mass distribution

As it was mentioned two leading groups of astronomers monitored bright IR stars at the Galactic Center for decades, see [8, 9, 10, 16, 13, 14, 15] and references therein. One of main object for observations of both teams was source 2 (or star S2 or S0-2) with period around 15 years, the pericenter distance around 120 AU, eccentricity $e \approx 0.89$, another star which may be used to improve the black hole parameters is S-38 with eccentricity $e \approx 0.8$ and orbital period around 19 years [16]. In paper [17] it was proposed to use observations for S55/S0-102 star to improve estimates for parameters of the black hole. In 2008 one period of observations for S2 has been finished. A famous relativistic advance (introduced by A. Einstein to explain the Mercury anomaly) for the object is around 1 mas [19, 21, 20] which is greater than a current accuracy for the orbit observations, however at the beginning the orbit reconstruction errors for orbit reconstruction were around 10 mas, therefore one should wait to observe the relativistic effect for S2 star. Sometimes, it is not necessary to monitor one revolution more for the orbit if an accuracy of orbit reconstruction is good enough.

In the case if there is only supermassive black hole and there is no bulk distribution of mass inside stellar orbit there is only relativistic advance (additional terms due to presence black hole spin in around 100 times smaller) [19, 21, 20] (see also recent discussions in [22, 23]). However, if a small fraction of mass distributed inside the orbit it should cause pericenter (and apocenter) shifts in the direction which is opposite to relativistic advance. Therefore, as it was shown in [20] even now one can constrain some theoretical models for dark matter distributions. Astroparticle physicists introduced such models to explain γ-ray flux from the Galactic Center with neutralino annihilation. In the future probably observations of bright stars will significantly constrain (or even rule out) a wide class of dark matter distributions at the Galactic Center [24].

3.2. Constraints on R^n theory

If us the conventional general relativity we unavoidably face with dark matter (DM) and dark energy (DE) problems. As it was noted in [25] famous French astronomer U. J. Le Verrier

IOP Conf. Series: Journal of Physics: Conf. Series **934** (2017) 012037 doi:10.1088/1742-6596/934/1/012037

proposed a way to resolve any anomaly, namely, the first option is to add additional component into model, the second option is to change a fundamental law (the Newtonian gravity law in the Le Verrier's case). To avoid DM and DE problems people suggested to introduce generalizations of Einstein – Hilbert Lagrangian and change scalar curvature R with another function, generally speaking with arbitrary function $f(R)$ [26, 27, 28]. Using such a generalizations of gravity one can explain a number of cosmological and astrophysical phenomena such as an accelerating expansion of the Universe, flat rotation curves for spiral galaxies etc. But, in the framework of these theories sometimes gravity law has no Newtonian limit in a weak gravitational field approximation and one should check the validity of gravity law for all scales. For instance, alternative theory proposed in [27] does not fit observational data in Solar system. Another class of theories with $f(R) = R^n$ (the case $n = 1$ corresponds to GR) can describe accelerating expansion of the Universe and flat rotation curves and in these cases parameter n has to be significantly differ from 1 ($n \in [1.5, 3]$), however as it was shown in [29] to fit observational data in Solar system $n \approx 1$ (it means that gravity theory should be very similar to the classical GR), therefore we have a contradiction with cosmological fits.

One can use observations of bright stars near the Galactic Center to test the alternative gravity law, and as it was shown in paper [30] analyzing observational data from Keck and VLT telescopes for S2 star orbit we should conclude that $n \approx 1$ and as it was pointed out earlier, it contradicts to suitable n values obtained from cosmological SNeIa data and rotation curves of spiral galaxies. In paper [31] we considered a central object and a stellar cluster in the framework of alternative theory of gravity and concluded that a presence of cluster does not change our previous conclusions that $n \approx 1$ and a gravity theory should be very close to GR.

3.3. Constraints on Yukawa gravity theory

As it was discussed in [33] $f(R)$ theories of gravity may have Yukawa limit in a weak gravitational field approximation and different authors successfully used the Yukawa like gravity to fit observational data for rotation curves of spiral galaxies and galactic clusters.

In paper [33], we used the Yukawa like gravity to constrain parameters of the potential with S2 star orbit data collected in observations of VLT and Keck telescopes. So, we considered the following potential

$$\Phi(r) = -\frac{GM}{(1+\delta)r}\left[1 + \delta e^{-\left(\frac{r}{\Lambda}\right)}\right], \tag{1}$$

where Λ and δ are parameters of the Yukawa potential.

In the paper [33] we showed that the preferable range of Λ parameter in the case of S2 star and it has to be $\Lambda \in [5000, 7000]$ AU and we found that it is very difficult hard to constrain on δ parameter due behavior of χ^2 function. We also found that the same the same $\delta = 1/3$ may be used to observational data for S2 star orbit and for galactic clusters and rotation curves of spiral galaxies.

3.4. Constraints on massive graviton theory

Many years ago Fierz and Pauli introduced a theory of massive gravity [34] (see a simple introduction in [35]). After that later a lot of defects have been found in massive theories of gravity such as a presence of ghosts, discontinuities for vanishing graviton mass. In last years, theorists overcame many technical problems [36, 37].

Yukawa modification of a Newtonian potential may be written in the following form [7, 35, 38]

$$V(r) = \frac{GM}{r}\exp(-r/\lambda_g). \tag{2}$$

C. Will discussed an opportunity to evaluate a graviton mass with an analysis of delays of gravitational waves in respect to electromagnetic radiation from supernovae or GRB. One can obtain also constrain graviton mass analyzing a structure of gravitational wave signal because if graviton has a mass then gravitons with higher masses propagate faster. This idea has been used by the LIGO-Virgo collaboration to evaluate a graviton mass [7, 38].

Using S2 orbit data obtained with VLT and Keck telescopes we found that with 90% confidence level we have $\lambda_g > 2900$ AU$= 4.3 \times 10^{11}$ km or $m_g < 2.9 \times 10^{-21}$ eV [39, 40, 41, 42].

In February 2016 the LIGO-Virgo collaboration reported about the discovery of gravitational waves from merger of binary black holes with masses about 30 $M_\odot$ [43], moreover, it was claimed that there is the graviton mass constraint 1.2×10^{-22} eV. So, the LIGO-Virgo collaboration not only discovered gravitational waves and an existence of binary black holes with high stellar masses but they obtained also the result of constraints on possible generalizations of GR considering massive gravity theories. In June 2017 the LIGO-Virgo collaboration reported about the discovery of the third GW event from mergers of the BHs with 31 and 19 solar masses at redshift z=0.19 and improved the graviton mass constraint $m_g < 7.7 \times 10^{-23}$ eV [44]. The LIGO graviton mass constraint is in around 40 times better than our bound obtained from an analysis of S2 star trajectory, but our estimate is independent and it may be improved with current and future observational data as it was noted in [12, 45].

4. Shadows for the black hole at the Galactic Center

In papers [46, 47] the authors considered a toy model for a shadow formation around a black hole. For the case of the Galactic Center the size of the shadow is around 50 μas. The shadow may be detectable at mm and sub-mm bands while in cm band scattering is spoiling black hole images [46, 47]. Consequent simulations and observations have basically confirmed these claims. It was significant progress to reach better angular resolution in mm-band for the Sgr A^* [48], and Doeleman et al. claimed that found the smallest spot with a size 37^{+16}_{-10} μas [48] and the size is comparable with shadow size evaluated in the framework of the simple model. The observations have been done with VLBI technique. In the future radioastronomers plan to create world wide VLBI network [4] and they call it as The Event Horizon Telescope because the angular resolution of the network will be better than the angular size of the event horizon for the Galactic Center or for the center of galaxy M87. One can find theoretical investigations sizes and structures of shadows in papers [49, 50, 51, 52, 53].

5. Conclusions

One can use precise observations of bright stars near the Galactic Center to check alternative theories of gravity and to investigate a presence of an extended mass distribution near the Galactic Center. One can the graviton mass constraint from an analysis of S2 star trajectory and the bound is consistent and comparable with the constraint presented recently by the LIGO collaboration.

Acknowledgements

The author thanks D. Borka, V. Borka Jovanović, F. De Paolis, G. Ingrosso, P. Jovanović, S. M. Kopeikin, A. A. Nucita, S.G. Rubin, B. Vlahovic for useful discussions. A. F. Z. thanks PIFI grant 2017VMA0014 of Chinese Academy of Sciences, the Strategic Priority Research Program of the Chinese Academy of Sciences (Grant No. XDB23040100), NSF (HRD-0833184) and NASA (NNX09AV07A) at NASA CADRE and NSF CREST Centers (NCCU, Durham, NC, USA) for a partial support. A. F. Z. also thanks the organizers of the Third Conference on Particle Physics and Astrophysics for their attention to the contribution.

References

[1] Eckart A, Schödel R and Straubmeier C 2005 *The Black Hole at the Center of the Milky Way* (London: Imperial College Press).
[2] Reid M 2009 *Intern. J. Mod. Phys.* D **18** 889
[3] Vincent F H, Meliani Z, Grandclément P *et al.* 2016 *Class. Quant. Grav.* **33** 105015 (19pp)
[4] Doeleman S, Agol E, Backer D *et al.* 2009 Imaging an Event Horizon: submm-VLBI of a Super Massive Black Hole *Preprint* 0906.3899
[5] Falcke H and Markoff S B 2013 *Class. Quan. Grav.* **30** 244003
[6] Johannsen T 2016 *Class. Quan. Grav.* **33** 113001
[7] Will C 2014 *Living Rev. Relat.* **17** 4
[8] Ghez A M, Morris M, Becklin E E, Tanner A and Kremenek T 2000 *Nature* **407** 349
[9] Meyer L, Ghez A M, Schödel R *et al.* 2012 *Science* **338** 84
[10] Morris M R, Meyer L and Ghez A M 2012 *Res. Astron. Astrophys.* **12** 995
[11] Boehle A, Ghez A M, Schödel R *et al.* 2016 *Astrophys. J.* 830:17 (23pp)
[12] Hees A, Do T, Ghez A M *et al.* 2017 *Phys. Rev. Lett.* **118** 211101
[13] Schödel R, Ott T, Genzel R *et al.* 2002 *Nature* **419**, 694 (2002).
[14] Gillessen S, Genzel R, Fritz T K *et al.* 2012 *Nature* **481** 51
[15] Gillessen S, Plewa P M, Eisenhauer F *et al.* 2017 *Astrophys. J.* **837**:30 (19pp)
[16] Boehle A, Ghez A M, Schdel R *et al.* 2016 *Astrophys. J.* **830**:17 (23pp)
[17] Parsa M, A. Eckart A, Shahzamanian B *et al.* 2017 *Astrophys. J.* **845**:22 (19pp)
[18] GRAVITY Collaboration: Abuter R, Accardo M, Amorim A *et al.* 2017 *Astron. & Astrophys.* **602** A94
[19] Rubilar G F and Eckart A 2001 *Astron. & Astrophys.* **374** 95
[20] Zakharov A F, Nucita A A, De Paolis F and Ingrosso G 2007 *Phys. Rev.* D **76** 062001
[21] Nucita A A, De Paolis F, Ingrosso G *et al.* 2007 *Proc. Astron. Soc. Pac.* **119** 349
[22] Saxton C J, Younsi Z and Wu K 2016 *Mon. Not. R. Astron. Soc.*, **461** 4295
[23] Dokuchaev V I and Eroshenko Yu N 2015 *Physics – Uspekhi* **58** 772
[24] Zakharov A F, Nucita A A, De Paolis F and Ingrosso G 2008 *J. Phys.: Conf. Series* **133** 012032
[25] Zakharov A F, Capozziello S, De Paolis F, Ingrosso G and Nucita A A 2009 *Space Sci. Rev.* **48** 301
[26] Capozziello S 2002 *Int. J. Mod. Phys.* D **11** 483
[27] Carroll S M, Duvvuri V, Trodden M and Turner M S, *Phys. Rev.* D **70** 043528
[28] Capozziello S, Cardone V F, Carloni S and Troisi A 2004 *Phys. Lett.* A **326** 292
[29] Zakharov A F, Nucita A A, De Paolis F and Ingrosso G 2006 *Phys. Rev.* D **74** 107101
[30] Borka D, Jovanović P, Borka Jovanović V and Zakharov A F 2012 *Phys. Rev.* D **85** 124004
[31] Zakharov A F, Borka D, Borka Jovanović V and Jovanović P 2014 *Adv. Space Res.* **54** 1108
[32] Capozziello S, Cardone V F and Troisi A 2007 *Phys. Rev.* D **76** 104019
[33] Borka B, Jovanović P, Borka Jovanović V *et al.*2013 *J. Cosm. Astropart. Phys.* JCAP **11** 2013 050
[34] Fierz M and Pauli W 1939 *Proc. Royal Soc. London* **A173** 211
[35] Visser M 1998 *Gen. Rel. Grav.* **30** 1717
[36] Rubakov V A and Tinyakov P G 2008 *Phys. – Usp.* **51** 759
[37] de Rham C, Deskins J T, Tolley A J and Zhou S - Y 2017 *Rev. Mod. Phys.* **89** 025004
[38] Will C 1998 *Phys. Rev.* D **57** 2061
[39] Zakharov A F, Jovanović P, Borka D *et al.* 2016 *J. Cosm. Astropart. Phys.* (JCAP) 05(2016)045.
[40] Zakharov A F, Jovanović P, Borka D and Borka Jovanović V 2016 *Eur. Phys. J. Web Conf.* **125** 01011
[41] Zakharov A F, Jovanović P, Borka D and Borka Jovanović V 2017 *J. Phys.: Conf. Series* **798** 012081
[42] Zakharov A F, Jovanović P, Borka D and Borka Jovanović V 2017 *Eur. Phys. J. Web of Conf.* **138** 010010
[43] Abbott B P *et al.* 2016 *Phys. Rev. Lett.* **116** 061102
[44] Abbott B P *et al.* 2017 *Phys. Rev. Lett.* **118** 221101
[45] Hees A, Ghez A M, Do T *et al.* 2017 Testing the gravitational theory with short-period stars around our Galactic Center *Preprint* 1705.10792
[46] Falcke H, Melia F and Agol E 2000 *Astrophys. J.* **528** L13
[47] Melia F and Falcke H 2001 *Ann. Rev. Astron. Astrophys.* **39** 309
[48] Doeleman S S *et al.* 2008 *Nature* **455** 78
[49] Zakharov A F, Nucita A A, De Paolis F and Ingrosso G 2005 *New Astron.* **10** 479
[50] Zakharov A F, Nucita A A, De Paolis F and Ingrosso G 2005 *Astron. & Astrophys.* **442** 795
[51] Zakharov A F, Nucita A A, De Paolis F and Ingrosso G 2012 *New Astron. Rev.* **56** 64
[52] Zakharov A F 2014 *Phys. Rev.* D **90** 062007
[53] Zakharov A F 2015 *J. Astrophys. Astron.* **36** 539

IOP Conf. Series: Journal of Physics: Conf. Series **934** (2017) 012038 doi:10.1088/1742-6596/934/1/012038

Positive spatial curvature does not falsify the landscape

B Horn
Manhattan College, Department of Physics, New York, NY 10471
E-mail: bhorn01@manhattan.edu

Abstract. We present a simple cosmological model where the quantum tunneling of a scalar field rearranges the energetics of the matter sector, sending a stable static ancestor vacuum with positive spatial curvature into an inflating solution with positive curvature. This serves as a proof of principle that an observation of positive spatial curvature does not falsify the hypothesis that our current observer patch originated from false vacuum tunneling in a string or field theoretic landscape. This poster submission is a summary of the work, and was presented at the 3rd annual ICPPA held in Moscow from October 2 to 5, 2017, by Prof. Rostislav Konoplich on behalf of the author.

1. Introduction: false vacuum tunneling and the landscape

Quantum tunneling and decay of a false vacuum is a means of transitioning between different low-energy vacua in a field or string theoretic landscape scenario. If the tunneling event leaves behind observable signatures, for instance positive spatial curvature which is observable in the cosmic microwave background radiation (CMB) or in the distribution of large scale structure (LSS), we may be able to test the hypothesis that our current observer patch originated in such a tunneling event. The large scale curvature of the spatial slices is quantified by the curvature parameter $\Omega_k = -3K/8\pi G_N H^2$, where K is the inverse square of the radius of curvature, and H is the Hubble parameter, all measured at the present time. A negative value for this parameter corresponds to positive curvature of the spatial slice, and vice versa. If measured, the curvature parameter will be one of the single most important parameters we can measure in cosmology, with potential implications for the existence of other vacua besides our own, and for the physics of the pre-inflationary era.

The case where false vacuum tunneling nucleates a bubble much smaller than the curvature scale was analyzed in the semiclassical approximation by [1]. The WKB amplitude for barrier tunneling can be related to the action of a solution to the equations of motion in Euclidean signature. Analytically continuing back to Lorentzian spacetime, the bubble expands after nucleation, and surfaces of constant time slicing inside the region of "true" vacuum have open or hyperbolic spatial curvature. If the inflationary epoch lasted just enough to solve the horizon and flatness problems, this curvature may be persist and be observable on the largest cosmological scales. It was argued in [4] that this may even be probabilistically favored in simple models of the landscape. This was further developed in [5] where it is argued that under fairly generic assumptions about the landscape, any spatial curvature observed today must be negative. This is often interpreted as a no-go theorem that an experimental detection of positive curvature would falsify the landscape. It is the purpose of the present work to present a counterexample to this statement, although we do not have the tools to decide whether our model can be likely in the context of a larger landscape scenario.

It should be noted that our model is not the first to feature positive curvature from a tunneling event (see e.g. [9, 11, 10, 18], and [19]), however, what makes the model here novel is that the tunneling process is described entirely within semiclassical field theory, and the tunneling rate is tunable and dominated by the tunneling of a single scalar field. This is accomplished using the tunneling of a scalar field to rearrange the energetic contributions of the matter sector between sources with different equations of state.

2. Model: Tunneling from the simple harmonic universe

Our starting point is the simple harmonic universe studied by [2, 6, 7, 8], which was shown in [2, 3] to be classically and quantum mechanically stable at the level of linearized perturbations. The simplest solution of this type consists of positive spatial curvature, a negative cosmological constant, and a matter source with $w = p/\rho = -2/3$. Adding an additional matter source with $w = p/\rho = -1/3$ tunes the energetics of the curvature contribution and helps control certain classes of perturbations. The metric is the FRW metric $ds^2 = -dt^2 + a(t)^2 ds^2_{S^3}$ and the Friedmann equation for the evolution of the scale factor is given by

$$\left(\frac{\dot{a}}{a}\right)^2 = \frac{K_{eff}}{a^2} + \frac{8\pi G_N}{3} V_{eff} = \frac{K_{eff}}{a^2} + \frac{8\pi G_N}{3}\left(\frac{\rho_0}{a} + \Lambda\right) \tag{1}$$

Here $K_{eff} > 0$ is the combination of curvature and matter with $w = -1/3$, and the term with ρ_0 is from the matter source with $w = -2/3$. It is simple to show that the scale factor oscillates sinusoidally in time. If the perturbations of the matter sector have positive speed of sound squared, it was shown in [2, 3] that there exists a parametric regime where the solution can be stable at the level of linearized perturbatioins. While there may be nonperturbative decay processes, the solution nevertheless can remain stable on exponentially long timescales.

We propose a model where the energetics of the matter sector depend on the vacuum expectation value of a scalar field ϕ, and the effective potential V_{eff} has multiple degenerate minima in ϕ with different equations of state. In particular, as ϕ tunnels from its initial to its final value, we have $\rho_0(\phi_i) > 0$, $\Lambda(\phi_i) < 0$, and $\rho_0(\phi_f) = 0$, $\Lambda(\phi_f) > 0$. The tunneling potential is constructed so that the tunneling is energetically possible only when the scale factor is at or near the minimum of the bounce. This must be true for the Euclidean as well as the Lorentzian solution, and so it is also necessary to make the tunneling transfer energy to a radiation sector as well, so that the tunneling will be energetically unfavorable for the classically forbidden region $a < a_{min}$ as well as for the classically allowed region $a > a_{min}$. The full potential for the matter sector is therefore given by

$$V_{eff}(\phi, a) = \frac{\rho_0(\phi)}{a} + \Lambda(\phi) + \frac{\rho_r(\phi)}{a^4}, \tag{2}$$

where the energy of the various matter sources depends on the vacuum expectation value of the scalar field so that $V_{eff}(\phi_i, a) \le V_{eff}(\phi_f, a)$, with equality only for $a = a_{min}$, and $\rho_0(\phi_i) > 0, \Lambda(\phi_i) < 0, \rho_r(\phi_i) = 0$; $\rho_0(\phi_f) = 0, \Lambda(\phi_f) > 0, \rho_r(\phi_f) > 0$. The scalar field tunnels over the entire volume of the ancestor at once, and is effectively a 0+1-dimensional process. As it tunnels through the barrier from the ancestor to the daughter vacuum, the universe transitions from the simple harmonic universe solution to an inflating solution with positive curvature. There is also a radiation bath present in the daughter vacuum, but this redshifts away within a few e-folds.

Since the energy is conserved, as long as the evolution of the scalar field in the Euclidean solution proceeds much faster than the evolution of the scale factor, gravitation effectively decouples and the scalar field evolves in a fixed background, with the transition taking place over the entire spherical spatial slice. On longer timescales, the Euclidean solution for the evolution of the scale factor can be analyzed with and without the presence of the scalar bounce, and it is found that the change in the Euclidean action is dominated by the scalar evolution. The scalar bounce evolution therefore decouples from gravity and proceeds without gravitational backreaction in this limit. Furthermore, there exists a parametric limit where the scalar field decay dominates the total decay rate over any purely global quantum gravitational processes (such as tunneling of the scale factor, as discussed in [3, 12, 17] which may be present.

Although gravity decouples from the homogeneous bounce solution, it is necessary to include backreaction effects when considering the second variation of the Euclidean action around the bounce solution. It was shown in [13], working separately with the eigenmodes of this operator which are odd/even in Euclidean time, that there cannot be more than one nonpositive eigenmode for the second variation if the Euclidean solution is to maximize the tunneling probability. This can be shown to be equivalent to finding the number of odd and even bound states of the potential $V_{eff}(\phi_{cl}(\tau))$ in flat space, which is replaced by a more complicated potential when gravitational backreaction is taken into account. For the bounce of a scalar field in flat space, it can be shown that there is a single even eigenmode with negative eigenvalue, and a single odd mode with zero eigenvalue. Here things are different because of the background evolution of the metric, and we find a cluster of odd negative eigenmodes closely spaced in energy, and a cluster of even negative eigenmodes closely spaced in energy. Within each cluster, however, these solutions for the wavefunction differ only in a small singular region of the effective potential, where the wavefunction undergoes rapid oscillations that may arise from quantum gravitational effects. This class of eigenvalues also arose in models studied in [14], and are believed to arise from quantum gravitational effects when the tunneling instanton includes most or all of the volume of the Universe. Although the interpretation of these eigenmodes is not fully understood, they do not seem to invalidate the interpretation of the Euclidean bounce solution as a tunneling process. The bounce action is not strictly speaking a decay rate when there are multiple negative eigenvalues present, but rather should be interpreted as a tunneling between degenerate minima.

3. Conclusion and experimental constraints

To conclude, we have presented a proof-of-principle model where an inflationary vacuum with positive spatial curvature is the endpoint of a scalar field tunneling event. The tunneling can be analyzed in the semiclassical regime and there exists a parametric limit where it proceeds self-consistently and dominates the tunneling rate. Whether such a process can be generic in the context of a larger landscape of possibilities will have to await further studies explicitly constructing the simple harmonic universe in the context of string or field theory, and it would be interesting to pursue this further. For the moment, however, we wish to make the point that positive curvature does not falsify the possibility that our present vacuum originated from tunneling in a landscape, and it is worth including positive values for the curvature in experimental searches. Together with the tensor to scalar ratio of primordial perturbations, if observed it is perhaps as significant as a single number can be.

The curvature parameter Ω_k must have an absolute value of at least 10^{-5} in order to be distinguishable from gauge-dependent primordial fluctuations of the same type that generated the seeds of structure visible in the CMB. Current observational limits from CMB and LSS surveys constrain the curvature parameter at the 1 per cent level, and upcoming experiments from will push this by at least a further order of magnitude and may even close the observationally interesting window entirely [15, 16].

References

[1] Coleman S and De Luccia F 1980 *Phys. Rev.* D **21** 3305
[2] Graham P *et al. JHEP* 1402 029
[3] Graham P *et al.* 2014 *JHEP* 1408 163
[4] Freivogel B *et al.* 2006 *JHEP* 0603 039
[5] Kleban M and Schillo M 2012 *JCAP* 1206 029
[6] Harrison E 1967 *MNRAS* 137 69
[7] Kardashev N 1990 *MNRAS* 243 252
[8] Dabrowski M 1996 *Annals Phys.* 248, 199
[9] Atkatz D and Pagels H 1982 *PRD* 25 2065
[10] Vilenkin A 1982 *Phys. Lett.* B **117** 25
[11] Bousso R and Linde A 1998 *PRD* 58 083503
[12] Mithani A and Vilenkin A 1998 *JCAP* 1201 028
[13] Coleman S 1988 *Nucl. Phys.* B **98** 178
[14] Lee H and Weinberg E 2012 *Phys. Rev.* D **90** 124002
[15] CMB S4 collaboration *Preprint* arXiv:1610.02743 [astro-ph.co]
[16] CORE collaboration *Preprint* arXiv:1612.08270 [astro-ph.co]
[17] Mithani A and Vilenkin A 1998 *JCAP* 1507 (2010) 010
[18] Vilenkin A 1983 *PRD* 27 2848
[19] Vilenkin A 1984 *PRD* 30 509

Relativistic anisotropic stars with the polytropic equation of state in general relativity

A A Isayev
Kharkov Institute of Physics and Technology, Academicheskaya Street 1, Kharkov, Ukraine
E-mail: isayev@kipt.kharkov.ua

Abstract. Spherically symmetric relativistic stars with the polytropic equation of state, which possess the local pressure anisotropy, are considered in the context of general relativity. The modified Lane-Emden equations are derived for the special ansatz for the anisotropy parameter Δ in the form of the differential relation between Δ and the metric function ν. The analytical solutions of the obtained equations are found for incompressible fluid stars. The dynamical stability of incompressible anisotropic fluid stars against radial oscillations is studied.

1. Introduction. Basic equations

It was suggested in Ref. [1] that, despite the spherically symmetric distribution of matter inside a compact stellar object, it can be characterized by the local pressure anisotropy. The analysis of the generalized equations of hydrostatic equilibrium, allowing for the pressure anisotropy, shows that anisotropy may have the substantial effect on the maximum equilibrium mass and gravitational surface redshift. The pressure anisotropy can be caused by different physical reasons, such as, e.g., availability of superfluid states with the finite orbital momentum of Cooper pairs [2, 3] or finite superfluid momentum [4, 5], or the presence of strong magnetic fields inside a star [6, 7, 8, 9, 10, 11, 12]. In the present work, we will study spherical relativistic anisotropic stars with the polytropic equation of state, aiming to obtain the modified Lane-Emden (LE) equations for the special ansatz for the anisotropy parameter Δ in the form of the differential relation between Δ and the metric function ν. In general case, the obtained LE equations can be integrated only numerically, but the analytic solutions can be found for incompressible fluid stars. Then we apply the Chandrasekhar variational procedure [13] to study the dynamical stability of incompressible anisotropic fluid stars with respect to radial oscillations.

For spherically symmetric stars, the line element is written in the form (in units with $c = 1$):

$$ds^2 = e^{\nu(r,t)}dt^2 - e^{\lambda(r,t)}dr^2 - r^2(d\theta^2 + \sin^2\theta d\varphi^2). \tag{1}$$

While the matter distribution inside a star is spherically symmetric, we allow the existence of the local pressure anisotropy in its interior with the different values of the radial p_r and transverse $p_\theta = p_\varphi \equiv p_t$ pressures. The energy–momentum tensor for a spherical anisotropic star reads

$$T_i^k = (\varepsilon + p_t)u_i u^k - p_t\delta_i^k + (p_r - p_t)s_i s^k, \tag{2}$$

where ε is the energy density of the system, $u^i = \frac{dx^i}{ds}$ is the fluid four-velocity, and s^i is the unit space-like vector with the properties $s^i u_i = 0,\ s^i s_i = -1$. For the motions in the radial direction, the four-vectors u^i and s^i have the structure $u^i = \left(\frac{e^{-\frac{\nu}{2}}}{\sqrt{1-v^2e^{\lambda-\nu}}}, \frac{ve^{-\frac{\nu}{2}}}{\sqrt{1-v^2e^{\lambda-\nu}}}, 0, 0\right)$, $s^i = \left(\frac{ve^{\frac{\lambda}{2}-\nu}}{\sqrt{1-v^2e^{\lambda-\nu}}}, \frac{e^{-\frac{\lambda}{2}}}{\sqrt{1-v^2e^{\lambda-\nu}}}, 0, 0\right)$,where $v = \frac{dr}{dt}$ is the radial velocity. In the spherically symmetric case, allowing for the motions in the radial direction, Einstein equations were written in Ref. [14]. The radial component of the equation $T^k_{i;k} = 0$ can be represented as

$$\dot{T}^0_1 + T^{1\,\prime}_1 + \frac{1}{2}T^0_1(\dot{\nu} + \dot{\lambda}) + \frac{\nu'}{2}(T^1_1 - T^0_0) + \frac{2}{r}(T^1_1 - T^2_2) = 0, \tag{3}$$

where $\dot{\nu} \equiv \frac{\partial \nu}{\partial t}$, $\nu' \equiv \frac{\partial \nu}{\partial r}$, etc. From Eq. (3), one can get in the static limit the equation for the hydrostatic equilibrium in the presence of the pressure anisotropy in the form

$$p_r' = -\frac{\nu'}{2}(\varepsilon + p_r) + \frac{2\Delta}{r}, \quad \Delta \equiv p_t - p_r, \tag{4}$$

where Δ is the anisotropy parameter. For the static configuration, the interior metric function λ reads [1]

$$e^{-\lambda(r)} = 1 - \frac{2G}{r}m(r), \quad r < R, \tag{5}$$

where R is the radial coordinate at the surface of a star, and $m(r) = 4\pi \int_0^r \varepsilon r^2 dr$ is the mass enclosed in the sphere of radius r. From Einstein equations in the static limit, one can find

$$\nu' = 2G\frac{m(r) + 4\pi p_r r^3}{r(r - 2Gm(r))}. \tag{6}$$

Hence, the equation of hydrostatic equilibrium for a spherical anisotropic star takes the form:

$$p_r' = -G\frac{(\varepsilon + p_r)(m(r) + 4\pi p_r r^3)}{r(r - 2Gm(r))} + \frac{2\Delta}{r}. \tag{7}$$

As the boundary condition to Eq. (7), we set the radial pressure at the center of a star $p_r(0) = p_{r0}$, and determine the radial coordinate R at the surface from the condition $p_r(R) = 0$. The total mass then can be calculated as $M = m(R)$. After finding the radial pressure distribution $p_r(r)$ (together with the mass distribution $m(r)$), the metric functions $\lambda(r), \nu(r)$ can be determined from Eqs. (5), (6). At the boundary $r = R$, the metric functions are matchable to the exterior vacuum Schwarzschild metric:

$$\lambda(R) = -\nu(R) = -\ln\left(1 - \frac{2GM}{R}\right). \tag{8}$$

In order to solve Eq. (7), it is necessary to set the equation of state (EoS) of the system. Further, as the EoS of the system, we choose the polytropic EoS in the form [15]:

$$p_r = K\varrho^\gamma \equiv K\varrho^{1+\frac{1}{n}}, \tag{9}$$

where ϱ is the mass density, K is some constant, γ is the polytropic exponent, n is the polytropic index. For the EoS (9) the energy density ε is related to the mass density ϱ and the radial pressure p_r by the equation $\varepsilon = \varrho + \frac{p_r}{\gamma-1}$ [15].

It is convenient to introduce the auxiliary dimensionless LE function θ according to equations:

$$p_r = p_{r0}\theta^{n+1}, \quad \varrho = \varrho_0\theta^n, \tag{10}$$

where ϱ_0 is the central mass density. It follows from the boundary conditions for the radial pressure p_r that

$$\theta(0) = 1, \quad \theta(R) = 0. \tag{11}$$

Then Eq. (4) of hydrostatic equilibrium can be rewritten as

$$2q_0(n+1)d\theta - \frac{4\Delta dr}{\varrho_0 r\theta^n} + \big(1 + (n+1)q_0\theta\big)d\nu = 0, \tag{12}$$

where $q_0 \equiv \frac{p_{r0}}{\varrho_0}$. In order to solve the equation of hydrostatic equilibrium, one needs to specify the anisotropy parameter Δ. We will suppose that the presence of the anisotropy parameter Δ doesn't change the general form of LE equations for relativistic isotropic stars [15], but only can change the coefficients in these equations. Specifically, we will assume that the parameter Δ and the metric function ν are related by the differential equation

$$-\frac{4\Delta dr}{\varrho_0 r\theta^n} + \big(1 + (n+1)q_0\theta\big)d\nu = (1 + \beta q_0\theta)d\nu, \tag{13}$$

where β is some real constant. Substituting Eq. (13) into Eq. (12) and integrating it, one can obtain the metric function $\nu(r)$ in the form

$$\nu(r) = \ln \frac{1 - \frac{2GM}{R}}{\big(1 + \beta q_0\theta\big)^{\frac{2(n+1)}{\beta}}}. \tag{14}$$

Let us define the auxiliary function

$$u(r) \equiv \frac{m(r)}{M} = \frac{r}{2GM}\left(1 - e^{-\lambda(r)}\right), \; u(0) = 0, u(R) = 1, \tag{15}$$

which satisfies the differential equation $Mu' = 4\pi\varepsilon r^2$. Introducing the dimensionless variable ξ and dimensionless function η by the equations

$$r = \alpha\xi, \quad \eta = \frac{M}{4\pi\varrho_0\alpha^3}u, \tag{16}$$

where $\alpha^2 = \frac{q_0(n+1)}{4\pi G\varrho_0}$, equations for the functions θ and η can be obtained in the form

$$\frac{\xi - 2(n+1)q_0\eta}{1 + \beta q_0\theta}\xi\frac{d\theta}{d\xi} + \eta + q_0\xi^3\theta^{n+1} = 0, \tag{17}$$

$$\frac{d\eta}{d\xi} = \xi^2\theta^n(1 + nq_0\theta). \tag{18}$$

Eqs. (17) and (18) represent the modified LE equations for relativistic anisotropic polytropes with the EoS (9) and ansatz (13) for the anisotropy parameter Δ, after solving which one can find from Eqs. (10), (16) the radial distribution of the radial pressure and mass in the interior of a spherical anisotropic star. One can see that the obtained LE equations formally look as in the isotropic case [15], but with that difference that the impact of the anisotropy parameter is reflected in the coefficient β (substituting the multiplier $(n+1)$). As follows from Eqs. (11), (16), the boundary conditions for the functions $\theta(\xi)$ and $\eta(\xi)$ read

$$\theta(0) = 1, \; \theta(\xi_R) = 0, \quad \xi_R \equiv R/\alpha \tag{19}$$

$$\eta(0) = 0, \; \eta(\xi_R) = \frac{M}{4\pi\varrho_0\alpha^3}. \tag{20}$$

In general case, the LE equations (17) and (18) can be integrated only numerically, but the analytical solutions can be found for incompressible anisotropic fluid stars, characterized by the constant density $\varrho = \text{const}$. At $n = 0$, solutions for the functions $\theta(\xi)$ and $\eta(\xi)$, satisfying the boundary conditions (19) and (20), are given by

$$\theta(\xi) = \frac{1}{q_0}\frac{(1+3q_0)\left(1-\frac{2q_0}{3}\xi^2\right)^{\frac{3-\beta}{4}} - (1+\beta q_0)}{3(1+\beta q_0) - \beta(1+3q_0)\left(1-\frac{2q_0}{3}\xi^2\right)^{\frac{3-\beta}{4}}}, \quad \eta(\xi) = \frac{\xi^3}{3}. \tag{21}$$

The positive root of the $\theta(\xi)$ is $\xi_R = \sqrt{\frac{3}{2q_0}\left[1 - \left(\frac{1+\beta q_0}{1+3q_0}\right)^{\frac{4}{3-\beta}}\right]}$.

2. Dynamical stability of incompressible anisotropic fluid stars

Let us consider the stability of spherically symmetric anisotropic stars with respect to radial oscillations, assuming that they do not violate the spherical symmetry. Further we will study the small radial oscillations and will represent the unknown quantities as $\varepsilon = \varepsilon^0 + \delta\varepsilon$, $p_r = p_r^0 + \delta p_r$, $p_t = p_t^0 + \delta p_t$, $\nu = \nu^0 + \delta\nu$, $\lambda = \lambda^0 + \delta\lambda$, where $\delta\varepsilon, \delta p_r, \delta p_t, \delta\nu, \delta\lambda$ are small perturbations with respect to the corresponding values at the state of hydrostatic equilibrium, denoted by the upper index "0". Also, for the small radial oscillations we will consider that $v \ll 1$. Following Ref. [13], it is convenient to introduce a "Lagrange displacement" ψ by the equation $v = \dot{\psi}$. We will assume that all perturbations depend on time only through the exponential factor $e^{i\omega t}$, where ω is the frequency of radial oscillations. Then, using the linearized form of Einstein equations [14], and equation for the conservation of the total baryon number, the variations $\delta\varepsilon, \delta p_r, \delta\nu, \delta\lambda$ can be expressed through the Lagrange displacement ψ. Substituting these expressions to the linearized form of Eq. (3), in the case of the polytropic EoS one gets

$$\begin{aligned}\omega^2 e^{\lambda^0-\nu^0}(\varepsilon^0 + p_r^0)\psi &= \frac{2\psi}{r}p_r^{0\prime} - \frac{2\psi}{r}\Big(\gamma\big(\nu^{0\prime} + \frac{\lambda^{0\prime}}{2} + \frac{2}{r}\big) + \frac{2}{r}\Big)(p_t^0 - p_r^0) \\ &+8\pi G e^{\lambda^0} p_t^0(\varepsilon^0 + p_r^0)\psi - \gamma\frac{d}{dr}\Big(\frac{2}{r}(p_t^0 - p_r^0)\psi\Big) - \frac{\psi}{\varepsilon^0 + p_r^0}\Big(p_r^{0\prime} - \frac{2}{r}(p_t^0 - p_r^0)\Big)^2 \\ &-\gamma e^{-(\nu^0 + \frac{\lambda^0}{2})}\frac{d}{dr}\Big(e^{\frac{3\nu^0+\lambda^0}{2}}\frac{p_r^0}{r^2}\frac{d}{dr}\big(r^2 e^{-\frac{\nu^0}{2}}\psi\big)\Big) - \frac{2}{r}\Big(\gamma p_r^0\frac{e^{\frac{\nu^0}{2}}}{r^2}\frac{d}{dr}\big(r^2 e^{-\frac{\nu^0}{2}}\psi\big) + \delta p_t\Big).\end{aligned} \tag{22}$$

Solutions of Eq. (22) for the frequencies of radial oscillations should be sought under the boundary conditions $\psi(r=0) = 0$, $\delta p_r(r=R) = 0$. In order to get the variational basis for finding the frequencies ω, let us multiply both parts of Eq. (22) on $r^2\psi\exp(\frac{\nu^0+\lambda^0}{2})$ and integrate over the range of r. We will write the corresponding equation already for incompressible fluid stars ($n = 0$), when the polytropic exponent $\gamma \to \infty$. Omitting the upper indexes zero, one gets

$$\begin{aligned}\omega^2\int_0^R e^{\frac{3\lambda-\nu}{2}}(\varepsilon + p_r)r^2\psi^2\,dr &= \gamma\int_0^R e^{\frac{\lambda+3\nu}{2}}\frac{p_r}{r^2}\Big(\frac{d}{dr}\big(r^2 e^{-\frac{\nu}{2}}\psi\big)\Big)^2 dr \\ -\gamma\int_0^R e^{\frac{\lambda+\nu}{2}}r^2\psi\frac{d}{dr}\Big(\frac{2}{r}(p_t - p_r)\psi\Big)\,dr &- 2\gamma\int_0^R e^{\frac{\lambda+\nu}{2}}r\psi^2\Big(\nu' + \frac{\lambda'}{2} + \frac{2}{r}\Big)(p_t - p_r) \\ &-2\gamma\int_0^R e^{\frac{\lambda}{2}+\nu}\psi\frac{p_r}{r}\frac{d}{dr}\big(r^2 e^{-\frac{\nu}{2}}\psi\big)\,dr.\end{aligned} \tag{23}$$

In the variational equation (23), the Lagrange displacement ψ should be chosen such that ω^2 is minimized. If all frequencies of radial oscillations are real, a spherical anisotropic star is dynamically stable; if some frequency appears to be imaginary, the configuration is unstable.

Table 1. The critical values of the parameter q_0 for the appearance of the dynamical instability of an incompressible anisotropic fluid star at different values of the parameter β.

β	q_{0c} evaluated with the trial function	
	$\chi_1 = e^{-\frac{\nu}{2}}\xi^2$	$\chi_2 = \xi$
0.2	2.069	-
0.4	2.986	-
0.6	5.189	-
0.8	16.421	-

A sufficient condition for the occurrence of the dynamical instability is vanishing of the r.-h. s. of Eq. (23) for some trial form of the Lagrange displacement ψ satisfying the boundary conditions.

Let us introduce, following Ref. [15], the auxiliary function $\chi = e^{-\frac{\nu}{2}}\psi$. After changing the integration variable in Eq. (23) according to Eq. (16), substituting $p_r = q_0\varrho_0\theta$, $\varepsilon = \varrho_0$, and using the anisotropy parameter Δ from Eq. (13) and expressions (14), (15) for the metric functions at $n = 0$, Eq. (23) takes the form

$$\frac{\omega^2}{\omega_0^2}\frac{1}{1-\frac{2GM}{R}}\int_0^{\xi_R}\frac{(1+q_0\theta)\xi^2\chi^2}{(1-\frac{2q_0\xi^2}{3})^{\frac{3}{2}}}\frac{d\xi}{(1+\beta q_0\theta)^{\frac{1}{\beta}}} = \gamma\int_0^{\xi_R}\frac{\theta\Big(\frac{d}{d\xi}(\xi^2\chi)\Big)^2}{\xi^2(1-\frac{2q_0\xi^2}{3})^{\frac{1}{2}}}\frac{d\xi}{(1+\beta q_0\theta)^{\frac{3}{\beta}}} \tag{24}$$
$$-\frac{\gamma(1-\beta)}{2}\int_0^{\xi_R}\frac{\xi^2\chi}{(1-\frac{2q_0\xi^2}{3})^{\frac{1}{2}}}\frac{d}{d\xi}\Big(\frac{\nu'\chi\theta}{(1+\beta q_0\theta)^{\frac{1}{\beta}}}\Big)\frac{d\xi}{(1+\beta q_0\theta)^{\frac{2}{\beta}}}$$
$$-\frac{\gamma(1-\beta)}{2}\int_0^{\xi_R}\frac{\xi^2\chi^2\nu'\theta(\nu'+\frac{\lambda'}{2}+\frac{2}{\xi})}{(1-\frac{2q_0\xi^2}{3})^{\frac{1}{2}}}\frac{d\xi}{(1+\beta q_0\theta)^{\frac{3}{\beta}}} - 2\gamma\int_0^{\xi_R}\frac{\chi\theta\frac{d}{d\xi}(\xi^2\chi)}{\xi(1-\frac{2q_0\xi^2}{3})^{\frac{1}{2}}}\frac{d\xi}{(1+\beta q_0\theta)^{\frac{3}{\beta}}},$$

where $\omega_0^2 = 4\pi\varrho_0 G$. Let us use the trial functions of the form $\chi_1 = e^{-\frac{\nu}{2}}\xi^2, \chi_2 = \xi$. Then for each given β we will try to find such q_{0c} at which the r.-h. s. of Eq. (24) vanishes, and, hence, the dynamical instability for an incompressible anisotropic fluid star occurs at $q_0 > q_{0c}$.

The results of calculations are presented in Table 1. The most important conclusion is that there are solutions for q_{0c} in the case of the trial function χ_1 at $\beta < 1$, i.e., for $\Delta = p_t - p_r > 0$ (and there are no solutions at $\beta > 1$). This means that the local pressure anisotropy with $p_t > p_r$ can affect the dynamical stability of spherical incompressible fluid stars, unlike to incompressible isotropic fluid stars with the polytropic EoS (9), which are stable against radial oscillations [15].

References

[1] Bowers R L and Liang E P T 1974 *Astrophys. J.* **188** 657
[2] Takatsuka T and Tamagaki R 1993 *Prog. Theor. Phys. Supplement* **112** 27
[3] Isayev A A and Röpke G 2002 *Phys. Rev.* C **66** 034315
[4] Isayev A A 2002 *Phys. Rev.* C **65** 031302
[5] Casalbuoni R and Nardulli G 2004 *Rev. Mod. Phys.* **76** 263
[6] Khalilov V R 2002 *Phys. Rev.* D **65** 056001
[7] Isayev A A and Yang J 2011 *Phys. Rev.* C **84** 065802
[8] Isayev A A and Yang J 2012 *Phys. Lett.* B **707** 163
[9] Isayev A A and Yang J 2013 *J. Phys.* G **40** 035105
[10] Isayev A A 2014 *Int. J. Mod. Phys.* A **29** 1450173
[11] Isayev A A 2015 *Phys. Rev.* C **91** 015208
[12] Chu P C, Wang X, Chen L W and Huang M 2015 *Phys. Rev.* D **91** 023003
[13] Chandrasekhar S 1964 *Astrophys. J.* **140** 417
[14] Landau L D and Lifshitz E M 2000 *The Classical Theory of Fields* (Oxford: Pergamon Press)
[15] Tooper R F 1965 *Astrophys. J.* **142** 1541

A mechanism for protogalaxies nuclei formation from primordial black holes clusters

V V Nikulin[1], A V Grobov[1,2] and S G Rubin[1]

[1] Department of elementary particles physics, National Research Nuclear University MEPhI, Moscow, Russia
[2] National Research Center Kurchatov Institute, Moscow, Russia

E-mail: N-Valer@yandex.ru

Abstract. A model describing the formation of protogalaxies is developed. Compact supermassive clusters of primordial black holes assumed to act as a nuclei for the galaxy formation. The mechanism of PBH formation based on a collapse of massive walls of scalar field due to second order phase transition. Mass spectra of PBH are obtained analytically and shown possibility of the formation of PBH clusters with a total mass of $\sim 10^5 - 10^8 M_\odot$, having a size of $\sim$ 10 parsec, in an amount of $\sim 10^{11}$, which corresponds to observational data on the value of galaxies on the visible Universe. The primary fractal structure of galaxies is naturally explained through the mechanism. Proposed approach is the cornerstone for a principally new scenario of the galaxy formation in the early Universe.

1. Introduction

Today, practically no one has any doubt that the centers of most galaxies contain supermassive black holes [1]. Nevertheless, the mechanism of galactic nuclei formation remains unclear. According to many studies, star formation and the formation of galactic nuclei were simultaneous [2]. The issue of the "pre-star" origin of supermassive black holes in the centers of galaxies remains open. In addition, unlike the "stellar" black holes, primordial black holes can have arbitrary mass.

One of the possibilities of how primrordial black holes can form is the phase transitions during cosmological inflation. During inflation, quantum fluctuations of various fields very quickly "swell" to macroscopic dimensions, forming primary inhomogeneities, including clusters of primordial black holes. These clusters may serve as a triggering mechanism for an accumulation baryons with further evolution in a protogalactic structure [3, 4].

2. Main Idea

Our calculations are based on the mechanism proposed in the works [5, 6]. We suppose that there is a field ϕ with the "Mexican hat"-like potential:

$$V(\phi) = \lambda(|\phi|^2 - f^2/2)^2 + \Lambda^4(1 - \cos\theta), \tag{1}$$

where $\phi = re^{i\theta}$, and $\Lambda^4(1 - \cos\theta)$ is the contribution of instanton effects in the renormalization of the Lagrangian. The potential is slightly inclined, so the minimum points are in $\theta = 2\pi n$.

It is assumed that during inflation this field is already in the ring minimum, and the radial component of the mass is quite large, so the field moves along the ring minimum.

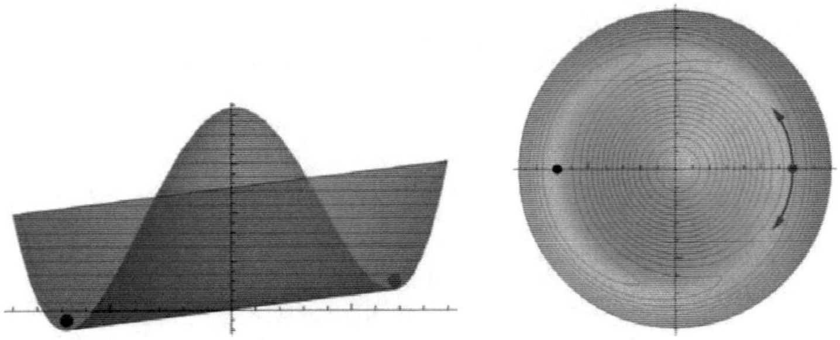

Figure 1. The potential defined by the formula (1) and its possible states.

Because of quantum fluctuations during inflation, random walks occur in the phase θ. We can obtain $f(\phi, t)$ - the distribution of the field ϕ, due to its quantum fluctuations. This distribution function is a solution of the Fokker-Planck equation [7]. Let the initial value of the field θ_U be slightly different from the minimum. After inflation, the classical evolution of field will roll it down to a minimum. Some fluctuations transit the field through $\theta = \pi$. It leads field to roll down to the minimum on the other side and forms space region of false vacuum state. A massive potential wall appears between the different vacuums, capable of collapsing into a black hole.

At the time of inflation, the size of the horizon H^{-1} is also the characteristic size of the fluctuations. After the end of inflation, a domain wall forms, separating the false vacuum. We can calculate the occurrence probability of critical fluctuations

$$P(t) = \int_{\pi}^{\infty} f(\theta, t) d\theta, \tag{2}$$

and, consequently, the rate at which the number of future domain walls increase:

$$\frac{dn}{dt} = \frac{dn_U}{dt} P(t) = 3He^{3Ht} P(t), \tag{3}$$

where dn_U/dt is the the rate of increasing in the number of cause-independent areas in the observable Universe, $Ht_{inf} = N$ the number of e-folds.

3. Mass spectra of PBH

Since we know the expansion rate of the Universe at the inflation stage, we can calculate the distribution of our regions in size because

$$\frac{dn}{dr} = \frac{dn}{dt} \frac{dt}{dr(t)}, \tag{4}$$

where $r(t)$ is the expansion law at the time of inflation. And therefore we can calculate a mass spectra because the mass of primordial black hole

$$m = 4\pi r^2 \rho_S, \tag{5}$$

where $\rho_S = 4\lambda^2 f$ is an energy density of the potential wall beetwen two vacuum states [5].

Omitting the long calculation, the mass sceptra of PBH:

$$\frac{dn}{dm} = \frac{3}{8} e^{3N} \sqrt[4]{\frac{\pi^3 \rho_S^3}{H^6 N^6 m^7}} \,\mathrm{Erfc}\left(\frac{2(\pi - \theta_U)\pi f}{H\sqrt{2N - \ln 2HN\sqrt{\frac{m}{4\pi\rho_S}}}} \right). \tag{6}$$

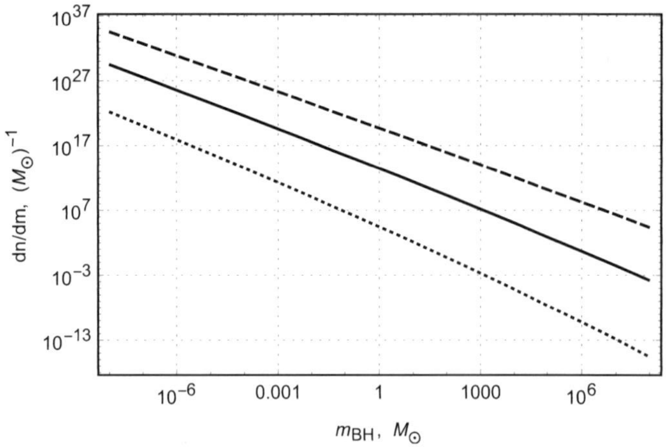

Figure 2. The differential mass spectrum of the primordial black holes:
– – – $f = 1.0H$,
—— $f = 1.6H$,
······ $f = 2.2H$.

Figure 2 shows the differential mass spectrum of the primordial black holes that appear in the model. We choose the following parameters of the model, consistent with the observed anisotropy of the background radiation: $H = 10^{13}$ GeV, $N = 60$, field parameter $\lambda = 5$ and the initial value $\theta_U = 0.05\pi$.

The number of PBHs in the observable Universe depends strongly on the value of f, so that three values are chosen: $f = 1.0H, f = 1.6H, f = 2.2H$. The model with $f = 1.6H$ has the best correspondence with observational data.

4. Clusters of PBHs

Fluctuations that bring the field value to a false minimum are multiple. Therefore, a region with other false minima can be formed with greater probability within a region whose field value is already close to critical. So, the PBHs form clusters that have a fractal structure. We can use the already obtained formula (6) with the minor changes to calculate the spectra of clusters.

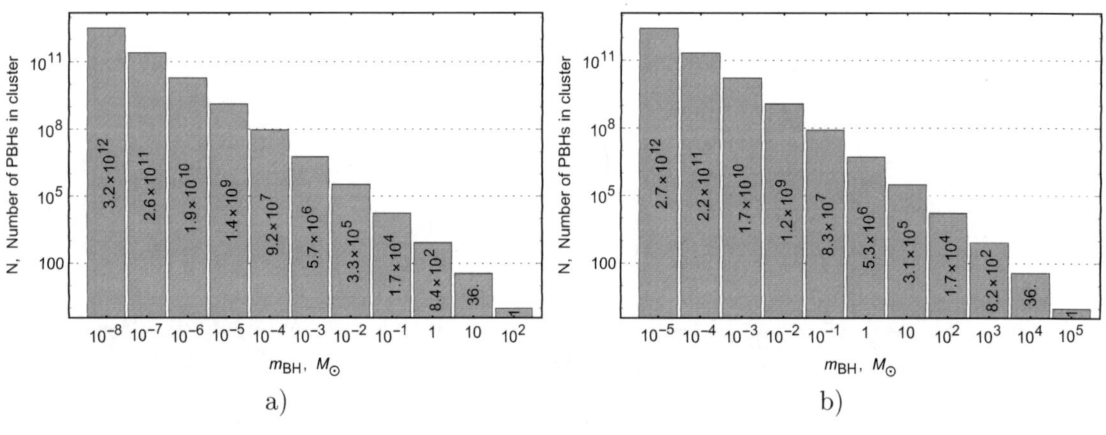

Figure 3. Mass spectra of PBHs clusters at $f = 1.8H$:
a) Cluster with central BH mass $10^2 M_\odot$, cluster size 4 pc, cluster mass $2.6 \cdot 10^5 M_\odot$, number of clusters in the Universe $4.1 \cdot 10^8$;
b) Cluster with central BH mass $10^5 M_\odot$, cluster size 40 pc, cluster mass $2.3 \cdot 10^8 M_\odot$, number of clusters in the Universe $8.1 \cdot 10^4$.

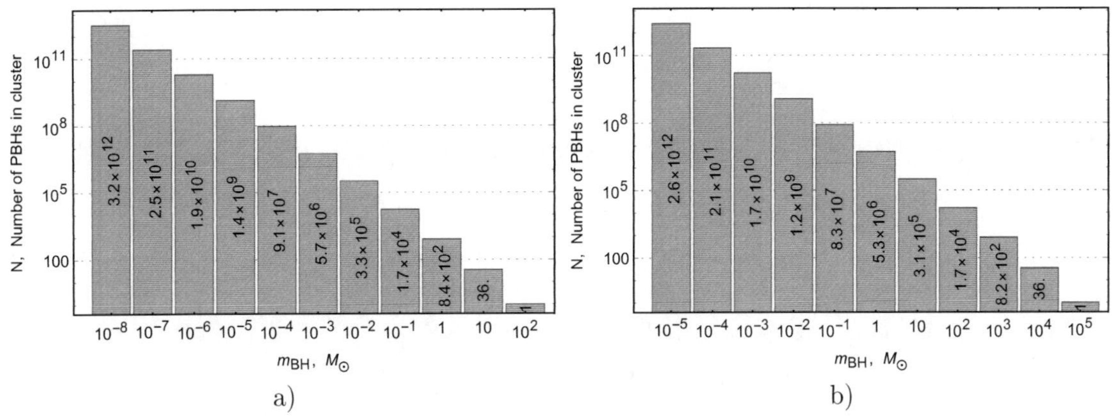

Figure 4. Mass spectra of PBHs clusters at $f = 1.6H$:
a) Cluster with central BH mass $10^2 M_\odot$, cluster size 4 pc, cluster mass $2.6 \cdot 10^5 M_\odot$, number of clusters in the Universe $2.9 \cdot 10^{11}$;
b) Cluster with central BH mass $10^5 M_\odot$, cluster size 40 pc, cluster mass $2.3 \cdot 10^8 M_\odot$, number of clusters in the Universe $1.1 \cdot 10^8$.

Figures 3 and 4 show the PBH mass spectra of the cluster for different model parameters. It is seen that at $f = 1.6H$ and central PBH mass $\sim 10^2 M_\odot$ the number of clusters with total mass $\sim 10^5 M_\odot$ corresponds to the number of galaxies in the observable Universe.

5. Conclusion

- Analytical mass spectra of primordial black holes for the observable Universe are obtained. With an appropriate choice of parameters, the model is consistent with the observational data.
- Primordial black holes form clusters have a fractal spatial structure. Most clusters are compact enough to form protogalactic nuclei.
- At $f = 1.6H$ and central PBH mass $\sim 10^2 M_\odot$ the number of clusters with total mass of $\sim 10^5 M_\odot$ is of $\sim 10^{11}$, so the model may explain forming the galactic nuclei in early Universe.

6. Forthcoming Research

It is planned to clarify in detail the spatial structure of the cluster (distribution with respect to r). In addition, we are going to simulate further movements of the PBHs in clusters to find out how they can form a supermassive black hole in the centers of future galaxies.

References

[1] Rosenberg D E and Rollino J 2000 *Preprint* astro-ph/0012023
[2] Veilleux S 2001 The starburst-agn connection *Springer Proceedings in Physics: Starburst Galaxies: Near and Far (Springer)* pp 8894
[3] Stiavelli M 1998 *ApJL* **495** L91
[4] Merrifield M, Forbes D A and Terlevich A 2000 *MNRAS* **313** L29L32
[5] Khlopov M Y, Konoplich R V, Rubin S G and Sakharov A S 2000 **G&C 6** 153156
[6] Dokuchaev V, Eroshenko Y N, Rubin S and Samarchenko D 2010 *Astron. Lett.* **36** 773779
[7] Linde A D 1982 *Phys. Lett.* B **116** 335339

What drives the kinematic evolution of star-forming galaxies?

Chao-Ling Hung
Manhattan College, 4513 Manhattan College Pkwy, Bronx, NY 10471, USA
E-mail: chaoling.hung@manhattan.edu

Abstract. One important result from recent large integral field spectrograph (IFS) surveys is that the intrinsic velocity dispersion of galaxies increases with redshift. Massive, rotation-dominated discs are already in place at $z \sim 2$, but they are dynamically hotter than spiral galaxies in the local Universe. Although several plausible mechanisms for this elevated velocity dispersion (e.g. star formation feedback, elevated gas supply, or more frequent galaxy interactions) have been proposed, the fundamental driver of the velocity dispersion enhancement at high redshift remains unclear. We investigate the origin of this kinematic evolution using a suite of cosmological simulations from the FIRE (Feedback In Realistic Environments) project. These simulations reproduce the observed trends between intrinsic velocity dispersion ($\sigma_{\rm intr}$), SFR, and z. In both the observed and simulated galaxies, $\sigma_{\rm intr}$ is positively correlated with SFR. $\sigma_{\rm intr}$ increases with redshift out to $z \sim 1$ and then flattens beyond that. In the FIRE simulations, $\sigma_{\rm intr}$ can vary significantly on timescales of $\lesssim 100$ Myr. These variations closely mirror the time evolution of the SFR and gas inflow rate ($\dot{M}_{\rm gas}$). By cross-correlating pairs of $\sigma_{\rm intr}$ $\dot{M}_{\rm gas}$, and SFR, we show that the increased gas inflow leads to subsequent enhanced star formation, and enhancements in $\sigma_{\rm intr}$ tend to temporally coincide with increases in $\dot{M}_{\rm gas}$ and SFR.

1. Introduction

The increasing capabilities of optical and near-infrared integral field spectrographs (IFSs) have revealed the internal dynamics of hundreds of star-forming galaxies out to $z \lesssim 3$ (see the review by Glazebrook K [1]). These IFS surveys at $1 \lesssim z \lesssim 3$ reveal that a significant fraction ($> 1/3$) of massive galaxies ($M_* \gtrsim 10^{10}{\rm M}_\odot$) exhibit smooth velocity gradients, indicative of rotating discs, whereas the rest exhibit irregular, merger-like kinematics or are dispersion-dominated (e.g. [2, 3]). However, a key difference between these massive rotating discs at $z \gtrsim 1$ and spiral galaxies in the local Universe is that intrinsic velocity dispersions traced by ionized gas (hereafter $\sigma_{\rm intr}$) of $z \gtrsim 1$ systems are significantly higher than those of their local counterparts.

The elevated $\sigma_{\rm intr}$ in high$-z$ galaxies has been associated with enhanced turbulent motions. However, the physical driver(s) of the enhanced turbulence (e.g. feedback or gravitational instability) remains unclear. Lehnert M D show that there is a positive correlation between $\sigma_{\rm intr}$ and star formation rate (SFR) surface density in $z \sim 2$, possibly suggesting that the enhanced $\sigma_{\rm intr}$ is driven by star formation feedback processes such as supernovae and radiation pressure [4, 5]. It has also been claimed that gravitational instabilities due to external sources, such as cosmological gas accretion and galaxy interactions, or internal dynamics, such as disc instabilities and clump-clump interactions, are responsible for the elevated $\sigma_{\rm intr}$ exhibited by high-z galaxies. Several works have found that the evolution of gas fraction (in a marginally stable disc) can explain the increase in $\sigma_{\rm intr}$ with z (e.g. [6, 3]).

In this study, we aim to explore the physical drivers of $\sigma_{\rm intr}$ in star-forming galaxies using a suite of cosmological simulations from the FIRE project [7]. In parallel, we gather measurements of $\sigma_{\rm intr}$ of star-forming galaxies at $0 \leq z \lesssim 3$ from IFS surveys presented in the literature.

2. A comparison with observations

We compare the kinematic properties of the simulated galaxies from the FIRE project with observations from the literature. Specifically, we examine whether the FIRE simulations exhibit the general trends in $\sigma_{\rm 1D}$, z and SFR present in the observations, and here $\sigma_{\rm 1D}$ is defined as the minimum value of the SFR-weighted standard deviation of the velocity distribution over different viewing angles in the simulations. Figure 1 shows $\sigma_{\rm 1D}$ as a function of redshift. The median $\sigma_{\rm 1D}$ values of the observed galaxies are generally $\sim 10-20$ km s^{-1} higher than those of the simulated galaxies. The possible origins of this systematic offset is still under investigation. Regardless, the observed and simulated samples follow a similar trend with redshift, in that the median $\sigma_{\rm 1D}$ increases from $z \sim 0$ to $z \sim 1$ and then remains approximately constant at $z \gtrsim 1$.

Figure 2 shows the distribution of $\sigma_{\rm 1D}$ and SFR for the observed and simulated galaxies. For the observed galaxies, a positive correlation is seen out to $z \lesssim 1$, consistent with the results of Green A W [8]. Although the dynamical range in SFR is smaller for the lower redshift galaxies compared with the higher redshift galaxies, no obvious offset is seen in the $\sigma_{\rm 1D}$-SFR relation within the redshift range probed. However, it becomes less clear whether the same positive correlation extends to $z \gtrsim 2$ galaxies owing to the small number of galaxies. The star-forming galaxies (galaxies that lie on or above the galaxy main sequence) from the FIRE simulations also exhibit a positive correlation between $\sigma_{\rm 1D}$ and SFR, consistent with the observed galaxies, and there is also no obvious offset in this $\sigma_{\rm 1D}$-SFR relations at different redshifts.

Although the FIRE simulations reproduce the observed trends on the $\sigma_{\rm 1D}$–z and $\sigma_{\rm 1D}$–SFR planes, understanding what physical processes are responsible for these trends remains challenging. First, the interpretation that velocity dispersion increases with redshift is tangled with the positive correlation between $\sigma_{\rm 1D}$ and SFR, given that galaxies targeted by IFS surveys at $z \gtrsim 1$ tend to have higher SFRs compared with galaxies targeted by lower-redshift IFS surveys. Second, the $\sigma_{\rm 1D}$ values of the observed galaxies exhibit large scatter even after taking into account the differences in SFR and removing galaxies that fall below the main sequence, suggesting that star formation is not the only relevant driver of velocity dispersion (although this interpretation can be complicated if the SFR and $\sigma_{\rm 1D}$ vary on short timescales and are not perfectly synchronised; this is explored in detail below).

3. Physical drivers of kinematic evolution

To gain further insight into the physical drivers of galaxy dynamics, we analyze the time evolution of various physical properties of the individual central galaxies of each halo. The time evolution of SFR and $\dot{M}_{\rm gas}$ most closely resemble the variations in $\sigma_{\rm 1D}$; all three properties vary significantly on timescales of < 100 Myr. By definition, $\sigma_{\rm 1D}$ approaches zero at low SFR (SFR $\lesssim 1$ M$_\odot$ yr^{-1}), since only few SPH gas particles with non-zero SFRs are available to trace the galaxy dynamics. In general, such time periods also correspond to when galaxies have lower gas inflow rates. Increases in $\sigma_{\rm 1D}$ occur near enhancements in $\dot{M}_{\rm gas}$ and SFR, although not exactly simultaneously. The approximate temporal correspondence of variations in $\sigma_{\rm 1D}$ and SFR is reflected in the positive SFR–$\sigma_{\rm 1D}$ correlation, and the fact that peaks do not exactly coincide in time may be responsible for the large scatter in this relation (figure 2). The similarities between the evolutions of SFR, $\sigma_{\rm 1D}$, and $\dot{M}_{\rm gas}$ suggest that variations in velocity dispersion, star formation (and thus prompt stellar feedback) and variations in gas inflow are all related.

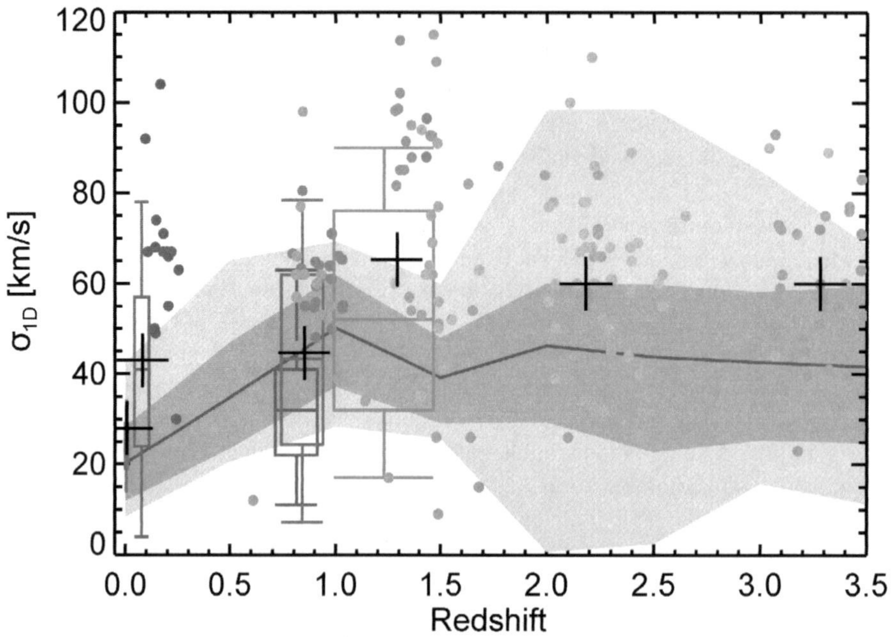

Figure 1. σ_{1D} as a function of redshift. The dark grey solid line shows the median value for the simulated galaxies that lie on or above the star-forming galaxy main sequence within redshift bins of width Δz =0.5, and the lighter (lightest) grey coloured area encloses 68 per cent (95 per cent) of the data. The coloured dots and the distribution bars summarize the observations described in Section 3.1; each color represents an IFS survey. In each distribution bar, the middle bar corresponds to the median σ_{1D}, and the boxes and vertical bars encompass 68 and 95 per cent of the data, respectively. The width of each distribution bar is proportional to the redshift coverage of the survey. The black crosses represent the median σ_{1D} of the observed galaxies that fall on or above the galaxy main sequence in six redshift bins: $z < 0.1$, $0.1 < z < 0.3$, $0.7 < z < 1.0$, $1.0 < z < 1.5$, $2.0 < z < 2.5$, and $3.0 < z < 3.5$. The simulated and observed galaxies exhibit the same qualitative trend, i.e. the median σ_{1D} increases from $z = 0$ to $z \sim 1$ and then remains constant, but at a given redshift, the simulated galaxies have median σ_{1D} values $\sim 10 - 20$ km s^{-1} less than the observed galaxies.

To examine whether the similarities in the temporal evolutions of SFR, σ_{1D}, and $\dot{M}_{gas}$ are indicative of any causal effects, we measure the time delays between pairs of these three quantities by cross-correlating their time series. For each simulation suite, we calculate temporal cross-correlation functions of pairs of time series [$(x, y) = (\dot{M}_{gas}$, SFR), ($\dot{M}_{gas}$,σ_{1D}), and (σ_{1D}, SFR)] that are normalised to have means of 0 and variances of 1, $\mathrm{CCF}(\tau_k) = \frac{1}{N}\sum_{i=1}^{N-k} x_i y_{i+k}$, where i indicates the time bin, N is the total number of time bins, τ_k is the lag corresponding to the k time bin, and k is varied to probe lags from -500 to 500 Myr.

In general, variations in $\dot{M}_{gas}$ occur prior to variations in the SFR, suggesting that inflows through 0.2 R_{vir}, due to cosmological gas inflow or galactic fountains, lead to subsequent enhanced star formation activity. As expected, the relation between $\dot{M}_{gas}$ and SFR becomes weaker if we define the inflow rate as the mass infalling through a shell with a radius of 1 R_{vir}. The negative signs of $\tau(\dot{M}_{gas}$-$\sigma_{1D})$ or $\tau(\sigma_{1D}$-SFR) are measured in a few simulation suites, suggesting that enhancements in velocity dispersion may occur after enhancements in the gas inflow rate at 0.2 R_{vir} but before the SFR is enhanced. However, these trends are less robust because positive or zero time delays are also detected for some simulation suites.

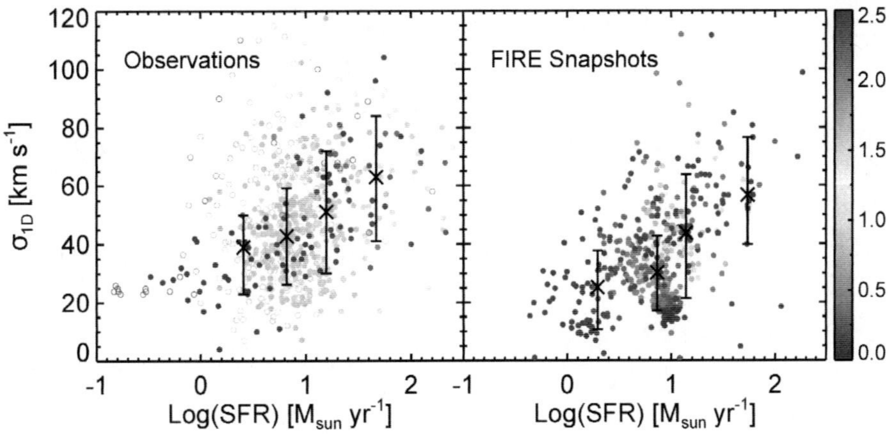

Figure 2. Distribution of observed and simulated galaxies on the σ_{1D}–SFR plane, colored-coded according to redshift. The open circles in the observations panel indicate galaxies that fall below the galaxy main sequence. The black x's and error bars overlaid on the data points represent the median σ_{1D} and 68 per cent distribution of the galaxies that lie on or above the main sequence in four log(SFR) bins. The simulations and observations both exhibit correlations between σ_{1D} and SFR, but at fixed SFR, except for the highest-SFR bin, the simulations have σ_{1D} values ~ 20 km s^{-1} less than the observed galaxies.

Given that the absolute values of $\tau(\dot{M}_{gas}\text{-}\sigma_{1D})$ and $\tau(\sigma_{1D}\text{-SFR})$ are typically smaller than the resolution of the times series, we can only conclude that variations in σ_{1D} temporally coincide with variations in $\dot{M}_{gas}$ and SFR.

References

[1] Glazebrook K 2013 *PASA* **30** 56
[2] Law D R *et al.* 2009 *ApJ* **697** 2057
[3] Wisnioski E *et al.* 2015 *ApJ* **799** 209
[4] Lehnert M D *et al.* 2009 *ApJ* **699** 1660
[5] Lehnert M D *et al.* 2013 *A&A* **555** A72
[6] Swinbank A M *et al.* 2012 *ApJ* **760** 130
[7] Hopkins P F, Keres D and Onorbe J 2014 *MNRAS* **445** 581
[8] Green A W, Glazebrook K and McGregor P J 2014 *MNRAS* **437** 581

Cosmological magnetic fields in turbulent matter

Maxim Dvornikov

Pushkov Institute of Terrestrial Magnetism, Ionosphere and Radiowave Propagation (IZMIRAN), 108840 Troitsk, Moscow, Russia
Physics Faculty, National Research Tomsk State University, 36 Lenin Avenue, 634050 Tomsk, Russia

E-mail: maxdvo@izmiran.ru

Abstract. We study the evolution of magnetic fields in turbulent hot plasma of the early Universe accounting for the chiral magnetic effect. The magnetohydrodynamic turbulence is modeled by replacing the matter velocity in the advection term in the Faraday equation with the Lorentz force. The system of the kinetic equations for the spectra of the densities of the magnetic helicity and the magnetic energy, as well as for the chiral imbalance, is derived. The amplification of the magnetic field is shown to result from the presence of the chiral magnetic effect solely. The system of the kinetic equations is solved numerically in primordial plasma after the electroweak phase transition. The influence of the matter turbulence on the magnetic field evolution is examined for different seed magnetic fields.

Magnetic fields are important for various physical processes including the cosmic rays propagation, the influence on the stellar and solar activities etc. Galactic magnetic fields, having the huge length scale comparable with a galaxy size, are supposed to be amplified from a seed magnetic field. The origin of this seed field is still an open problem in astrophysics and cosmology [1]. There is a possibility that seed magnetic fields are generated in the epoch of the structure formation. Another option, which is assumed in the present work, is that seed fields arise in the early Univese before the recombination.

The evolution of cosmic magnetic fields in primordial plasma, having the asymmetry of right and left particles, and the leptogenesis in (hyper-)magnetic fields before and after the electroweak phase transition (EWPT) were studied in [2–4]. The instability of the magnetic field in this primordial plasma, leading to the field generation, is owing to the chiral magnetic effect (CME) [5].

In this work, we shall discuss the influence of the matter turbulent motion on the generation of magnetic fields driven by the CME [6]. The advection term in the Faraday equation was neglected in [2]. The MHD turbulence only was accounted for in [7] while describing the evolution of magnetic fields in the early Universe. Recently the effect of the CME on the magnetic fields evolution in the turbulent primordial plasma was estimated numerically in [8].

We start with the system of the MHD equations in the single fluid approximation for the hot relativistic plasma in the early Universe after EWPT with the temperatures $10\,\text{MeV} < T < T_{\text{EWPT}} \simeq 100\,\text{GeV}$. For plasma obeying the equation of state $P = \rho/3$, these

IOP Conf. Series: Journal of Physics: Conf. Series **934** (2017) 012042 doi:10.1088/1742-6596/934/1/012042

equations have the form,

$$\partial_t \rho + \frac{4}{3}\nabla \cdot (\rho \mathbf{v}) = 0, \tag{1}$$

$$\frac{4}{3}\partial_t(\rho \mathbf{v}) - \frac{4}{3}\rho \mathbf{v} \times (\nabla \times \mathbf{v}) = (\mathbf{J} \times \mathbf{B}) - \nabla P + \frac{4}{3}\rho\nu\nabla^2 \mathbf{v}, \tag{2}$$

$$\partial_t \mathbf{B} = -\nabla \times \mathbf{E}, \tag{3}$$

where ρ is the energy density of the fluid, P is the pressure, $\mathbf{B}$ is the magnetic field, $\mathbf{E}$ is the electric field, $\mathbf{v}$ is the fluid velocity, $\mathbf{J} = (\nabla \times \mathbf{B})$ is the electric current, and ν is the kinematic viscosity.

Since charged particles in plasma are ultrarelativistic, we have to account for the anomalous CME current $\mathbf{J}_{\text{CME}} = (2\alpha_{\text{em}}\mu_5/\pi)\mathbf{B}$ along with the usual ohmic current $\mathbf{J}_{\text{Ohm}} = \sigma_{\text{cond}}[\mathbf{E} + (\mathbf{v} \times \mathbf{B})]$ in equations (1)-(3). Using the MHD approximation, we derive the Faraday equation modified by the CME contribution,

$$\partial_t \mathbf{B} = \nabla \times (\mathbf{v} \times \mathbf{B}) + \eta_m \nabla^2 \mathbf{B} + \frac{2\alpha_{\text{em}}\mu_5}{\pi\sigma_{\text{cond}}}\nabla \times \mathbf{B}. \tag{4}$$

Here $\alpha_{\text{em}} \approx 7.3 \times 10^{-3}$ is the fine structure constant, $\eta_m = \sigma_{\text{cond}}^{-1}$ is the magnetic diffusion coefficient, $\sigma_{\text{cond}} = \sigma_c T$ is the hot plasma conductivity with $\sigma_c \simeq 100$, $\mu_5 = (\mu_{\text{R}} - \mu_{\text{L}})/2$ is the chiral imbalance, and $\mu_{\text{R,L}}$ are the chemical potentials of right and left electrons.

To proceed with the analysis of equation (4) we shall model the fluid velocity $\mathbf{v}$, obeying the Navier-Stokes equation (2), which is rather difficult to solve, with the Lorentz force $\mathbf{F}_{\text{L}} \sim (\mathbf{J} \times \mathbf{B})$ [6],

$$\mathbf{v} = \frac{\tau_d}{P + \rho}(\mathbf{J} \times \mathbf{B}), \tag{5}$$

accompanied with the phenomenological drag time parameter $\tau_d \sim \alpha_{\text{em}}^{-2}/T$ [9].

Basing on the master equations (4) and (5), we derive the kinetic equations for the spectra of the magnetic energy density $\mathcal{E}_{\text{B}} = \mathcal{E}_{\text{B}}(k,t)$ and the density of the magnetic helicity $\mathcal{H}_{\text{B}} = \mathcal{H}_{\text{B}}(k,t)$ [6],

$$\frac{\partial \mathcal{E}_{\text{B}}}{\partial t} = -2k^2\eta_{\text{eff}}\mathcal{E}_{\text{B}} + \alpha_+ k^2 \mathcal{H}_{\text{B}}, \quad \frac{\partial \mathcal{H}_{\text{B}}}{\partial t} = -2k^2\eta_{\text{eff}}\mathcal{H}_{\text{B}} + 4\alpha_- \mathcal{E}_{\text{B}}, \tag{6}$$

where

$$\eta_{\text{eff}} = \sigma_{\text{cond}}^{-1} + \frac{4}{3}\frac{\tau_d}{P+\rho}\int \mathrm{d}p \mathcal{E}_{\text{B}}, \quad \alpha_d = \frac{2}{3}\frac{\tau_d}{P+\rho}\int \mathrm{d}p p^2 \mathcal{H}_{\text{B}},$$

$$\alpha_\pm = \alpha_{\text{CME}} \mp \alpha_d, \quad \alpha_{\text{CME}} = \frac{\Pi(t)}{\sigma_{\text{cond}}}, \quad \Pi(t) = \frac{2\alpha_{\text{em}}}{\pi}\mu_5(t). \tag{7}$$

The spectra $\mathcal{E}_{\text{B}}$ and $\mathcal{H}_{\text{B}}$ are related to the density of the magnetic energy and of the magnetic helicity by the following relations:

$$E_{\text{B}}(t) = \frac{B^2}{2} = \int \mathrm{d}k\, \mathcal{E}_{\text{B}}(k,t), \quad H_{\text{B}}(t) = \frac{1}{V}\int \mathrm{d}^3x(\mathbf{A} \cdot \mathbf{B}) = \int \mathrm{d}k\, \mathcal{H}_{\text{B}}(k,t), \tag{8}$$

where V is the normalization volume and $\mathbf{A}$ is the vector potential. The details of the derivation of equations (6) and (7) are provided in [6].

IOP Conf. Series: Journal of Physics: Conf. Series **934** (2017) 012042 doi:10.1088/1742-6596/934/1/012042

The difference of our results from findings of [2] is seen from the second non-linear terms in equation (7) which contain the drag time τ_d when we take into account the turbulent motion $\sim \mathbf{v}$. Note that the effective magnetic diffusion coefficient η_{eff} in equation (7) coincides with that in [7] (accounting also for the factor $P + \rho$ in the denominator missed there). The analog of the α-dynamo parameter, $\alpha_\pm$, in equation (7) differs from that derived in [7] mainly because of the absence of the CME term there, and due to the different signs ($\pm$) in turbulent contributions for the evolution of the spectra $\mathcal{E}_\text{B}$ and $\mathcal{H}_\text{B}$ instead of the same sign (+) in both equations.

When we study the evolution of magnetic fields in a hot plasma in the expanding universe, it is convenient to rewrite equations (6) and (7) using the conformal dimensionless variables. They are introduced in the following way: $t \to \eta = M_0/T$ and $\tilde{k} = ak$, where $a = 1/T$, $M_0 = M_{\text{Pl}}/1.66\sqrt{g^*}$, $M_{\text{Pl}} = 1.2 \times 10^{19}$ GeV is the Plank mass, and $g^* = 106.75$ is the number of the relativistic degrees of freedom. In these variables, equation (6) takes the form,

$$\frac{\partial \tilde{\mathcal{E}}_\text{B}}{\partial \eta} = -2\tilde{k}^2 \tilde{\eta}_{\text{eff}} \tilde{\mathcal{E}}_\text{B} + \tilde{\alpha}_+ \tilde{k}^2 \tilde{\mathcal{H}}_\text{B}, \quad \frac{\partial \tilde{\mathcal{H}}_\text{B}}{\partial \eta} = -2\tilde{k}^2 \tilde{\eta}_{\text{eff}} \tilde{\mathcal{H}}_\text{B} + 4\tilde{\alpha}_- \tilde{\mathcal{E}}_\text{B}, \tag{9}$$

where $\tilde{\mathcal{E}}_\text{B} = \tilde{\mathcal{E}}_\text{B}(\tilde{k}, \eta)$ and $\tilde{\mathcal{H}}_\text{B} = \tilde{\mathcal{H}}_\text{B}(\tilde{k}, \eta)$ are the conformal spectra, as well as $\tilde{\eta}_{\text{eff}}$ and $\tilde{\alpha}_\pm$ are the effective magnetic diffusion and effective α-dynamo parameters expressed in the conformal variables [6]. The evolution equation for the chiral imbalance $\tilde{\mu}_5 = \tilde{\mu}_5(\eta)$ has the form,

$$\frac{\mathrm{d}\tilde{\mu}_5}{\mathrm{d}\eta} = -\frac{6\alpha_{\text{em}}}{\pi} \int \mathrm{d}\tilde{k} \frac{\partial \tilde{\mathcal{H}}_\text{B}}{\partial \eta} - \tilde{\Gamma}_f \tilde{\mu}_5, \quad \tilde{\Gamma}_f = \alpha_{\text{em}}^2 \left(\frac{m_e}{3M_0} \right)^2 \eta^2, \tag{10}$$

where take into account the helicity flip rate $\tilde{\Gamma}_f$ in a hot relativistic plasma [2].

We shall use the initial energy spectrum in the form, $\tilde{\mathcal{E}}_\text{B}(\tilde{k}, \eta_0) = \mathcal{C}\tilde{k}^{\nu_\text{B}}$. The factor $\mathcal{C}$ can be found from the equation (8), rewritten in the conformal variables, by setting there $\tilde{B} = \tilde{B}_0$, where $\tilde{B}_0 = \tilde{B}(\eta = \eta_0)$ is the seed magnetic field. If we use the Batchelor initial spectrum with $\nu_\text{B} = 4$ and $0 < \tilde{k} < \tilde{k}_{\max} = 10^{-6}$, then it is convenient to introduce the following dimensionless variables:

$$\mathcal{H}(\kappa, \tau) = \frac{12\alpha_{\text{em}}^2}{\pi^2} \tilde{\mathcal{H}}_\text{B}(\tilde{k}, \eta), \quad \mathcal{R}(\kappa, \tau) = \frac{24\alpha_{\text{em}}^2}{\pi^2 \tilde{k}_{\max}} \tilde{\mathcal{E}}_\text{B}(\tilde{k}, \eta), \quad \mathcal{M}(\tau) = \frac{2\alpha_{\text{em}}}{\pi \tilde{k}_{\max}} \tilde{\mu}_5(\eta),$$
$$\tau = \frac{2\tilde{k}_{\max}^2}{\sigma_c} \eta, \quad \kappa = \frac{\tilde{k}}{\tilde{k}_{\max}}, \quad \mathcal{G} = \frac{\sigma_c}{2\tilde{k}_{\max}^2} \tilde{\Gamma}_f. \tag{11}$$

Using these variables, the system of kinetic equations takes the form,

$$\frac{\partial \mathcal{H}}{\partial \tau} = -\kappa^2 \mathcal{H} \left[1 + K_d \int_0^1 \mathrm{d}\kappa' \mathcal{R}(\kappa', \tau) \right] + \mathcal{R} \left[\mathcal{M} + K_d \int_0^1 \mathrm{d}\kappa' \kappa'^2 \mathcal{H}(\kappa', \tau) \right], \tag{12}$$

$$\frac{\partial \mathcal{R}}{\partial \tau} = -\kappa^2 \mathcal{R} \left[1 + K_d \int_0^1 \mathrm{d}\kappa' \mathcal{R}(\kappa', \tau) \right] + \kappa^2 \mathcal{H} \left[\mathcal{M} - K_d \int_0^1 \mathrm{d}\kappa' \kappa'^2 \mathcal{H}(\kappa', \tau) \right], \tag{13}$$

$$\frac{\mathrm{d}\mathcal{M}}{\mathrm{d}\tau} = \int_0^1 \mathrm{d}\kappa \left(\kappa^2 \mathcal{H} - \mathcal{R}\mathcal{M} \right) - \mathcal{G}\mathcal{M}, \tag{14}$$

where $K_d = 5\sigma_c \tilde{k}_{\max}^2 / 4\alpha_{\text{em}}^4 g^*$. It is interesting to note that the contribution of the turbulent terms cancels out in equation (14). Nevertheless there is a turbulence contribution in

IOP Conf. Series: Journal of Physics: Conf. Series **934** (2017) 012042 doi:10.1088/1742-6596/934/1/012042

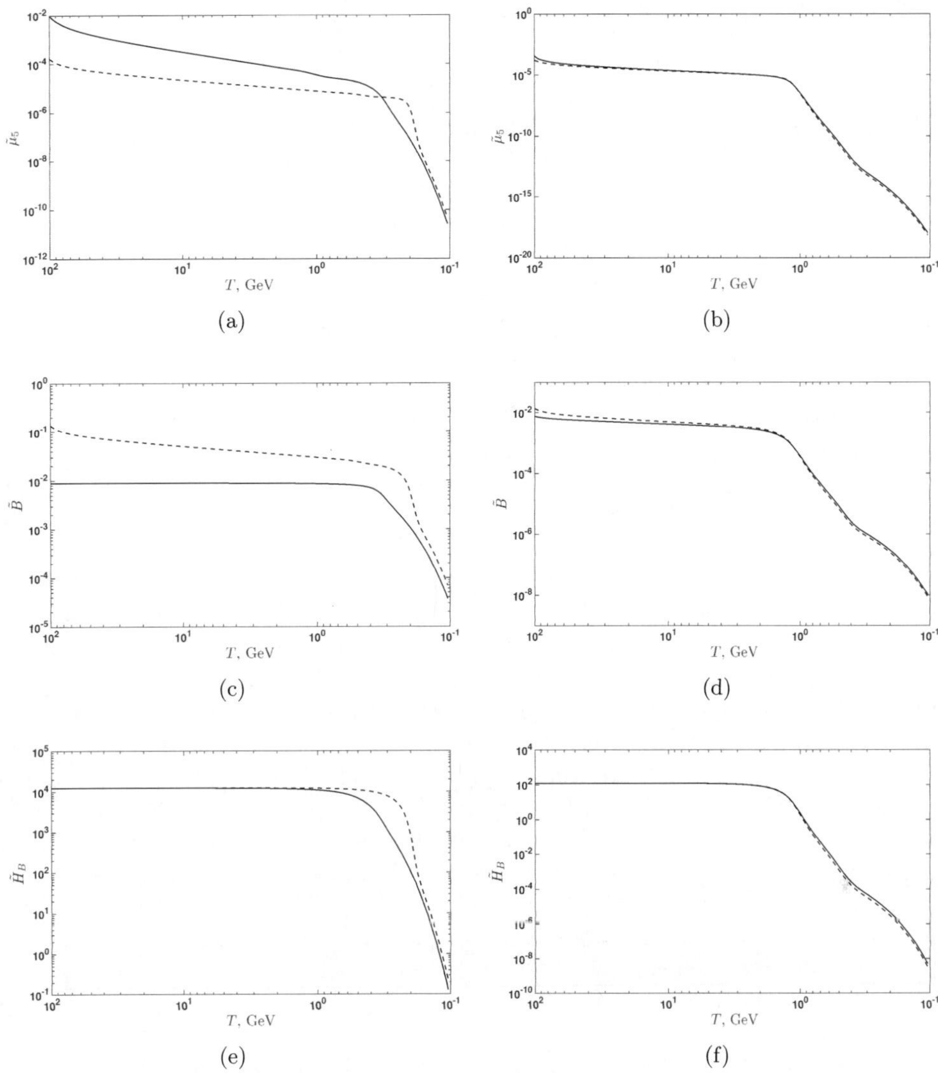

Figure 1. The evolution of the chiral imbalance, the magnetic energy density and the helicity density in plasma of the early universe at $10^2\,\text{MeV} < T < 10^2\,\text{GeV}$. (a) and (b) The evolution of the chiral imbalance. (c) and (d) The evolution of the magnetic field. (e) and (f) The evolution magnetic helicity density. Panels (a), (c), and (e) correspond to $\tilde{B}_0 = 10^{-1}$ whereas panels (b), (d), and (f) to $\tilde{B}_0 = 10^{-2}$. Solid lines show the evolution accounting for both the turbulence effects ($K_d \neq 0$) and the CME, whereas dashed lines are for the CME case only ($K_d = 0$).

equation (12). The initial values of the functions $\mathcal{R}$ and $\mathcal{H}$ are $\mathcal{R}(\kappa, \tau_0) = \mathcal{R}_0 \kappa^{\nu_\text{B}}$ and $\mathcal{H}(\kappa, \tau_0) = q\mathcal{R}_0 \kappa^{\nu_\text{B}-1}$, where $\mathcal{R}_0 = 12\alpha_\text{em}^2 \tilde{B}_0^2(\nu_\text{B}+1)/\pi^2\tilde{k}_\text{max}^2$ and the parameter $0 \leq q \leq 1$ fixes the initial magnetic helicity.

We solve kinetic equations (12)-(14) numerically. The influence of the turbulent matter motion $\sim \mathbf{v}$ on the MHD characteristics, such as the magnetic field strength and the magnetic helicity, as well as on the chiral asymmetry parameter $\mu_5(t)$ in a hot plasma in the broken phase of early Universe is illustrated in figure 1. The solid lines correspond to the case where both effects, i.e. the CME and the turbulent motion of matter $\mathbf{v} \sim \tau_d$, are taken into account, while the dash lines correspond to the CME effect only applied in [2]. Note that we present the numerical solutions of equations (12)-(14) for the maximum helicity parameter $q = 1$ only. In figures 1(a), 1(c) and 1(e) we show results for the maximum initial magnetic field $\tilde{B}_0 = 0.1$ still obeying the BBN bound on the magnetic field [10]. In figures 1(c), 1(d) and 1(f) we show the results for a smaller seed field $\tilde{B}_0 = 10^{-2}$.

One can see in figure 1 that the stronger the initial magnetic field is, the more noticeable the difference between the turbulent and nonturbulent cases is. The dependence of μ_5 on the turbulence parameter K_d is hidden non-trivially in equation (14). Such a hidden dependence comes rather from the magnetic field characteristics $\mathcal{H}$ and $\mathcal{R}$, which evidently depend on that parameter as seen in equations (12) and (13). While the diffusion terms for the magnetic helicity $\mathcal{H}$ and the magnetic energy densities spectra $\mathcal{R}$ are both enhanced by turbulent motions $\sim K_d$, the instability (generation) terms $\sim \mathcal{M} \sim \tilde{\mu}_5$ are supplemented differently through the same parameter K_d. The magnetic helicity $\mathcal{H}$ is supported by turbulent motions even for a decreasing chiral anomaly μ_5; cf. figures 1(e) and 1(f). The magnetic energy $\mathcal{R}$ reduces additionally through the turbulent parameter K_d. This is a reason why the solid curve for magnetic field strength in figures 1(c) and 1(d) occurs below the dash curves corresponding to the pure CME effect.

Let us stress that such opposite contributions of the turbulent motion $\sim \mathbf{v}$ to the evolution of $\mathcal{H}$ and $\mathcal{R}$ comes directly from different signs of the parameter α_d in equation (7) as we found in contrast to the results in [7]. Another important result obtained in [6] is the examination of the possibility for the plasma turbulence to drive the magnetic field instability. In our work we have approximated the plasma velocity by the Lorentz force. We can see that, if one accounts for only the plasma turbulence contribution, i.e. assuming that $\alpha_d \neq 0$ and $\alpha_{\text{CME}} = 0$, the initial magnetic field cannot be amplified. This our new finding confronts the statement of [7], where it was claimed that plasma turbulence described within the chosen model can provide the enhancement of a seed magnetic field.

Acknowledgments

I am thankful to the Tomsk State University Competitiveness Improvement Program and RFBR (research project No. 15-02-00293) for a partial support.

References

[1] Durrer R and Neronov A 2013 *Astron. Astrophys. Rev.* **21** 62 (*Preprint* arXiv:1303.7121)
[2] Boyarsky A, Fröhlich J and Ruchayskiy O 2012 *Phys. Rev. Lett.* **108** 031301 (*Preprint* arXiv:1109.3350)
[3] Dvornikov M and Semikoz V B 2012 *J. Cosmol. Astropart. Phys.* JCAP **02** 040 (*Preprint* arXiv:1111.6876)
[4] Dvornikov M and Semikoz V B 2013 *Phys. Rev.* D **87** 025023 (*Preprint* arXiv:1212.1416)
[5] Vilenkin A 1980 *Phys. Rev.* D **22** 3080–4
[6] Dvornikov M and Semikoz V B 2017 *Phys. Rev.* D **95** 043538 (*Preprint* arXiv:1612.05897)
[7] Campanelli L 2007 *Phys. Rev. Lett.* **98** 251302 (*Preprint* arXiv:0705.2308)
[8] Brandenburg A *et al.* 2017 *Astrophys. J. Lett.* **845** L21 (*Preprint* arXiv:1707.03385)
[9] Cornwall J M 1997 *Phys. Rev.* D **56** 6146–54 (*Preprint* hep-th/9704022)
[10] Cheng B, Schramm D N and Truran J W 1994 *Phys. Rev.* D **49** 5006–18 (*Preprint* astro-ph/9308041)

Black-holes-hedgehogs in the false vacuum and a new physics beyond the Standard Model

C R Das[1], L V Laperashvili[2], B G Sidharth[3], H B Nielsen[4]

[1]Bogoliubov Laboratory of Theoretical Physics, Joint Institute for Nuclear Research, International Intergovernmental Organization, Joliot-Curie 6, 141980 Dubna, Moscow region, Russia
[2]The Institute of Theoretical and Experimental Physics, National Research Center "Kurchatov Institute", Bolshaya Cheremushkinskaya, 25, 117218 Moscow, Russia
[3]International Institute of Applicable Mathematics and Information Sciences, B.M. Birla Science Centre, Adarsh Nagar, 500063 Hyderabad, India
[4]Niels Bohr Institute, Blegdamsvej, 17-21, DK 2100 Copenhagen, Denmark

E-mail: [1]das@theor.jinr.ru, [2]laper@itep.ru, [3]iiamisbgs@yahoo.co.in, [4]hbech@nbi.dk

Abstract. In the present talk, we consider the existence of the two degenerate universal vacua: a) the first Electroweak vacuum at $v = 246$ GeV - "true vacuum", and b) the second Planck scale "false vacuum" at $v_2 \sim 10^{18}$ GeV. In these vacua, we investigated the different topological defects. The main aim of this paper is an investigation of the hedgehog's configurations as defects of the false vacuum. In the framework of the $f(R)$ gravity, suggested by authors in their Gravi-Weak Unification model, we obtained a black hole solution, which corresponds to a "hedgehog" - global monopole, "swallowed" by a black-hole with mass $\sim 10^{19}$ GeV. These black-holes form a lattice-like structure of the vacuum at the Planck scale. Considering the results of the hedgehog lattice theory in the framework of the $SU(2)$ Yang-Mills gauge-invariant theory with hedgehogs in the Wilson loops, we have used the critical value of temperature for the hedgehog's confinement phase. This result gave us the possibility to conclude that there exist triplet Higgs fields which can contribute to the SM at the energy scale $\simeq 10^4 \sim 10^5$ GeV. Showing a new physics at the scale $10 \div 100$ TeV, these triplet Higgs particles can provide the stability of the EW-vacuum of the SM.

1. The investigation

The talk is devoted to studying of topological defects of the universal vacua and to existence of two degenerate vacua in our Universe. The main ideas are published in Refs. [1, 2].

During the expansion after the Planck era, the early Universe underwent a series of phase transitions as a result of which there were arisen such vacuum topological defects (widely discussed in literature) as monopoles or hedgehogs (point defects), strings (line defects), bubbles and domain walls (sheet defects). These topological defects appeared due to the breakdown of local or global gauge symmetries.

This work is essentially based on the discovery that a cosmological constant of our Universe is extremely small, almost zero [3, 4, 5]. We considered a Multiple Point Principle (MPP) first suggested by D.L. Bennett and H.B. Nielsen [6], which predicts the existence in Nature of several degenerate vacua with very small energy density (cosmological constants).

The model developed in this article confirms the existence of the two degenerate vacua of the Universe: The first ("true") Electroweak (EW) vacuum with VEV $v_1 \approx 246$ GeV, and the second ("false") Planck scale vacuum with VEV $v_2 \sim 10^{18}$ GeV.

The main idea of this investigation is the hedgehog's configurations [7, 8] as defects of the false vacuum. We have shown that at superhigh (Planck scale) energies the black-holes-hedgehogs are responsible for the creation of the false vacuum of the Universe. In the framework of the $f(R)$ gravity, we have obtained a solution for a global monopole, which is a black-hole-hedgehog at the Planck scale. Here we have used the $f(R)$ gravity predicted by the Gravi-Weak unification model previously developed by authors in papers [9, 10, 11, 12, 13].

Using the results of Refs.[14, 15] obtained for the $SU(2)$ Yang-Mills theory of the gauge-invariant hedgehog-like structures in the Wilson loops, we have considered the lattice theory giving the critical value of temperature for the hedgehogs confinement phase. Considering the hedgehog lattice theory, we have concluded that hedgehogs can exist only at the energy scale $\mu \gtrsim 10^4$ GeV. Triplet Higgs fields Φ^a (with $a = 1, 2, 3$), which are responsible for the formation of hedgehogs, can show a new physics at the scale ~ 10 TeV.

We reviewed the Multiple Point Principle (MPP) suggested by D.L. Bennett and H.B. Nielsen [6]. In the assumption of the existence of the two degenerate vacua (Electroweak vacuum at $v_1 \approx 246$ GeV, and Planck scale one at $v_2 \sim 10^{18}$ GeV), Froggatt and Nielsen [16] obtained the first prediction of the top-quark and Higgs boson masses, which was further improved by several authors in the next approximations. We devoted to the general properties of topological defects of the universal vacua. We considered topological defects in the "false vacuum", which is presented as a spherical bubble spontaneously produced in the de Sitter-like universe. The space-time inside the bubble, which we refer to as a "true vacuum", has the geometry of an open Friedmann-Lemaitre-Robertson-Walker (FLRW) universe. The Gravi-Weak unification (GWU) model [9, 10, 11, 12, 13] as an example of the $f(R)$ gravity. We considered the existence of the de Sitter solutions in the Planck phase. In our calculations of parameters of the GWU-model, where we predicted the Planck scale false vacuum VEV equal to $v_2 \approx 6.28 \times 10^{18}$ GeV. We have investigated the hedgehog's configurations as defects of the false vacuum and we obtained a solution for a black-hole in the framework of the $f(R)$ gravity, which corresponds to a global monopole "swallowed" by a black-hole. The metric around of the global monopole was considered. The mass M_{BH}, radius δ and "horizon radius" r_h of the black-hole-hedgehog were estimated. The lattice-like structure of the false vacuum which is described by a non-differentiable space-time: by a foam of black-holes, having lattice-like structure, in which sites are black-holes with "hedgehog" monopoles inside them. This manifold is described by a non-commutative geometry predicted an almost zero cosmological constant. The phase transition from the "false vacuum" to the "true vacuum", where it was shown that the Electroweak spontaneous breakdown of symmetry $SU(2)_L \times U(1)_Y \to U(1)_{el.mag}$ created new topological defects of EW vacuum: the Abrikosov-Nielsen-Olesen closed magnetic vortices ("ANO strings") of the Abelian Higgs model and Sidharth's Compton phase objects. Then the "true vacuum" (EW-vacuum) again presents the non-differentiable manifold with non-commutative geometry, and again has an almost zero cosmological constant. Here we estimated the black-hole-hedgehog's mass and radius: $M_{BH} \approx 3.65 \times 10^{18}$ GeV and $\delta \approx 0.29\lambda_{Pl} \approx 10^{-21}$ GeV^{-1} near the second vacuum v_2. We emphasize that due to the energy conservation law, the vacuum density before the phase transition is equal to the vacuum density after the phase transition, and we have

$$\rho_{vac}(\text{at Planck scale}) = \rho_{vac}(\text{at EW scale}).$$

Therefore, we confirmed the Multiple Point Principle: we have two degenerate vacua v_1 and v_2 with an almost zero vacuum energy (cosmological constants). This means that our EW-vacuum, in which we live, is stable. The Planck scale vacuum cannot be negative:

IOP Conf. Series: Journal of Physics: Conf. Series **934** (2017) 012043 doi:10.1088/1742-6596/934/1/012043

$V_{eff}(min1) = V_{eff}(min2)$, these potentials are equal exactly. Hedgehogs in Wilson loops of the $SU(2)$ Yang-Mills theory, and phase transitions in this theory were investigated using the results of Refs. [14, 15]. Their lattice results gave the critical value of the temperature for the hedgehog's confinement phase: $\beta_{crit} \approx 2.5$, and this result gives the value of critical temperature $T_c \sim 10^{18}$ GeV. We show that the hedgehog's confinement happens at energy ~ 10 TeV, which is a threshold energy of the production of a pair of the $SU(2)$-triplet Higgs bosons. We reviewed the problem of the vacuum stability in the Standard Model. We show that hedgehogs can contribute at energy scale $\mu > 10^4$ GeV. Therefore, a triplet Higgs field Φ^a provides a new physics at the scale ~ 10 TeV. We predict an exact stability of the EW-vacuum and the accuracy of the MPP.

2. The results

(i) In this investigation, we were based on the discovery that a cosmological constant of our Universe is extremely small, almost zero, and assumed a new law of Nature which was named as a Multiple Point Principle (MPP). The MPP postulates: *There are two vacua in the SM with the same energy density, or cosmological constant, and both cosmological constants are zero, or approximately zero.* We considered the existence of the following two degenerate vacua in the SM: a) the first Electroweak vacuum at $v_1 = 246$ GeV, which is a "true" vacuum, and b) the second "false" vacuum at the Planck scale with VEV $v_2 \sim 10^{18}$ GeV.

(ii) The bubble, which we refer to as "the false vacuum", is a de Sitter space with its constant expansion rate H_F. The initial radius of this bubble is close to the de Sitter horizon, which corresponds to the Universe radius. The space-time inside the bubble, which we refer to as "the true vacuum", has the geometry of an open FLRW universe.

(iii) We investigated the topological structure of the universal vacua. Different phase transitions, which were resulted during the expansion of the early Universe after the Planck era, produced the formation of the various kind of topological defects. The aim of this investigation is the consideration of the hedgehog configurations as defects in the false vacuum. We have obtained a solution for a black-hole in the region which contains a global monopole in the framework of the $f(R)$ gravity, where $f(R)$ is a function of the Ricci scalar R. Here we have used the results of the Gravi-Weak unification (GWU) model. The gravitational field, isovector scalar Φ^a with $a = 1, 2, 3$, produced by a spherically symmetric configuration in the scalar field theory, is pointing radially: Φ^a is parallel to $\hat{r}$ – the unit vector in the radial direction. In this GWU approach, we obtained a "hedgehog" solution (in Alexander Polyakov's terminology). We also showed that this is a black-hole solution, corresponding to a global monopole that has been "swallowed" by a black-hole.

(iv) We estimated all parameters of the Gravi-Weak unification model, which gave the prediction of the Planck scale false vacuum VEV equal to $v = 2\sqrt{2}M_{Pl}^{red} \approx 6.28 \times 10^{18}$ GeV.

(v) We have shown, that the Planck scale Universe vacuum is described by a non-differentiable space-time: by a foam of black-holes, or by lattice-like structure, where sites are black-holes with the "hedgehog" monopoles inside them. This manifold is described by a non-commutative geometry, leading to a tiny value of cosmological constant $\Lambda \approx 0$.

(vi) Taking into account that the phase transition from the "false vacuum" to the "true vacuum" is a consequence of the electroweak spontaneous breakdown of symmetry $SU(2)_L \times U(1)_Y \to U(1)_{el.mag}$, we considered topological defects of EW-vacuum: the Abrikosov-Nielsen-Olesen closed magnetic vortices ("ANO strings") of the Abelian Higgs model and Sidharth's Compton phase objects. We showed that the "true vacuum" (EW-vacuum) again is presented by the non-differentiable manifold with non-commutative geometry leading to an almost zero cosmological constant.

(vii) By solving the gravitational field equations we estimated the black-hole-hedgehog's mass, radius and horizon radius are $M_h \approx 3.65 \times 10^{18}$ GeV, $R_h \sim 10^{-21}$ GeV^{-1} and $r_h \approx 2.29 R_h$ respectively.

(viii) We considered that due to the energy conservation law, the vacuum energy density before the phase transition is equal to the vacuum energy density after the phase transition: ρ_{vac}(at Planck scale) $=$ ρ_{vac}(at EW scale). This result confirms the Multiple Point Principle: we have two degenerate vacua v_1 and v_2 with an almost zero vacuum energy density (cosmological constants). By these considerations we confirmed the vacuum stability of the EW-vacuum, in which we live. The Planck scale vacuum cannot be negative because of the exact equality $V_{eff}(min_1) = V_{eff}(min_2)$.

(ix) Hedgehogs in the Wilson loops of the $SU(2)$ Yang-Mills theory, and phase transitions in this theory were investigated revising the results of Refs. [14, 15]. Using their lattice result for the critical value of the temperature of hedgehog's confinement phase: $\beta_{crit} \approx 2.5$, we predicted the production of the $SU(2)$-triplet Higgs bosons at LHC at energy scale $\mu \sim 10$ TeV, providing a new physics in the SM.

(x) We considered an additional confirmation of the vacuum stability and accuracy of the MPP taking into account that hedgehog fields Φ^a produce a new physics at the scale ~ 10 TeV, and calculating at high energies the contribution of the black-hole-hedgehog corrections to the effective Higgs potential. This result essentially depends on the hedgehog field parameters: mass, radius and mixing coupling constant λ_{hH} of the interaction of hedgehogs with the SM doublet Higgs fields H.

(xi) Please see ref. [17] for detail investigation.

Acknowledgments

LVL greatly thanks to the B.M. Birla Science Centre (Hyderabad, India) and personally Prof. B.G. Sidharth, for hospitality, collaboration and financial support. HBN wishes to thank the Niels Bohr Institute for the status of professor emeritus and corresponding support. CRD is thankful to Prof. D.I. Kazakov for support.

References

[1] Sidharth B G, Das C R, Laperashvili L V and Nielsen H B 2017 Black Holes-Hedgehogs and Strings as Defects of the Universal Vacua *Preprint* arXiv:1703.05594
[2] Das C R, Laperashvili L V, Nielsen H B and Sidharth B G 2017 Black Holes-Hedgehogs in the False Vacuum and a New Physics beyond the Standard Model *3rd Int. Conf. on particle physics and astrophysics* (Moscow: National Research Nuclear University MEPhI)
[3] Sidharth B G 1997 *Proc. of the 8th Marcell Grossmann Meeting on General Relativity* (Jerusalem: World Scientific, Singapore) pp. 476-479
[4] Sidharth B G 1998 *Int. J. Mod. Phys.* A**13** 2599 (*Preprint* quant-ph/9808031)
[5] Perlmutter S, Schmidt B and Riess A 1998 *Nature* **391** 51.
[6] Bennett D L and Nielsen H B 1994 *Int. J. Mod. Phys.* A**9** 5155 (*Preprint* arXiv:hep-ph/9311321)
[7] t'Hooft G 1974 *Nucl. Phys.* B**79** 276
[8] Polyakov A M 1974 *Pisma Zh. Eksp. Teor. Fiz.* **20** 43 [*JETP Lett.* **20** 194]
[9] Das C R, Laperashvili L V and Tureanu A 2013 *Int. J. Mod. Phys.* A**28** 1350085 (*Preprint* arXiv:1304.3069)
[10] Froggatt C D, Das C R, Laperashvili L V, Nielsen H B and Tureanu A 2013 Gravi-Weak Unification and Multiple Point Principle *Preprint* arXiv:1311.4413
[11] Laperashvili L V, Nielsen H B and Tureanu A 2015 *Int. J. Mod. Phys.* A**30** no.09 1550044 (*Preprint* arXiv:1411.6456)
[12] Das C R, Laperashvili L V, Nielsen H B, Tureanu A and Froggatt C D 2015 *Phys. Atom. Nucl.* **78** 3 440
[13] Laperashvili L V, Nielsen H B and Sidharth B G, Planck Scale Physics 2015 Gravi-Weak Unification and the Higgs Inflation *Preprint* arXiv:1503.03911.
[14] Belavin V A, Chernodub M N and Kozlov I E 2000 *Nucl. Phys.* B**748** 524 (*Preprint* hep-lat/0512030)
[15] Chernodub M N 2006 *Phys. Lett.* **634** 255 (*Preprint* hep-lat/0503018)
[16] Froggatt C D and Nielsen H B 1996 *Phys. Lett.* B**368** 96 (*Preprint* hep-ph/9511371)

IOP Conf. Series: Journal of Physics: Conf. Series **934** (2017) 012043 doi:10.1088/1742-6596/934/1/012043

[17] Sidharth B G, Das C R, Laperashvili L V and Nielsen H B 2017 *Int. J. Mod. Phys.* D**27** 1850022 (*Preprint* arXiv:1703.05594)

IOP Conf. Series: Journal of Physics: Conf. Series **934** (2017) 012044 doi:10.1088/1742-6596/934/1/012044

Star motion around rotating black hole in the Galactic Center in real time

Vyacheslav Dokuchaev[1,2] and Natalia Nazarova[2]
[1]Institute for Nuclear Research of the Russian Academy of Sciences, Prospekt 60-letiya Oktyabrya 7a, Moscow, 117312, Rissia
[2]National Research Nuclear University "MEPhI" (Moscow Engineering Physics Institute) Kashirskoe shosse 31, Moscow, 115409, Russia
E-mail: dokuchaev@inr.ac.ru, nazarova.mephi@gmail.com

Abstract. The Event Horizon Telescope team intends by the 2020 to resolve the shadow of supermassive black hole SgrA* in the Galactic Center. It would be the first attempt for direct identification of the enigmatic black hole. In other words, it would be the first experimental verification of the General Relativity in the strong field limit. There is a chance to find a star moving on the relativistic orbit close to this black hole. We present the animated numerical model of the gravitational lensing of a star (or any other lighting probe), moving around rotating Kerr black hole in the Galactic Center and viewed by the distant observer.

It is awaited that the submillimeter interferometer "The Event Horizon Telescope" [1]-[8] in the nearest three years would be possible to measure the form of the shadow [9]-[17] of the supermassive black hole SgrA* in the Galactic Center. The next step would be the detailed investigations of the form of this black hole shadow and also the specific features of the relativistic motion of the ambient objects [19]-[28].

We made numerical calculations of the gravitational lensing the finite size star moving on the circular relativistic orbit in the Kerr metric. As an representative example we choose the circular equatorial orbit of star with the orbital period $T = 3.22$ hrs rotating around the supermassive black hole SgrA*. The detailed observation of this gravitational lensing is possible in principle with the help of the space interferometer of the type of the Russian project Millimetron [18]. Our numerical calculations are based on the approach and formalism developed by A. G. Polnarev [29] and [30, 31]. We find the error in the integral (20) in [31]. The correct expression for this integral, which is related with observed energy flux from the lensed star image, is

$$\int \frac{r[r^3 - q^2(r-2)]}{[r^4 + r^2 + 2r - 4r\lambda - r(r-2)\lambda^2 - (r-1)^2 q^2]^{3/2}} dr. \tag{1}$$

By using, in particular, this expression we recalculated the observed light curve from the lensed star images.

The lensed star (or other probe) is moving in the equatorial plane of the Kerr Black hole, $\theta_s = \pi/2$, along the circular orbit with radius $r_s = 20GM/c^2$, where M is black hole mass, G is the Newton constant, c is the velocity of star. It supposed that lensed star is corotating with respect to the direction of the black hole rotation. It is considered the case of the example of the black hole with the mass $M = (4.2 \pm 0.2)\, 10^6 M_\odot$ and dimensionless spin (Kerr rotation

IOP Conf. Series: Journal of Physics: Conf. Series **934** (2017) 012044 doi:10.1088/1742-6596/934/1/012044

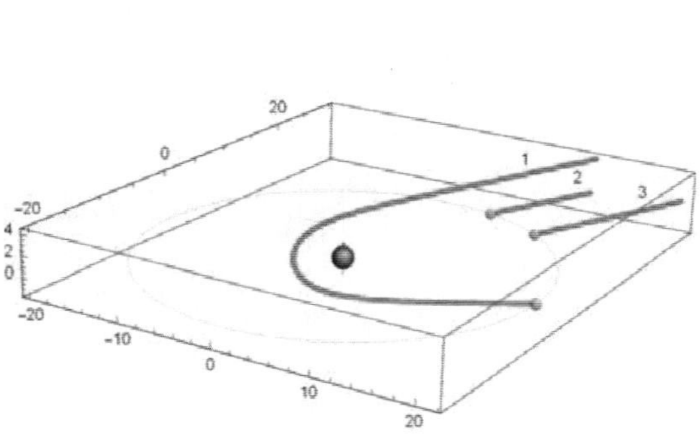

Figure 1. Examples of $3D$ photon trajectory for the first light echo with $\lambda = -6.891$, $q = 1.741$ (curve 1) and the photon trajectories for the direct images with $\lambda = -20.95$, $q = 2.112$ (curve 2) and $\lambda = -0.012$, $q = 2.005$ (curve 3), respectively.

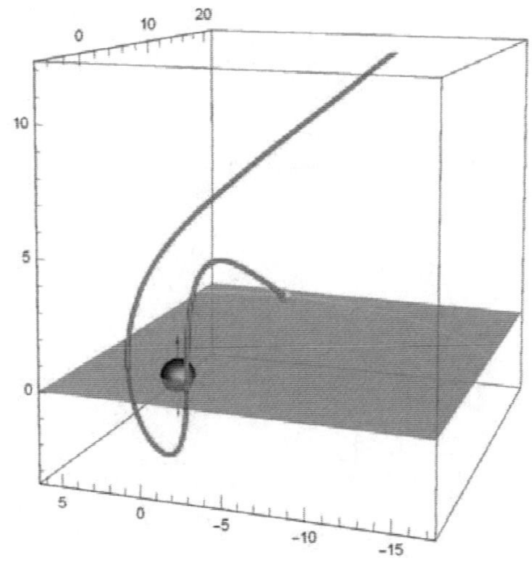

Figure 2. Example of the $3D$ photon trajectory for the second light echo with $\lambda = -1.784$, $q = 5.206$.

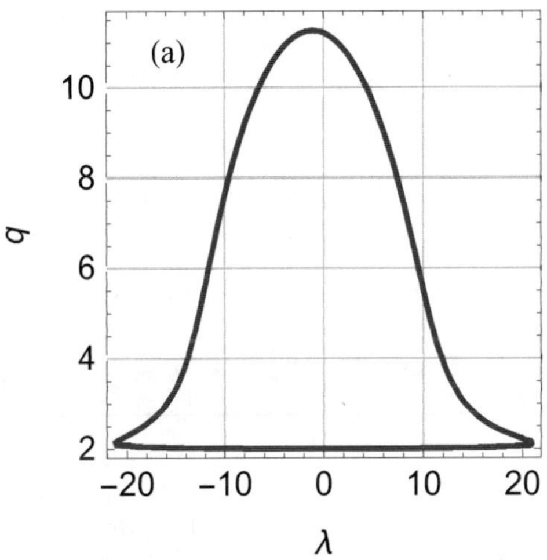

Figure 3. Parameters of photon trajectories (λ, q) for direct images (a).

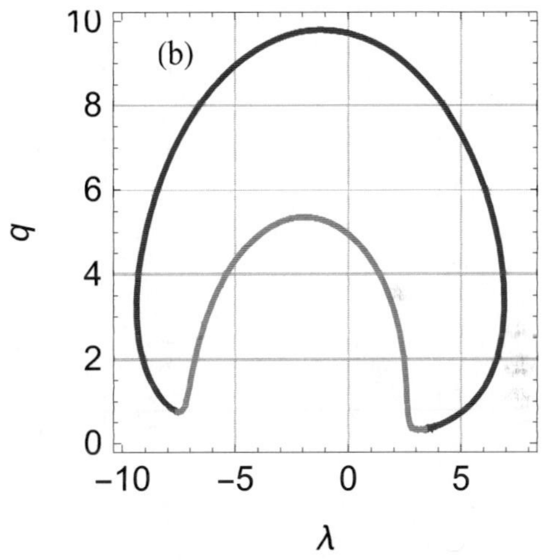

Figure 4. Parameters of photon trajectories (λ, q) for the first light echoes (b).

parameter) $a = 0.998$. The star orbital velocity $v \simeq 0.22c$ and orbital period $T \simeq 3.22\,\text{hrs}$, correspondingly.

The distant observer (telescope) is placed very far from the black hole in the fixed point with the radial distance from the black hole $r_0 \gg GM/c^2$, polar angle $\cos\theta_0 = 0.1$, $\theta_0 \simeq 84^\circ\!.24$ and azimuthal angle $\varphi_0 = 0$. It is supposed also that the lensed star is spherical with the own radius $d \ll r_s$ and is shining isotropically the thermal radiation with the total luminosity L. In the Newtinian limit the corresponding observed energy flux from this equals $F_N = L/4\pi r_0^2$. The radial distance r is normalized in units GM/c^2 in the following figures.

The photon trajectories are defined by two dimensionless parameters: the azimuthal angular momentum λ Carter constant q, which is related with the nonequatorial photon motion. These dimensionless parameters λ and q are called also the Chandrasekhar constants [10].

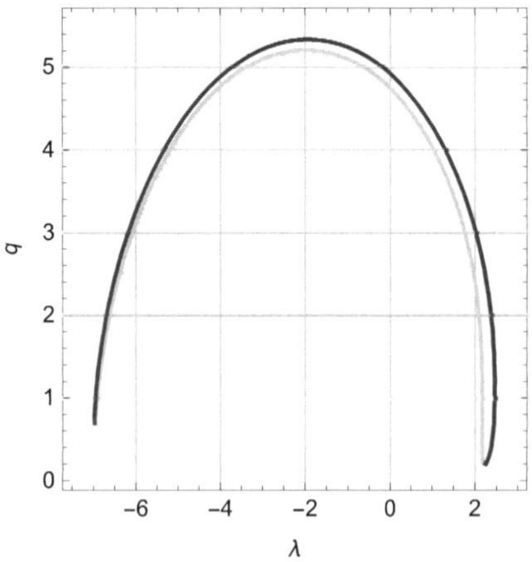

Figure 5. Parameters of photon trajectories (λ, q) for the first light echoes.

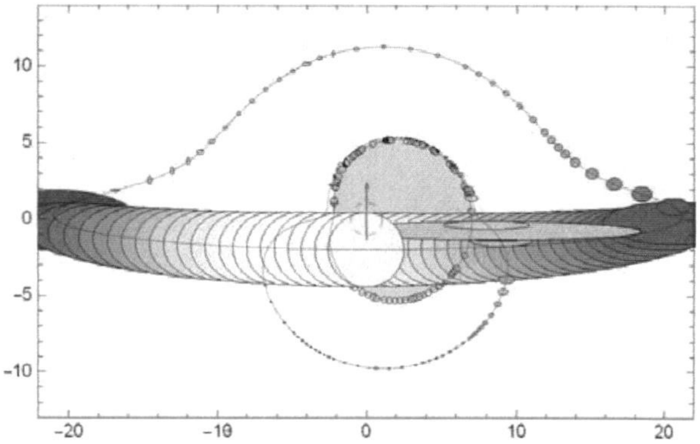

Figure 6. Invisible black hole event horizon (dashed circle); black hole rotation axes (vertical arrow); orbit of the direct images (the most extended closed curve); orbits of the first and second light echoes and the last is the most closed to the black hole shadow (filled region) and the instant images of the lensed star on these three orbits. Instant direct images, the first and second light echoes are shown in equal time intervals. Images of the second light echoes are enlarged to be visible.

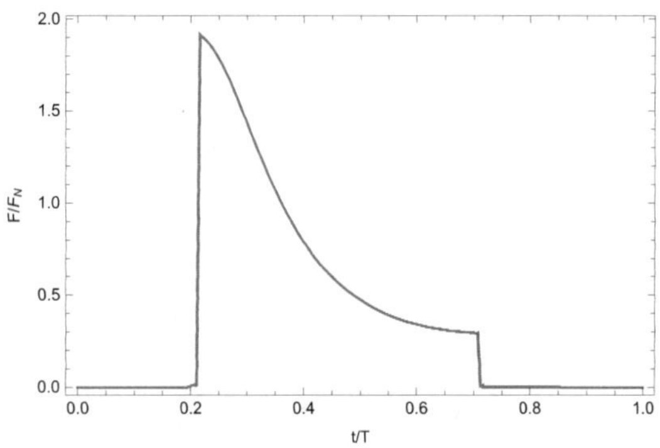

Figure 7. Light curve of the direct image.

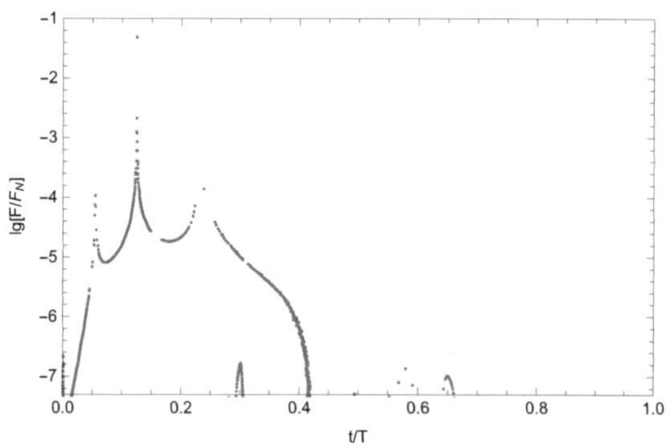

Figure 8. Light curve of the first light echo.

The distant observer formally, may see the infinite number of instant star images. We provided calculation only for the three different types of images: the direct image and the first and second light echoes (see for details [32]). The direct image of the lensed star is produced by the photons, which do not intersect the black hole equatorial plane. The first light echo is produced by photons, which intersect the equatorial plane only one time (these are "one-orbit photons" according the classification scheme from [31]). Respectively the second light echo is produced by the photons, which intersect the equatorial plane two times ("two-orbit photons"). See in figures 1 and 2 the corresponding $3D$ photon trajectories. The deformation of star images from the circular form were calculated in first approximation for the deflection of light rays with respect to the central ray in this approximation these deflections have elliptical forms.

See in figures 9 and 10 the frequency shift $g(t)$ along the trajectories of direct images and

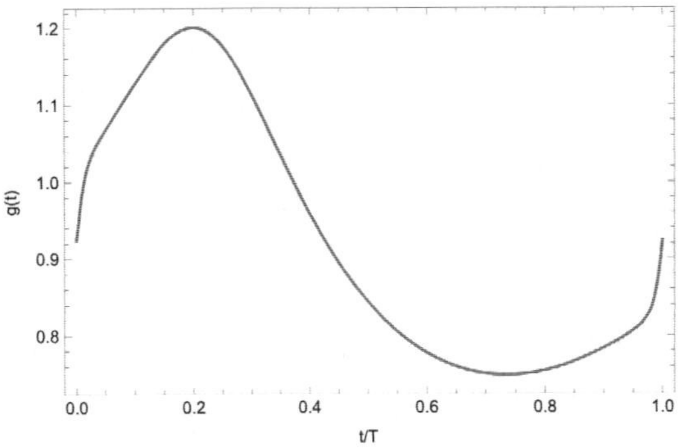

Figure 9. Frequency shift $g(t)$ for direct images.

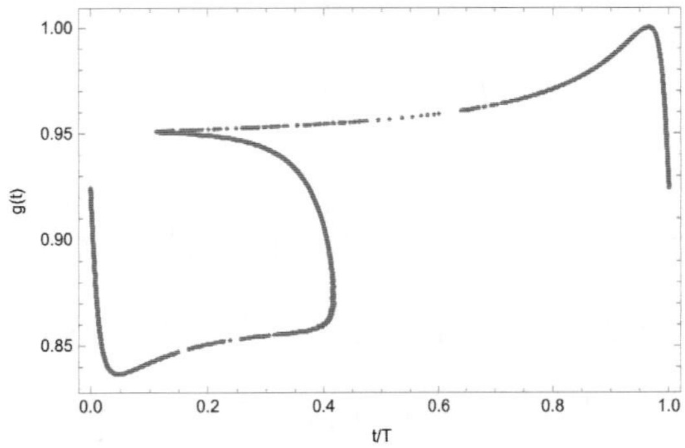

Figure 10. Frequency shift $g(t)$ for the first light echoes.

the first light echoes. The frequency shifts $g(t)$ is defined as the ratio of the observed light frequency ν_0 to the corresponding one in rest frame of the star. Function $g(t)$ takes into account as the gravitational redshift and also the Doppler effect due to star motion. See time evolution animation of the lensed star images in [33].

References

[1] Fish V L et al [Event Horizon Telescope Collaboration] 2016 *Galaxies* **4** 54
[2] Lacroix T and Silk J 2013 *Astron. Astrophys.* **554** A36
[3] Johannsen T et al 2016 *Phys. Rev. Lett.* **116** 031101
[4] Broderick A E et al 2016 *Astrophys. J.* **820** 137
[5] Chaelet A A et al 2016 *Astrophys. J.* **829** 11
[6] Kim J, Marrone D P, Chan C, Medeiros L, Özel F and Psaltis D 2016 *Astrophys. J.* **832** 156
[7] Roelofs F, Johnson M D, Shiokawa H, Doeleman S S and Falcke H 2017 *Astrophys. J.* **847** 55
[8] Doeleman S 2017 *Nature Astron.* **1** 646
[9] Bardeen J M 1973 in *Black Holes* ed DeWitt C and DeWitt B S (New York: Gordon and Breach) p. 215
[10] Chandrasekhar S 1983 *The Mathematical Theory of Black Holes* (Oxford: Clarendon Press)
[11] Falcke H, Melia F and Agol E 2000 *Astrophys. J.* **528** L13
[12] Takahashi R 2004 *Astrophys. J.* **611** 996
[13] Falcke H and Markoff S 2013 *Class. Quantum Grav.* **30** 244003
[14] Li Z and Bambi C 2014 *JCAP* **01** 041
[15] Cunha P V P, Herdeiro C A R, Radu E and Runarsson H F 2015 *Phys. Rev. Lett.* **115** 211102
[16] Abdujabbarov A A, Rezzolla L and Ahmedov B J 2015 *Mon. Not. R. Astron. Soc.* **454** 2423
[17] Younsi Z, Zhidenko A, Rezzolla L, Konoplya R and Mizuno Y 2016 *Phys. Rev.* D **94** 084025
[18] Kardashev N S et al 2014 *Phys. Usp.* **57** 1199
[19] Shatskiy A A, Novikov I D and Kardashev N S 2008 *Phys. Usp.* **51** 457
[20] Babichev E O, Dokuchaev V I and Eroshenko Yu N 2013 *Phys. Usp.* **56** 1155
[21] Dokuchaev V I 2014 *Gen. Relativ. Grav.* **46** 1832
[22] Dokuchaev V I and Eroshenko Yu N 2015 *JETP Lett.* **101** 777
[23] Dokuchaev V I and Eroshenko Yu N 2015 *Phys. Usp.* **58** 772
[24] Herrera-Aguilar A and Nucamendi U 2015 *Phys. Rev.* D **92** 045024
[25] Becerril R, Valdez-Alvarado S and Nucamendi U 2016 *Phys. Rev.* D **94** 124024
[26] Will C M and Maitra M 2017 *Phys. Rev.* D **95** 064003
[27] Ferrer F, da Rosa A M and Will C M 2017 *Phys. Rev.* D **96** 083014
[28] Goddi C et al 2017 *Int. J. Mod. Phys.* D **26** 1730001
[29] Polnarev A G 1972 *Astrofizika* **8** 273
[30] Cunnungham C T and Bardeen J M 1972 *Astrophys. J.* **173** L137
[31] Cunnungham C T and Bardeen J M 1973 *Astrophys. J.* **183** 237
[32] Dokuchaev V I and Nazarova N O 2017 *Pisma ZhETF* **106** 609
[33] `https://www.youtube.com/watch?v=x0a41_yFnEQ`

IOP Conf. Series: Journal of Physics: Conf. Series **934** (2017) 012045 doi:10.1088/1742-6596/934/1/012045

High temperature limit of the Standard Model due to gauge groups contraction

N A Gromov

Institute of Physics and Mathematics, Komi Science Center UrD RAS, Kommunisticheskaya st., 24, Syktyvkar, 167982, Russia

E-mail: gromov@dm.komisc.ru

Abstract. The high temperature (high energy) limit of the Standard Model is developed with the help of contractions its gauge groups. The elementary particles evolution in the early Universe from Plank time up to several milliseconds is deduced from this limit theory. Particle properties at the infinite temperature look very unusual: all particles are massless, only neutral Z-bosons, u-quarks, neutrinos and photons are survived in this limit. The weak interactions become long-range and are mediated by neutral currents, quarks have only one color degree of freedom.

1. Introduction

The modern theory of elementary particles known as Standard Model consist of Electroweak Model, which unified electromagnetic and weak interactions, as well as Quantum Chromodynamics (QCD), describing their strong interactions. The Standard Model gives a good description of the experimental data and has been recently confirmed by the discover of Higgs boson at LHC. No new physics was observed in the experiments up to now. So today the Standard Model is the only theory of elementary particles and if we are interested in particle properties in the early Universe we need use its high temperature (high energy) limit.

Standard Model is a gauge theory with the direct product of a simple groups $SU(3) \times SU(2) \times U(1)$ as a gauge group. The well known in physics operation of group contraction (or transition) [1] transforms initial group into some more simple limit group. For a symmetric physical system the contraction of its symmetry group means a transition to some limit state. We use this method in order to re-establish the evolution of elementary particles and their interactions in the early Universe. We are based on the modern knowledge of the particle world which is concentrated in Standard Model. For this we investigate the high-temperature (high-energy) limit of Standard Model generated by appropriate contractions of the gauge groups $SU(2)$ and $SU(3)$ [2–4].

2. High-temperature lagrangian of Electroweak Model

The contracted group $SU(2;\epsilon)$ is obtained [3] by *the consistent rescaling* of the fundamental representation of $SU(2)$ and the space $\mathbf{C}_2$ as follows

$$z'(\epsilon) = \begin{pmatrix} z_1' \\ \epsilon z_2' \end{pmatrix} = \begin{pmatrix} \alpha & \epsilon\beta \\ -\epsilon\bar{\beta} & \bar{\alpha} \end{pmatrix} \begin{pmatrix} z_1 \\ \epsilon z_2 \end{pmatrix} = u(\epsilon)z(\epsilon),$$

$$\det u(\epsilon) = |\alpha|^2 + \epsilon^2|\beta|^2 = 1, \quad u(\epsilon)u^\dagger(\epsilon) = 1. \tag{1}$$

This transformation induces the following rescaling of the standard boson fields as well as the left lepton and quark fields:

$$W_\mu^\pm \to \epsilon W_\mu^\pm, \;\; Z_\mu \to Z_\mu, \;\; A_\mu \to A_\mu,$$

$$e_l \to \epsilon e_l, \quad d_l \to \epsilon d_l, \quad \nu_l \to \nu_l, \quad u_l \to u_l. \tag{2}$$

From the special mechanism of spontaneous symmetry breaking, which is used to generate mass of vector bosons and other elementary particles of the model, the following transformation rule for Higgs boson field χ and masses of all particles

$$\chi \to \epsilon\chi, \quad v \to \epsilon v, \quad m_p \to \epsilon m_p, \quad p = \chi, W, Z, e, u, d \tag{3}$$

is obtained.

After transformations (2), (3) the Lagrangian of the Electroweak Model can be represented in the form

$$L(\epsilon) = L_\infty + \epsilon L_1 + \epsilon^2 L_2 + \epsilon^3 L_3 + \epsilon^4 L_4. \tag{4}$$

The explicit form of itermediate parts of Lagrangian L_k was given in [3]. In particular the infinite temperature limit ($\epsilon = 0$) Lagrangian is equal to

$$L_\infty = -\frac{1}{4}\mathcal{Z}_{\mu\nu}^2 - \frac{1}{4}\mathcal{F}_{\mu\nu}^2 + \nu_l^\dagger i\tilde{\tau}_\mu\partial_\mu\nu_l + u_l^\dagger i\tilde{\tau}_\mu\partial_\mu u_l +$$

$$+ e_r^\dagger i\tau_\mu\partial_\mu e_r + d_r^\dagger i\tau_\mu\partial_\mu d_r + u_r^\dagger i\tau_\mu\partial_\mu u_r + L_\infty^{int}(A_\mu, Z_\mu). \tag{5}$$

The contraction parameter is monotonous function $\epsilon(T)$ of the temperature of a system under consideration with the property $\epsilon(T) \to 0$ for $T \to \infty$. Very high temperatures can exist in the early Universe just after its creation on the first stages of the Hot Big Bang [5] in pre-electroweak epoch.

3. Quantum Chromodynamics at high temperature

Like the Electroweak Model QCD is a gauge theory based on the local color degrees of freedom [6]. The QCD gauge group is $SU(3)$, acting in three dimensional complex space of color quark states $q^t = (q_1, q_2, q_3)^t \equiv (q_R, q_G, q_B)^t \in \mathbf{C}_3$. The $SU(3)$ gauge bosons (gluons) are the force carrier of the theory between the quarks. The QCD Lagrangian is taken in the form

$$\mathcal{L} = \sum_q \bar{q}^i(i\gamma^\mu)(D_\mu)_{ij}q^j - \frac{1}{4}\sum_{\alpha=1}^{8} F_{\mu\nu}^\alpha F^{\mu\nu\,\alpha}, \tag{6}$$

where $D_\mu q$ are covariant derivatives of quark fields

$$D_\mu q = \left(\partial_\mu - ig_s\left(\frac{\lambda^\alpha}{2}\right)A_\mu^\alpha\right)q, \tag{7}$$

g_S is the strong coupling constant, $t^a = \lambda^a/2$ are generators of $SU(3)$, λ^a are Gell-Mann matrices. Gluon stress tensor is introduced as usual

$$F_{\mu\nu}^\alpha = \partial_\mu A_\nu^\alpha - \partial_\nu A_\mu^\alpha + g_s f^{\alpha\beta\gamma}A_\mu^\beta A_\nu^\gamma, \tag{8}$$

where $[t^\alpha, t^\beta] = if^{\alpha\beta\gamma}t^\gamma$, $\alpha, \beta, \gamma = 1, \ldots, 8$. Mass terms are not included as far as they are present in the electroweak Lagrangian.

The contracted special unitary group $SU(3;\kappa)$ is defined by the action

$$q'(\kappa) = \begin{pmatrix} q_1' \\ \kappa_1 q_2' \\ \kappa_1\kappa_2 q_3' \end{pmatrix} = \begin{pmatrix} u_{11} & \kappa_1 u_{12} & \kappa_1\kappa_2 u_{13} \\ \kappa_1 u_{21} & u_{22} & \kappa_2 u_{23} \\ \kappa_1\kappa_2 u_{31} & \kappa_2 u_{32} & u_{33} \end{pmatrix} \begin{pmatrix} q_1 \\ \kappa_1 q_2 \\ \kappa_1\kappa_2 q_3 \end{pmatrix} = U(\kappa)q(\kappa),$$

$$\det U(\kappa) = 1, \quad U(\kappa)U^\dagger(\kappa) = 1 \tag{9}$$

on the complex space $\mathbf{C}_3(\kappa)$ in such a way that the hermitian form

$$q^\dagger(\kappa)q(\kappa) = |q_1|^2 + \kappa_1^2\left(|q_2|^2 + \kappa_2^2|q_3|^2\right) \tag{10}$$

remains invariant, when the contraction parameters tend to zero: $\kappa_1, \kappa_2 \to 0$. We take these parameters identical to the contraction parameter ϵ of the Electroweak Model: $\kappa_1 = \kappa_2 = \kappa = \epsilon$, so the limit $\kappa = \epsilon \to 0$ corresponds to the infinite temperature limit.

Transition from the classical group $SU(3)$ and space $\mathbf{C}_3$ to the group $SU(3;\epsilon)$ and space $\mathbf{C}_3(\epsilon)$ is given by the substitution

$$q_1 \to q_1, \quad q_2 \to \epsilon q_2, \quad q_3 \to \epsilon^2 q_3,$$

$$A_\mu^{GR} \to \epsilon A_\mu^{GR}, \quad A_\mu^{BG} \to \epsilon A_\mu^{BG}, \quad A_\mu^{BR} \to \epsilon^2 A_\mu^{BR}, \tag{11}$$

and diagonal gauge fields $A_\mu^{RR}, A_\mu^{GG}, A_\mu^{BB}$ remain unchanged. Then from (6) we obtain the quark part of Lagrangian

$$\mathcal{L}_q(\epsilon) = \sum_q \Biggl\{ i\bar{q}_1\gamma^\mu\partial_\mu q_1 + \frac{g_s}{2}|q_1|^2\gamma^\mu A_\mu^{RR} +$$

$$+\epsilon^2\left[i\bar{q}_2\gamma^\mu\partial_\mu q_2 + \frac{g_s}{2}\left(|q_2|^2\gamma^\mu A_\mu^{GG} + q_1\bar{q}_2\gamma^\mu A_\mu^{GR} + \bar{q}_1 q_2\gamma^\mu \bar{A}_\mu^{GR}\right)\right] +$$

$$+\epsilon^4\Biggl[i\bar{q}_3\gamma^\mu\partial_\mu q_3 + \frac{g_s}{2}\Bigl(|q_3|^2\gamma^\mu A_\mu^{BB} + q_1\bar{q}_3\gamma^\mu A_\mu^{BR} + \bar{q}_1 q_3\gamma^\mu \bar{A}_\mu^{BR} +$$

$$+q_2\bar{q}_3\gamma^\mu A_\mu^{BG} + \bar{q}_2 q_3\gamma^\mu \bar{A}_\mu^{BG}\Bigr)\Biggr]\Biggr\} = L_q^\infty + \epsilon^2 L_q^{(2)} + \epsilon^4 L_q^{(4)}. \tag{12}$$

The gluon part of (6) is cumbersome, but can be represented as a sum of terms [4]

$$\mathcal{L}_{gl}(\epsilon) = L_{gl}^\infty + \epsilon^2 L_{gl}^{(2)} + \epsilon^4 L_{gl}^{(4)} + \epsilon^6 L_{gl}^{(6)} + \epsilon^8 L_{gl}^{(8)}. \tag{13}$$

In the infinite temperature limit the most parts of gluon tensor components are equal to zero, so we can write out the QCD Lagrangian $\mathcal{L}_\infty$ explicitly

$$\mathcal{L}_\infty = L_q^\infty + L_{gl}^\infty = \sum_q \left\{ i\bar{q}_R\gamma^\mu\partial_\mu q_R + \frac{g_s}{2}|q_R|^2\gamma^\mu A_\mu^{RR}\right\} -$$

$$-\frac{1}{4}\left(F_{\mu\nu}^{RR}\right)^2 - \frac{1}{4}\left(F_{\mu\nu}^{GG}\right)^2 - \frac{1}{4}F_{\mu\nu}^{RR}F_{\mu\nu}^{GG}. \tag{14}$$

$$L_{gl}^{(6)} = -\frac{g_s^2}{16}\Bigl\{\left([4,7]-[5,6]\right)^2 + \left([4,6]+[5,7]\right)^2 +$$

$$+\Big([2,4]-[1,5]\Big)^2+\Big([1,4]+[2,5]\Big)^2+4\Big([1,2]+[6,7]\Big)[4,5]\Big\},$$

,

$$L_{gl}^{(8)}=-\frac{g_s^2}{4}[4,5]^2. \quad (15)$$

Let us introduce the notations

$$\partial A^k \equiv \partial_\mu A_\nu^k - \partial_\nu A_\mu^k, \quad [k,m]\equiv A_\mu^k A_\nu^m - A_\mu^m A_\nu^k, \quad (16)$$

The Standard Model passes in this limit through several stages, which are distinguished by the powers of contraction parameter, i.e. by powers of the temperature. Starting with $L_q(\epsilon)$ (4), $\mathcal{L}_q(\epsilon)$ (12) and $\mathcal{L}_{gl}(\epsilon)$ (13) one can construct a number of intermediate models, namely, one model with explicit Lagrangian for each stage [4]. These models describe the restoration of particles properties in the Universe evolution.

4. Conclusion

The contraction of gauge group of the Standard Model gives an opportunity to order in time different stages of its development, but does not make it possible to bear their absolute dates.

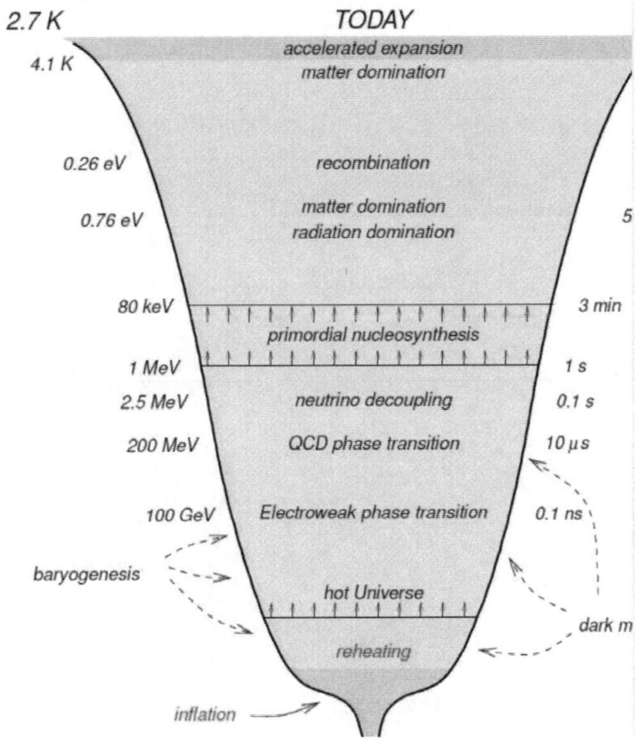

Figure 1. History of the Universe

These dates can be estimated with the help of additional assumptions. We use the fact that the electroweak epoch starts at the temperature $T_4 = 100$ GeV (1 GeV=10^{13} K) whereas the QCD epoch begins at $T_8 = 0.2$ GeV (Figure 1), suppose that the contraction parameter inversely depends on temperature $\epsilon(T) = AT^{-1}$ and use cutoff level Δ for $\epsilon^k L_k$, i.e. for $\epsilon^k < \Delta$ all the terms L_k are negligible quantities in Lagrangian. Simple calculations [4] give the following boundary values of the temperature in the early Universe (GeV): $T_1 = 10^{18}$, $T_2 = 10^7$, $T_3 = 10^3$, $T_4 = 10^2$, $T_6 = 1$, $T_8 = 2 \cdot 10^{-1}$. The obtained estimation for "infinity" temperature $T_1 \approx 10^{18}$ GeV is comparable with Planck energy $\approx 10^{19}$ GeV, where the gravitation effects are important. So the developed evolution of the elementary particles does not exceed the range of the problems described by electroweak and strong interactions.

At the infinite temperature limit ($T > 10^{18}$ GeV) all particles including vector bosons lose their masses and electroweak interactions become long-range. Monochromatic massless quarks exchange by only one sort of R-gluons. It follows from the explicit form of Lagrangians $L_{\infty}^{int}(A_\mu, Z_\mu)$ (5) and $\mathcal{L}_\infty$ (14) that only the particles of the same sort interact with each other. Particles of different sorts do not interact. It looks like some stratification of leptons and quark-gluon plasma with only one sort of particles in each stratum.

At the level of classical gauge fields it is already possible to give some conclusions on the stages of elementary particles mass appearance in the Universe evolution. In particular we can conclude that half of quarks ($\approx \epsilon$, 10^{18} GeV $> T > 10^7$ GeV) first restore they mass. Then Z-bosons, electrons and other quarks become massive ($\approx \epsilon^2$, 10^7 GeV $> T > 10^3$ GeV). Finally Higgs boson χ and charged $W^{\pm}$-bosons restore their masses ($\approx \epsilon^4$, $T \leq 10^2$ GeV).

Acknowledgments

The author is thankful to V. V. Kuratov and V. I. Kostyakov for helpful discussions. The study is supported by Program of UrD RAS, project No 15-16-1-3.

References

[1] Inönü E and Wigner E P 1953 *Proc. Nat. Acad. Sci. USA* **39** 510
[2] Gromov N A 2012 *Contractions of Classical and Quantum Groups* (Moscow: Fizmatlit) [in Russian]
[3] Gromov N A 2015 *Physica Scripta* **90** 074009
[4] Gromov N A 2016 *J. Cosmol. Astroparticle Phys.* **3** 053
[5] Gorbunov D S and Rubakov V A 2011 *Introduction to the Theory of the Early Universe: Hot Big Bang Theory* (Singapure: World Scientific)
[6] Emel'yanov V M 2007 *Standard Model and its Expansion* (Moscow: Fizmatlit) [in Russian]

IOP Conf. Series: Journal of Physics: Conf. Series **934** (2017) 012046 doi:10.1088/1742-6596/934/1/012046

Transitions between topologically non-trivial configurations

V A Gani[1,2], A A Kirillov[1,3] and S G Rubin[1,4]

[1]National Research Nuclear University MEPhI (Moscow Engineering Physics Institute), 115409 Moscow, Russia

[3]National Research Center Kurchatov Institute, Institute for Theoretical and Experimental Physics, 117218 Moscow, Russia

[3]Department of Theoretical Physics, Yaroslavl State P.G. Demidov University, 150003 Yaroslavl, Russia

N.I. Lobachevsky Institute of Mathematics and Mechanics, Kazan Federal University, 420008 Kazan, Russia

E-mail: vagani@mephi.ru

Abstract. We study formation and evolution of solitons within a model with two real scalar fields with the potential having a saddle point. The set of these configurations can be split into disjoint equivalence classes. We give a simple expression for the winding number of an arbitrary closed loop in the field space and discuss the evolution scenarios that change the winding number. These non-trivial field configurations lead to formation of the domain walls in the three-dimensional physical space.

1. Introduction and Motivation

Recently, the inflationary models with several fields have become the subject of a growing interest [1]. Many of these models involves complicated potentials, which arise in various supersymmetric models [2], in the brane approach [3,4], and in the landscape paradigm [5,6]. The inflation within the multi-field models has been thoroughly elaborated, its effect on fluctuations of the cosmic microwave background was analyzed [7]. One of the important features of the multi-field inflationary models is a potential with a number of minima [8]. A field-theoretical system with a potential with many minima can possess topologically non-trivial scalar field configurations — solitons [9]. Such configurations could form domains with extra energy density or collapse into black holes after the end of the inflation. It is important that non-trivial field configurations could be formed even if the potential of the model has only one minimum. In this paper we study the conditions at which these solitons can be formed.

In this paper we consider a system of two real scalar fields, φ and χ. As we show, the domain wall can be formed after the inflation even in the case of only one potential minimum. It is demonstrated that non-trivial field configurations can appear if the fields tend to the same minimum of the potential at different spatial asymptotics. At the same time, the existence of a saddle point of the potential is necessary for this phenomenon to occur. We introduce the topological classification, which allows us to divide non-trivial configurations into disjoint (homotopic) classes labeled by *winding number*. We also find the conditions at which the winding number can change, i.e. the configuration can move from one homotopic class to another. It is important to note that such transitions are not possible if the fields tend to different potential minima at different spatial asymptotics. Only transitions due to quantum tunneling are possible in this case [10].

The fields φ and χ live in the (3+1)-dimensional space-time. Nevertheless, we will study one-dimensional configurations. Thus the multi-field inflationary models are connected with (1+1)-dimensional field-theoretical models, where many important results have been obtained recently, in particular, related to soliton interactions [11–19], soliton stability [20], bubble decay [21, 22], and planar domain walls [23–28], see also book [9] and review [29].

2. The Model

Consider a model with two real scalar fields φ, χ in $(3+1)$ space-time dimensions, with its dynamics determined by the Lagrangian density

$$\mathcal{L} = \frac{1}{2} g^{\mu\nu} \left(\partial_\mu\varphi\partial_\nu\varphi + \partial_\mu\chi\partial_\nu\chi\right) - V(\varphi, \chi), \tag{1}$$

with the Minkowski metric tensor. The equations of motion following from the Lagrangian (1) are

$$\begin{cases} \varphi_{tt} + 3H\varphi_t - \varphi_{rr} - \dfrac{2}{r}\varphi_r = -\dfrac{\partial V}{\partial\varphi}, \\ \chi_{tt} + 3H\chi_t - \chi_{rr} - \dfrac{2}{r}\chi_r = -\dfrac{\partial V}{\partial\chi}, \end{cases} \tag{2}$$

where $H = \dot{a}/a$ is the Hubble parameter, which is small after the end of the inflation. The friction term ensures the dumping of long-period oscillations during the reheating stage. Let we have a domain $\mathcal{B}$, which is surrounded by the border $\partial\mathcal{B}$. In this border, the fields are at the saddle point (φ_s, χ_s) of the potential $V(\varphi, \chi)$. Below we demonstrate that a non-trivial field configuration can be formed even in the case of the same vacuum state at both sides of the border, i.e. inside and outside of the domain $\mathcal{B}$.

Below we assume that the cosmological expansion is small, and we use the physical distances $R = a(t)r$. Besides that, we study configurations of the type of planar domain walls, because for an observer near the border $\partial\mathcal{B}$ the wall is almost flat. According to this, the equations (2) become

$$\begin{cases} \varphi_{tt} + 3H\varphi_t - \varphi_{xx} = -\dfrac{\partial V}{\partial\varphi}, \\ \chi_{tt} + 3H\chi_t - \chi_{xx} = -\dfrac{\partial V}{\partial\chi}. \end{cases} \tag{3}$$

For our numerical simulations we use the following potential $V(\varphi, \chi)$ with a saddle point:

$$V(\varphi, \chi) = d\,(\varphi^2 + \chi^2) + a\,\exp\left[-b\,(\varphi - \varphi_0)^2 - c\,(\chi - \chi_0)^2\right], \tag{4}$$

where a, b, c, d are positive parameters of the model, and the point (φ_0, χ_0) is the local maximum of the potential (4). In our simulations we used the following values of the parameters: $a = 5\cdot 10^2$, $b = 1$, $c = 1$, $d = 1$, $\varphi_0 = -5$, $\chi_0 = 0$.

3. Topological classification and numerical simulations

We can split the set of all closed curves in the (φ, χ) plane into different equivalence classes (homotopic classes). In the spirit of this splitting, the equivalent curves have the same winding (or topological) number N, which is the number of turns of the curve around the local maximum of the potential, i.e. around the point (φ_0, χ_0). Then we use the residue theorem in order to obtain an explicit expression for N. According to this theorem, the line integral of some function

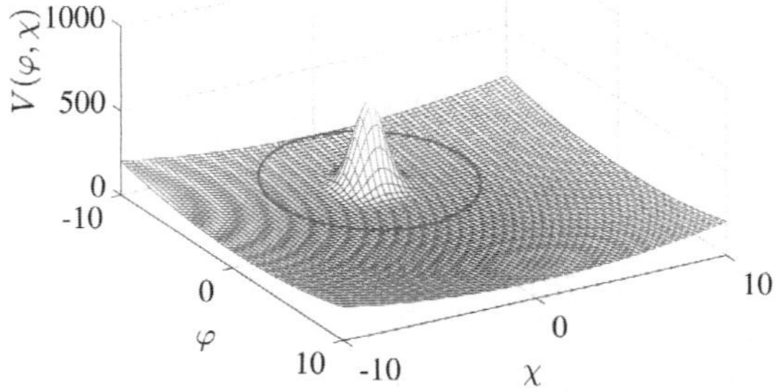

Figure 1. The potential (4) with the initial field configuration (7), (8).

$f(\zeta)$ of the complex variable $\zeta = \varphi + i\chi$ around the closed curve in the (φ, χ) plane is equal to the sum of residues of $f(\zeta)$ at isolated singular points (surely, each counted as many times as the curve winds around the point) multiplied by $2\pi i$. We take the function $f(\zeta) = \frac{1}{\zeta - \zeta_0}$. After some simple algebra, we obtain:

$$N[\varphi, \chi] = \frac{1}{2\pi} \int_{-\infty}^{+\infty} \frac{(\varphi - \varphi_0)\chi_x - (\chi - \chi_0)\varphi_x}{(\varphi - \varphi_0)^2 + (\chi - \chi_0)^2} \, dx. \tag{5}$$

Thus we have split all closed curves in the (φ, χ) plane into homotopic (equivalence) classes, depending on the number of windings of the curve around the local maximum of the potential (φ_0, χ_0).

In our numerical simulations we start from an initial field configuration, which is a closed loop in the (φ, χ) plane. According to this, the boundary conditions are

$$\begin{cases} \varphi(-\infty, t) = \varphi(+\infty, t), \\ \varphi_x(-\infty, t) = \varphi_x(+\infty, t), \\ \chi(-\infty, t) = \chi(+\infty, t), \\ \chi_x(-\infty, t) = \chi_x(+\infty, t). \end{cases} \tag{6}$$

Consider the initial configuration, which encircles the potential maximum:

$$\begin{cases} \varphi(x) = \varphi_0 + R\cos\theta(x), \\ \chi(x) = \chi_0 + R\sin\theta(x), \end{cases} \tag{7}$$

where $\theta(x)$ is

$$\theta(x) = N\pi\,(1 + \tanh x)\,. \tag{8}$$

Here $R > 0$ is the radius of the circle, and $0 \le \theta(x) \le 2\pi N$ at $-\infty \le x \le +\infty$, see fig. 1.

4. Results

We studied the evolution of the initial configuration (7), (8) for winding numbers $N = 1, 2, 3$ numerically. At all these winding numbers we observed the tightening of the closed loop around the potential maximum (φ_0, χ_0). This result means that a space trajectory connecting two points inside and outside of the domain $\mathcal{B}$ necessarily goes through a domain wall or through a series of domain walls, at $N = 1$ and at $N \ge 2$, respectively, see fig. 2.

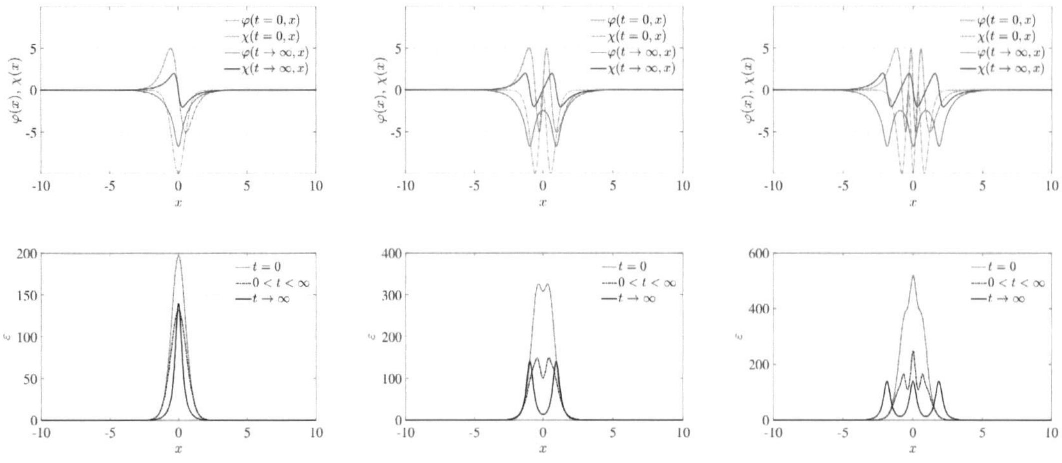

Figure 2. Top: spatial distribution of the fields at $t = 0$ and at $t \to +\infty$ for the potential (4) for various winding numbers $N = 1, 2, 3$. Bottom: the corresponding energy density ε of the solitons as a function of the spatial coordinate x.

5. Conclusion

A finite height of the maximum allows the change of the homotopic class of a given field configuration, along with the stepwise decrease of the winding number. The realisation of such changes depends on the parameters of the potential and on those of the initial field configuration. We have quantitatively confirmed these qualitative features of the transitions that change the winding number.

Field configurations similar to those considered in this work could be formed during the inflationary stage, and lead to strong inhomogeneities in the CMB radiation temperature fluctuations.

This new type of solitons that has been studied in this paper may be also useful as a mechanism of BHs with complex mass spectra production [30–33] and could be used for solution of a number of cosmological and astrophysical problems, e.g., high-z quasars [34] and supermassive black holes within galactic nuclei [31, 35–37], dark matter [32, 33, 38–42], unidentified sources of gamma-radiation [43, 44], reionization [45–47], the suppression of intermediate-mass black holes production [37, 48].

Acknowledgments

The authors would like to thank Dr. K. M. Belotsky for useful discussions. The work of A. A. K. was supported by Russian Science Foundation, Project № 15-12-10039. The work of S. G. R. was supported by the Ministry of Education and Science of the Russian Federation, Project № 3.4970.2017/BY. The research fulfilled in the framework of the MEPhI Academic Excellence Project (contract № 02.a03.21.0005, 27.08.2013) and according to the Russian Government Program of Competitive Growth of Kazan Federal University.

References

[1] Abbyazov R R and Chervon S V 2013 *Mod. Phys. Lett.* A **28** 1350024 (*Preprint* 1404.5836)
[2] Kawai S and Kim J 2016 *Phys. Rev.* D **93** 065023 (*Preprint* 1512.05861)
[3] Dias M, Frazer J and Liddle A R 2012 *JCAP* **06** 020 (*Preprint* 1203.3792)
[4] Lüst D 2009 *JHEP* **03** 149 (*Preprint* 0904.4601)
[5] Susskind L 2003 *The Davis Meeting On Cosmic Inflation* (Davis CA.) p 26 (*Preprint* hep-th/0302219)

[6] Masoumi A, Vilenkin A and Yamada M 2017 *JCAP* **05** 053 (*Preprint* 1612.03960)
[7] Frazer J and Liddle A R 2011 *JCAP* **02** 026 (*Preprint* 1101.1619)
[8] Liu J, Wang Y and Zhou S 2015 *Phys. Rev.* D **91** 103525(*Preprint* 1501.06785)
[9] Manton N and Sutcliffe P 2004 *Topological Solitons* (Cambridge: Cambridge University Press)
[10] Dine M and Paban S 2015 *JHEP* **10** 088 (*Preprint* 1506.06428)
[11] Gani V A and Kudryavtsev A E 1999 *Phys. Rev.* E **60** 3305
[12] Gani V A, Kudryavtsev A E and Lizunova M A 2014 *Phys. Rev.* D **89** 125009
[13] Moradi Marjaneh A, Gani V, Saadatmand D, Dmitriev S and Javidan K 2017 *JHEP* **07** 028
[14] Gani V A, Lensky V, Lizunova M A and Mrozovskaya E V 2016 *J. Phys.: Conf. Ser.* **675** 012019
[15] Gani V A, Lensky V and Lizunova M A 2015 *JHEP* **08** 147
[16] Belendryasova E and Gani V A *Preprint* 1708.00403
[17] Bazeia D, Belendryasova E and Gani V A 2017 *Preprint* 1710.04993
[18] Gani V A, Moradi Marjaneh A, Askari A, Belendryasova E and Saadatmand D 2017 *Preprint* 1711.01918
[19] Radomskiy R V, Mrozovskaya E V, Gani V A and Christov I C 2017 *J. Phys.: Conf. Ser.* **798** 012087
[20] Gani V A, Konyukhova N B, Kurochkin S V and Lensky V A 2004 *Comput. Math. Math. Phys.* **44** 1968 (*Preprint* 0710.2975)
[21] Gani V A, Kudryavtsev A E, Belova T L and Druzhinin B L 1999 *Phys. Atom. Nucl.* **62** 895 (*Preprint* hep-ph/9712526)
[22] Belova T I, Gani V A and Kudyavtsev A E 2001 *Phys. Atom. Nucl.* **64** 140 (*Preprint* hep-ph/0003308)
[23] Gani V A and Kudryavtsev A E 2001 *Phys. Atom. Nucl.* **64** 2043
[24] Gani V A, Ksenzov V G and Kudryavtsev A E 2010 *Phys. Atom. Nucl.* **73** 1889
[25] Gani V A, Ksenzov V G and Kudryavtsev A E 2011 *Phys. Atom. Nucl.* **74** 771
[26] Lensky V A, Gani V A and Kudryavtsev A E 2001 *Sov. Phys. JETP* **93** 677
[27] Gani V A, Lizunova M A and Radomskiy R V 2016 *J. Phys.: Conf. Ser.* **675** 012020
[28] Gani V A, Lizunova M A and Radomskiy R V 2016 *JHEP* **04** 043
[29] Belova T I and Kudryavtsev A E 1997 *Phys. Usp.* **40** 359
[30] Rubin S G, Khlopov M Yu and Sakharov A S 2000 *Grav. Cosmol. Suppl.* **6** 51 (*Preprint* hep-ph/0005271)
[31] Rubin S G, Khlopov M Yu and Sakharov A S 2005 *Astropart. Phys.* **23** 265 (*Preprint* astro-ph/0401532)
[32] Belotsky K M, Kirillov A A and Rubin S G 2015 *Int. J. Mod. Phys.* D **24** 1545005
[33] Belotsky K M et al. 2014 *Mod. Phys.Lett.* A **29** 1440005 (*Preprint* 1410.0203)
[34] Dokuchaev V I, Eroshenko Yu N and Rubin S G 2005 *Grav. Cosmol.* **11** 99 (*Preprint* astro-ph/0412418)
[35] Rubin S G, Sakharov A S and Khlopov M Yu 2001 *J. Exp. Theor. Phys.* **92** 921 (*Preprint* hep-ph/0106187)
[36] Dokuchaev V I, Eroshenko Yu N and Rubin S G 2008 *Astron. Rep.* **52** 779 (*Preprint* 0801.0885)
[37] Grobov A V, Rubin S G, Samarchenko D A and Zhizhin E D 2011 *Grav. Cosmol.* **17** 181
[38] Carr B, Kühnel F and Sandstad M 2016 *Phys. Rev.* D **94** 083504 (*Preprint* 1607.06077)
[39] Dolgov A D 2017 *EPJ Web Conf.* **142** 01012 (*Preprint* arXiv:1701.05774)
[40] Grobov A V, Rubin S G and Shalamova V Yu 2014 *Adv. High Energy Phys.* **2014** 319218
[41] Gani V A, Dmitriev A E and Rubin S G 2015 *Int. J. Mod. Phys.* D **24** 1545001 (*Preprint* 1411.4828)
[42] Gani V A, Dmitriev A E and Rubin S G 2015 *Phys. Proc.* **74** 32 (*Preprint* 1511.01869)
[43] Belotsky K M, Berkov A V, Kirillov A A and Rubin S G 2011 *Grav. Cosmol.* **17** 27
[44] Belotsky K M, Berkov A V, Kirillov A A and Rubin S G 2011 *Astropart. Phys.* **35** 28 (*Preprint* 1212.2524)
[45] Belotsky K M, Kirillov A A and Rubin S G 2015 *Phys. Atom. Nucl.* **78** 387
[46] Belotsky K M and Kirillov A A 2015 *JCAP* **01** 041 (*Preprint* 1409.8601)
[47] Belotsky K M, Kirillov A A, Nazarova N O and Rubin S G 2017 *Int. J. Mod. Phys.* D **26** 1750102 (*Preprint* 1702.06338)
[48] Dokuchaev V I, Eroshenko Yu N, Rubin S G and Samarchenko D A 2010 *Astron. Lett.* **36** 773(*Preprint* 1010.5325)

Topological geons with self-gravitating phantom scalar field

P V Kratovitch, I M Potashov, Ju V Tchemarina and A N Tsirulev
Faculty of Mathematics, Tver State University, 35 Sadovyi, Tver, Russia, 170002
E-mail: kratovitch.pv@tversu.ru, chemarina.yv@tversu.ru, potashov.im@tversu.ru, tsirulev.an@tversu.ru

Abstract. A topological geon is the quotient manifold M/Z_2 where M is a static spherically symmetric wormhole having the reflection symmetry with respect to its throat. We distinguish such asymptotically flat solutions of the Einstein equations according to the form of the time-time metric function by using the quadrature formulas of the so-called inverse problem for self-gravitating spherically symmetric scalar fields. We distinguish three types of geon spacetimes and illustrate them by simple examples. We also study possible observational effects associated with bounded geodesic motion near topological geons.

1. Introduction

A (topological) geon is an orientable asymptotically flat spacetime with topology $\mathbb{R}\times\mathbb{R}P^3\backslash\{p\}$ [1]. The 'spatial' factor is the punctured real projective space and the omitted point is normally associated with the asymptotic infinity. This 3D manifold is orientable and can be obtained by removing an open ball from $\mathbb{R}^3$ and identifying antipodal points on the boundary sphere. In Einstein gravity, geons can exist only in the presence of exotic matter violating the null energy condition. For example, a self-interacting phantom scalar field has nowadays become a useful mathematical tool in models of astrophysics and cosmology. In particular, some models of dark energy and dark matter are based on phantom scalar fields with special self-interaction potentials.

In what follows, we will restrict our consideration only to static spherically symmetric geons that are supported by a phantom scalar field and do not have an event horizon. In this case, there exists a more convenient presentation of a geon as the quotient by the group $\mathbb{Z}_2$ of a spherically symmetric traversable wormhole [2, 3, 4, 5, 6] that possesses the reflectional symmetry with respect to its throat. The nonidentical $\mathbb{Z}_2$-transformation in spherical coordinates has the form $(t, r, \vartheta, \varphi) \to (t, -r, \pi - \vartheta, \varphi + \pi)$, $r \in \mathbb{R}$. Formally, such a wormhole and the corresponding geon are interchangeable (and equal for $r > 0$) solutions of the Einstein-Klein-Gordon equations but have the different topologies and physical interpretations. Wormholes are assumed to have emerged as global primordial structures that connect two large regions of the Universe, while geons are particle-like objects which could be created in the early Universe [7] and now could be of the various size from the Planck length up to the galaxy scale. For this reason, it seems preferable to regard geons as potentially observable astrophysical objects [8].

There is a natural question whether supermassive black holes and not geons (or wormholes) are placed in the centers of massive galaxies, or in more general context, how we can distinguish geons from black holes [9]. In this paper, we study the distinctive features of the motion of massive test particles in geon spacetimes of various types. In Section 2 we introduce the appropriate mathematical technique for describing geons and give some concrete examples. Section 3 is devoted to studying the bounded orbits in the spirit of the Kepler problem.

2. Basic equations and types of topological geons

Mathematically, self-gravitating phantom scalar fields are described by the action (in geometric units, $G = c = 1$)

$$\Sigma = \frac{1}{8\pi}\int\left(-\frac{1}{2}S - \langle d\phi, d\phi\rangle - 2V(\phi)\right)\sqrt{|g|}\,d^4x\,, \tag{1}$$

where S is the scalar curvature, ϕ the scalar field, $V(\phi)$ the self-interaction potential, and the sign of the scalar field kinetic term is negative. It is convenient, for our purposes, to write the metric of a spherically symmetric spacetime in the form

$$ds^2 = Adt^2 - \frac{dr^2}{A} - C^2\left(d\theta^2 + \sin^2\theta\,d\varphi^2\right)\,, \tag{2}$$

where A, C and ϕ are functions of the radial coordinate r with the range from $-\infty$ to ∞. Then, in the orthonormal basis associated with the metric (2), the independent Einstein-Klein-Gordon equations for the action (1) are

$$2A\,\frac{C''}{C} + A'\,\frac{C'^2}{C^2} + A'\,\frac{C'}{C} - \frac{1}{C^2} = A\phi'^2 - 2V\,, \tag{3}$$

$$A'\,\frac{C'}{C} + A\frac{C'^2}{C^2} - \frac{1}{C^2} = -A\phi'^2 - 2V\,, \tag{4}$$

$$A\phi'' + \phi'\left(A' + 2A\,\frac{C'}{C}\right) + \frac{dV}{d\phi} = 0\,, \tag{5}$$

where a prime denotes differentiation with respect to r.

From a physical point of view, a scalar field can be considered both as a fundamental natural field and as a phenomenological construction. In any case, we do not know the explicit form of the self-interaction potential. Therefore we assume that the scalar field is nonlinear with an arbitrary potential and have to study geons and their properties using the so-called 'inverse problem method for self-gravitating spherically symmetric scalar fields' developed in Refs. [10, 11]. This method allows one to construct a general solution of the Einstein-Klein-Gordon equations; it can be represented as quadrature formulae [12, 13, 14]. Here we use the quadrature formulae in the form [14]

$$\phi' = \sqrt{C''/C}\,, \quad A = 2C^2\int_r^\infty \frac{r}{C^4}\,dr\,, \quad \widetilde{V}(r) = \frac{1}{2C^2}\left(1 - 3C'^2A - CC''A + 2r\,\frac{C'}{C}\right). \tag{6}$$

It can be directly verified that equations (3)–(5) hold identically for ϕ', A, and $\widetilde{V}$ defined by (6).

These quadrature formulae give us rich possibilities to model scalar hairy configurations in terms of the area metric function $C(r)$ with the asymptotic expansion $C(r) = r + 3m + o(1)$. Choosing $C(r)$ to be even and positive, and such that $C'' > 0$ for all r, one can find $\phi(r)$, $A(r)$, and $\widetilde{V}(r)$; then, if $\phi(r)$ is monotonic, one finds the potential $V(\phi)$; however, each solution to the system (3)–(5) satisfies (6) regardless of the monotonicity of $\phi(r)$. The requirement for $C(r)$ to be even eliminates the ambiguity in the choice of the radial coordinate so that m is the Schwarzschild mass. The formulae (3)–(6) are invariant under the scale transformation $r \to r/\alpha$, $C \to C/\alpha$, $V \to \alpha V$, so that every time we can choose an arbitrary scale (e.g., take $m = 1$).

There exist, as opposed to black holes, a variety of geons with qualitatively different behaviors of the metric function $A(r)$ near the geon surface (or the wormhole throat). In order to discuss these principal differences, we consider the three illustrative examples labeled by I, II, and III:

$$\text{(I) } C=(r^4+2r^2+a^4)^{1/4}, \quad \text{(II) } C=(r^2+4)^{1/2}+3\,, \quad \text{(III) } C=\left(r^6+r^4+\frac{r^2}{16}+\frac{1}{64}\right)^{1/6}. \quad (7)$$

The geons I and III have a zero Schwarzschild mass, while for the geon II $m = 1$. For the cases I and II, the function $C(r)$ has the same form, $C = \sqrt{r^2+a^2} + 3m$, so from (6) we find $A = 1 - 2m/C$. Note that for any admissible choice of $C(r)$, we have $A = 1 - 2m/r + o(1/r)$ as $r \to \infty$ and the potential is positive on the geon surface ($r = 0$) and negative at spatial infinity, $V \sim -3m/r^3$ as $r \to \infty$. For the cases I, II, and III, the functions $C(r)$ and $A(r)$ are plotted in Figure 1.

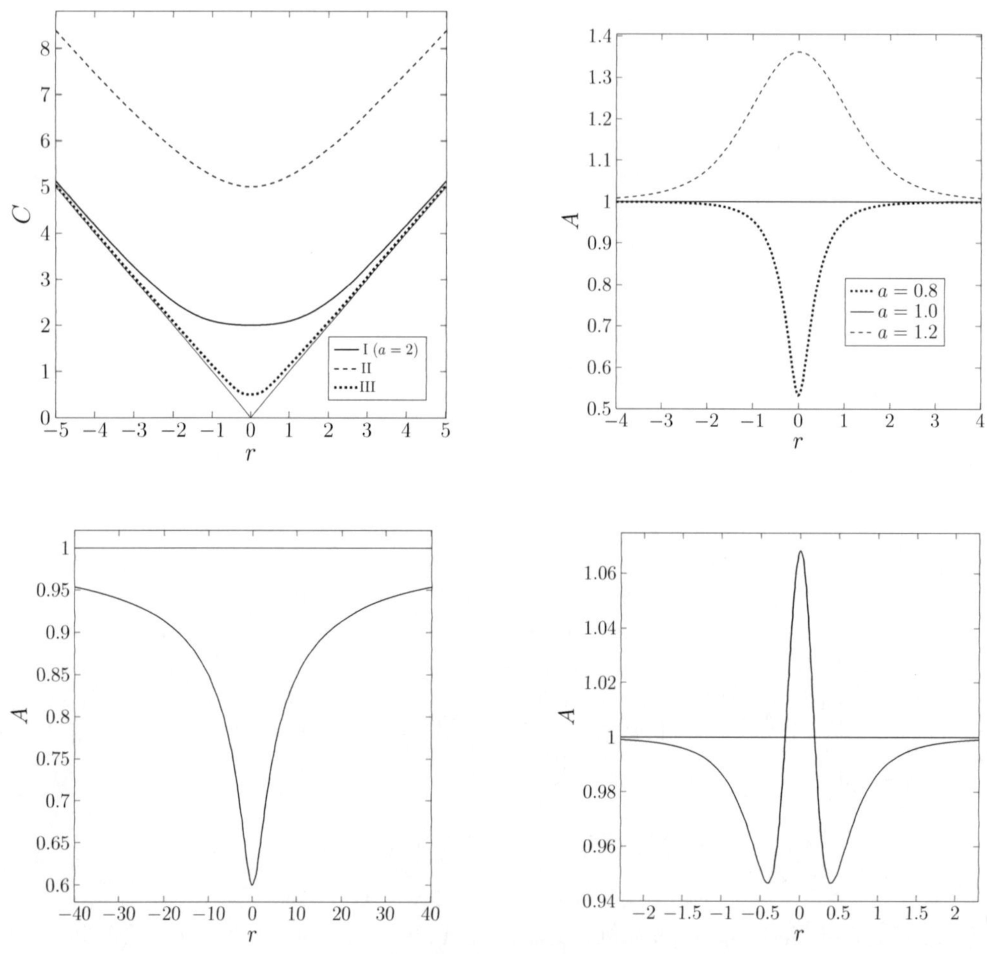

Figure 1. Top-left plot: the function C for the cases I, II, and III in (7). Top-right plot: $A(r)$ for the case I; the value $a = 1.0$ corresponds to the Ellis solution [16]. Bottom plots: $A(r)$ for the cases II (left) and III (right)

We see that there exist at least three types of geons distinguish by the form of A near the geon surface. The metric function A of the first or the second type has, respectively, only one

IOP Conf. Series: Journal of Physics: Conf. Series **934** (2017) 012047 doi:10.1088/1742-6596/934/1/012047

minimum or maximum that placed on the geon surface (or the wormhole throat); A of the third type has a maximum at $r = 0$ and at least one minimum at some point $r = r* > 0$.

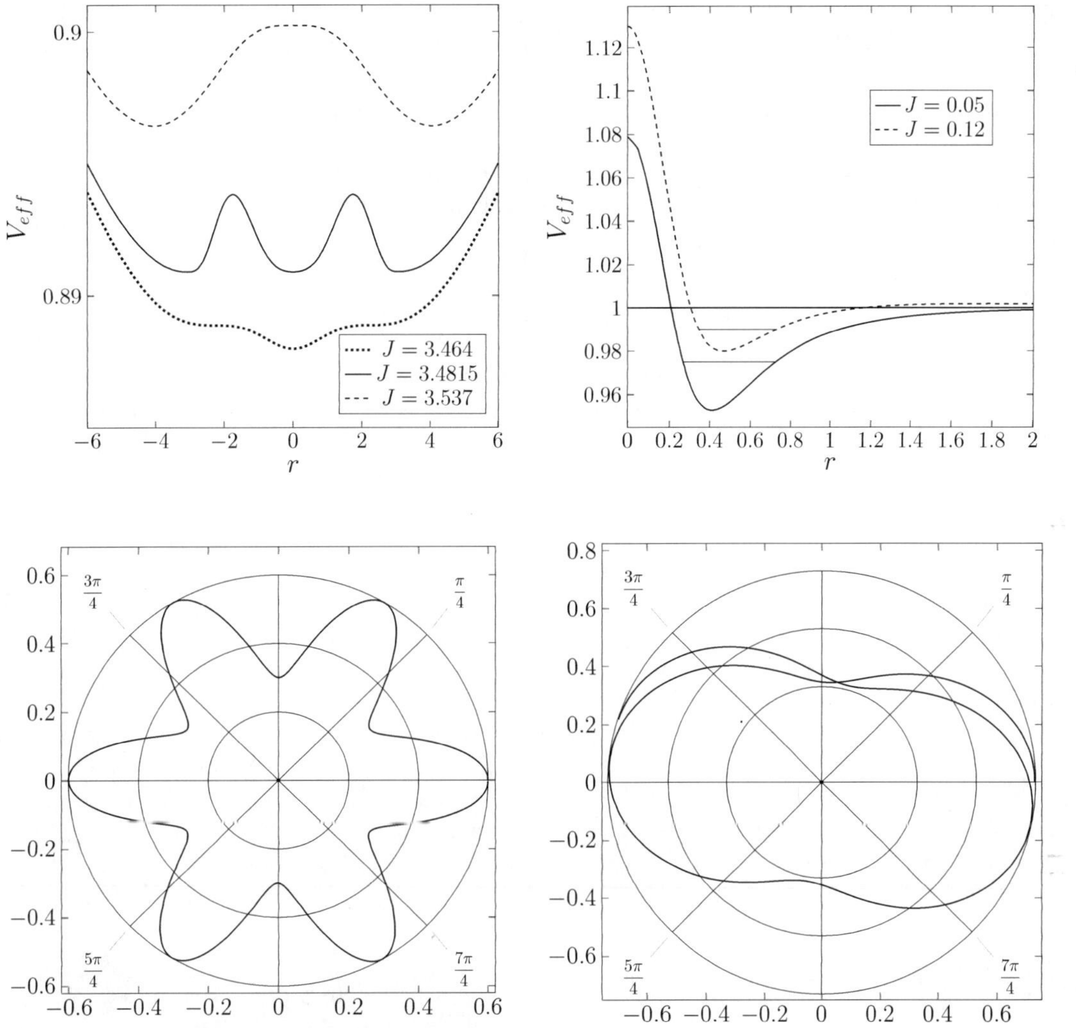

Figure 2. Top-left plot: the effective potential for the case II. Top-right plot: the effective potential for the case III. Bottom plots: two trajectories with $E^2 = 0.965$, $J = 0.05$ (left) and $E^2 = 0.99$, $J = 0.12$ (right) for the case III.

3. Bounded orbits around topological geons

The objects of our primary interest are minimums of the metric function A because the first integrals and the effective potential for a massive test particle are [15]

$$\frac{dt}{ds} = \frac{E}{A}, \quad \frac{d\varphi}{ds} = \frac{J}{C^2}, \quad \left(\frac{dr}{ds}\right)^2 = E^2 - V_{eff}, \quad V_{eff} = A\left(1 + \frac{J^2}{C^2}\right), \tag{8}$$

where E and J denote, respectively, the 'specific energy' and the 'specific angular momentum' of the particle. A minimum of $A(r)$ guarantees that there exist bounded orbits and stable circular orbits in its neighborhood, including the stable orbits with $J = 0$ for particles at rest. Figure 2 shows, first, that the effective potential of a geon with $m > 0$ can have a minimum on the geon surface and at least one minimum at some point $r = r* > 0$ (in the bounded interval, $3.464 < J < 3.537$). Second, there exist orbits that are impossible in neighborhoods of black holes. The angular velocities of test particles in the circular orbits for the cases II and III, as measured by a static observer at infinity, are given in the table below. If a test particle is in a potential well with nonzero radius $r = r^* > 0$ and $E^2 = V_{eff}(r^*)$, then the particle has a zero angular velocity (formally, in the stable circular 'orbit'). The similar potential wells were also found in other gravitating systems [17, 18]. On the contrary, a test particle can have an arbitrary angular velocity in the stable circular orbit on the geon surface.

Table 1. The angular velocity ω of a test particle in the circular orbits.

	The case II			The case III		
r	0	0	3.072	4.052	0.412	0.468
J	3.464	3.4815	3.4815	3.537	0.05	0.12
ω	0.089	0.089	0.058	0.049	0.124	0.259

4. Conclusions

We show that in neighborhoods of topological geons, test particles can accumulate in spherical layers around potential wells (minimums of the time-time metric function). Near the geon surface, the particles can have arbitrary velocities and the corresponding gas can be at the very high temperature, while non-central potential wells attractive particles with zero or small velocities, thus forming cold spherical layers around the geon. This is a crucial difference from the case of black holes. In the latter case, accretion disks with the innermost stable circular orbit must be observed. We can expect that a massive or supermassive geon can be revealed and identified by the orbital motion of test particles (planets or stars) and the specific behavior of clouds of gas in its neighborhood.

References

[1] Sorkin R D 1986 *Proc. NATO Advanced Study Institute on Topological Properties and Global Structure of Space-Time (Erice, 1985)* ed Bergmann P G and De Sabbata V (New York: Plenum) p 249
[2] Bronnikov K A and Fabris J C 2006 *Phys. Rev. Lett.* **96** 251101 (*Preprint* gr-qc/0511109)
[3] Bronnikov K A, Dehnen H and Melnikov V N 2007 *Gen. Rel. Grav.* **39** 973 (*Preprint* gr-qc/0611022)
[4] Bronnikov K A and Sushkov 2010 *Class. Quantum Grav.* **27** 095022 (*Preprint* 1001.3511)
[5] Bolokhov S V, Bronnikov K A and Skvortsova M V 2012 *Class. Quantum Grav.* **29** 245006
[6] Bronnikov K A and Korolyov P A 2017 *Grav. Cosmol.* **23** 273 (*Preprint* 1705.05906)
[7] Dowker H F and Sorkin R D 1998 *Class. Quantum Grav.* **15** 1153 (*Preprint* gr-qc/9609064)
[8] Kardashev N S 2009 *Phys. Usp.* **52** 1127
[9] Li Z, Bambi C 2014 *Phys. Rev.* D **90** 024071
[10] Bechmann O and Lechtenfeld O 1995 *Class. Quantum Grav.* **12** 1473 (*Preprint* gr-qc/9502011)
[11] Bronnikov K A and Shikin G N 2002 *Grav. Cosmol.* **8** 107 (*Preprint* gr-qc/0109027)
[12] Tchemarina Ju V and Tsirulev A N 2009 *Grav. Cosmol.* **15** 94
[13] Azreg-Aïnou M 2010 *Gen. Rel. Grav.* **42** 1427 (*Preprint* gr-qc/0912.1722)
[14] Solovyev D A and Tsirulev A N 2012 *Class. Quantum Grav.* **29** 055013
[15] Nikonov V V, Potashov I M and Tsirulev A N 2016 Circular orbits around static self-gravitating scalar field configurations *Math. Model. Geom.* MMG04(2016)02
[16] Ellis H G 1974 *J. Math. Phys.* **15** 520, 1973 *J. Math. Phys.* **14** 103
[17] Pugliese D, Quevedo H, and Ruffini R 2011 *Phys. Rev.* D **83** 024021 (*Preprint* 1003.2687)
[18] Vieira R S S, Schee J, Kluźniak W, Stuchlík Z and Abramowicz M 2014 *Phys. Rev.* D **90** 024035

IOP Conf. Series: Journal of Physics: Conf. Series **934** (2017) 012048 doi:10.1088/1742-6596/934/1/012048

The using of non-standard according to Hans Reichenbach's closk synchronization in the integral covariant formulation of the laws of conservation

Valery Egorowich Stepanov

Physics and Industry Faculty, North Eastern Federal University, Yakutsk, 677000, Russia

E-mail: tritiy51@mail.ru

Abstract. The non-standard Reichenbach's synchronization is necessary for the coincidence of the numerical values of the integrals of the conserved quantities in two different inertial frames of reference, thereby ensuring an integral covariant formulation of the laws of conservation of physical quantities. The inclined axes of time in Minkowski's space have a physical meaning of time with non-standard clock synchronization according to Reichenbach.

1. Introduction

In quantum electrodynamics, the law of conservation of the electric charge is proved by the Noether's theorem, using condition of the invariance of the action for the Dirac's spinor field according to the one-parameter group of unitary transformations. Locally, this law corresponds to the equality to zero of the four-dimensional divergence of the electric current density vector. For a system of charges located in a bounded volume and vanishing at spatial infinity, the integral over the three-dimensional hypersurface of the space is equal to the total charge of the system and its magnitude is conserved in time as discussed by Michkewich [1].

The laws of conservation of energy-impulse and impulse momentum are formulated in the form of three-dimensional volume integrals connected by the Noether's theorem with the ten parametric Poincaré's group of continuous symmetry transformations of Minkowski's four-dimensional space-time as explained in [1]. Integrals over the spatial three-dimensional hypersurface from the densities of conserved quantities are constant magnitudes in time. Thus, integral conserved quantities are necessary parts in the theory of interactions of a system of particles and physical fields. However, the relativity of the simultaneity of different events, resulting from the Lorentz's transformations, raises the question of the compatibility of integral conservation laws with the principle of relativity of the special theory of relativity. We know the extreme point of view, which affirms the exclusively local (in the form of a four-dimensional differential equation of continuity) character of the conservation laws. For example, as explained by Feynman [2], where he considered a cylinder of finite length. In the own reference frame for a cylinder, an electron and a positron simultaneously emitted from the ends of the cylinder. In its own system, the full charge of the cylinder does not change, but in a different inertial reference frame the total charge varies at different moments of time on the clock at rest because of the relativity of the simultaneity of the different events. In this non-local incompatibility of the conservation law with the principle of relativity is expressed the Feynman's paradox.

Here is considered the tensor formulation of the integral law of conservation of the total charge of a system bounded in space with is compatible respect to the Lorentz transformations.

2. Materials and Methods

2.1. Non-standard clock synchronization in distant simultaneity

The concept of time in the special theory of relativity is based on the dates of the space separated clocks with the same speed of motion, launched simultaneously. Simultaneous start-up of the clock is made by taking into consideration the time spent by the light on the path from the initial to the end points. In his famous work [3], Einstein proposed a condition about the equality of the speeds of light in the forward and backward directions and established the method of clock synchronization in the special theory of relativity. Einstein's synchronization will be called standard.

In the famous Michelson-Morley's experiment on an interferometer, light passes along a closed path, the absence of an ether wind is determined by the constancy of the average speed of light as it travels along a closed path C, the possible inequality of the light velocities in individual sections does not affect on the experimental result as discussed by Ugarov [4].

Reichenbach generalized Einstein's standard simultaneity to a topological one, compatible with the principle of causality and observability of the average speed of light along a closed route as explained in [5]. Nowadays in the scientific literature there is a discussion about the conventional status of the numerical value of the speed of light in one direction as discussed by Grunbaum [6].

Without entering into this discussion, we are going to prove here that by using non-standard clock synchronization according to Hans Reichenbach, we can formulate the law of charge conservation in a relativistically invariant manner, as the constancy of conserved quantities in their own frame of reference.

L_{AB}, at the instant t_{a0} the light signal is emitted to the point B, which arrives at the moment t_{b0} and returns back to the time t_{a1} (Figure 1). With the standard synchronization according to A. Einstein, the moment of arrival at point B is at the middle of the interval of the total time of the trip back and forth, which is seen in formula (1).

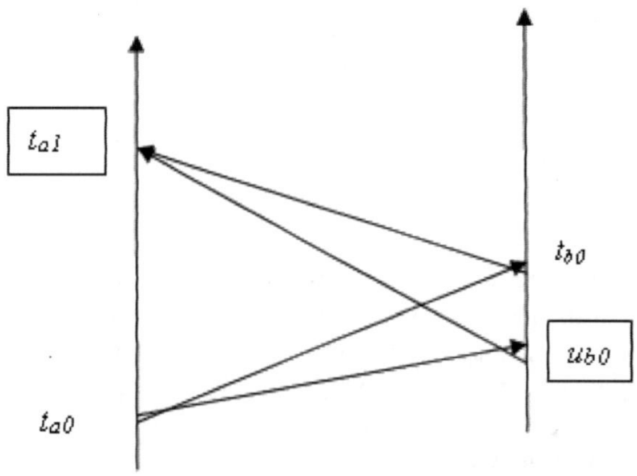

Figure 1. Temporal diagram of the clock A and B

$$t_{b0} = t_{a0} + \tfrac{1}{2}(t_{a1} - t_{a0}). \qquad (1)$$

With nonstandard synchronization, the 1/2 factor is replaced by the Reichenbach parameter ϵ_{AB}, varying from zero to one

$$u_{b0} = t_{a0} + \epsilon_{AB}(t_{a1} - t_{a0}) \qquad . \qquad (2)$$

The speed of light in one direction C_{AB} is related with parameter of Reichenbach's formula

$$C_{AB} = \frac{L_{AB}}{u_{bo} - t_{ao}} = \frac{c}{2\epsilon_{AB}} \ . \tag{3}$$

If at point B we shift back the time of starting the clock to an interval equal to $t_{bo} - u_{bo}$, we will get an apparent increasing speed of light there, corresponding to formula (3). It is easy to see that

$$u_{bo} = t_{bo} + u_{bo} - t_{bo} = t_{bo} + (\epsilon_{AB} - \frac{1}{2})\frac{2L_{AB}}{c} \ . \tag{4}$$

The origin of the Cartesian X-coordinate system is compatible with point A and put an arbitrary point B on the X-axis, this point will have the coordinate x. Then the time u of a reference frame with non-standard synchronization will be related with the time of the reference frame with Einstein's synchronization t by the relation

$$u = t + (\epsilon_{AB} - \frac{1}{2})\frac{2x}{c} \ . \tag{5}$$

Let us consider a reference frame $\acute{K}$ moving with a constant velocity v along the X axis. The stationary laboratory frame of reference is denoted by the symbol K. These two systems are connected by special Lorentz's transformations. We show that due to the shift the start of the remoted clocks the Feynman paradox can be resolved. Let it be a cylinder moving with velocity v, the axis of symmetry of which is aligned with the axis x . Let the electron emerge at the time $\acute{t}_1$ from the point with the coordinate $\acute{x}_1$ of the cylinder, and the positron emerge from the point $\acute{x}_2$. These two events in the fixed system K have coordinates (t_1, x_1) and (t_2, x_2) . By changing the clock synchronization in the laboratory system using formula (5) so that these two events are simultaneous in time u, and also using special Lorentz' transformations, we get that the Reichenbach's parameter is determined by the condition

$$\epsilon_{AB} = \frac{1}{2} - \frac{v}{2c}, \tag{6}$$

$$u = t - \frac{v}{c^2}x. \tag{7}$$

Consequently, due to the transition to non-standard clock synchronization, it is possible to resolve the Feynman's paradox in the sense that for both inertial reference systems the law of conservation of the total charge of the cylinder is simultaneously fulfilled in its self and in laboratory frames of reference.

2.2. The transition to non-standard synchronization leads to an oblique rectilinear time coordinate axis in the Minkowski's space

In Minkowski's space, the time coordinate x^0having the dimension of length is equal to the product of the time t by the speed of light c, the spatial coordinates following after it have the numbers 1, 2, 3. In the space-time plane, the coordinates u^0, u^1 for the laboratory inertial reference frame with non-standard clock synchronization are related to the old coordinates with Einstein's synchronization by the following transformations

$$u^0 = x^0 - \frac{v}{c}x^1, \tag{8}$$

$$u^1 = x^1. \tag{9}$$

A set of simultaneous events forms a three-dimensional hypersurface in the Minkowski's space that intersects the time axis orthogonally at the considered moment. When the reference frame $\acute{K}$ moves along the axis x^1 . the hypersurface of simultaneity of this proper reference frame of a charged cylinder cuts a line in the plane (x^0, x^1), which is defined by equation

$$u^0 = x^0 - \frac{v}{c}x^1 = const. \tag{10}$$

Differentiation of this equation leads to the result

$$\text{th}\,\theta = \frac{dx^0}{dx^1} = \frac{v}{c}. \tag{11}$$

The time axis u^0 is orthogonal to the line defined by equation (9) (figure 2).

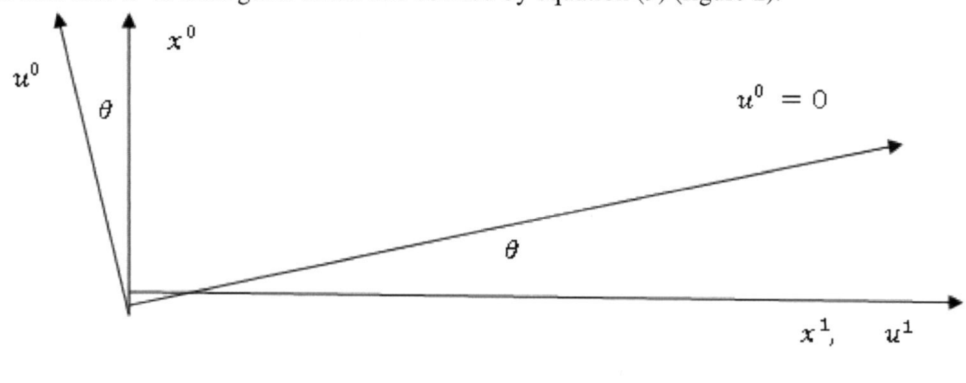

Figure 2. Slope of the time axis of non-standard synchronization

Thus, the coordinate axes (u^0, u^1) form an angle $\pi/2 + \theta$. The inverse formulas to (9) for the transition to new coordinates have the form

$$x^0 = u^0 + \frac{v}{c}u^1,$$
$$x^1 = u^1. \tag{12}$$

In the skew coordinate system, the nondiagonal component of the metric tensor differs from zero

$$g_{01} = \mu_{00}\frac{\partial x^0}{\partial u^0}\frac{\partial x^0}{\partial u^1} = -\frac{v}{c} \quad . \tag{13}$$

In addition, it turns out that

$$g_{00} = \mu_{00}\frac{\partial x^0}{\partial u^0}\frac{\partial x^0}{\partial u^0} = 1 \quad , \tag{14}$$

$$g_{11} = -1 + \frac{v^2}{c^2} \quad . \tag{15}$$

Thus, in the special theory of relativity, skew rectangular coordinate systems are uniquely associated with the shifts of the clock starting times equivalent to the non-standard clock synchronization according to Reichenbach.

2.3. *Integral covariant formulation of conservation laws*

Let us consider an insular system of mutually motionless particles. The law of conservation of charge for this system in the differential form is expressed as the vanishing of the divergence of the four-dimensional vector of the volumetric current density. To calculate the integral formulation of the integral over the four-dimensional tube world lines of particles insular system, forming a four-dimensional cylinder. This cylinder ends of two perpendicular to the time world lines of the surfaces of simultaneous events belonging to future and past moments of time of the observer in his own frame of reference insular system. According to the four-dimensional Gauss's theorem the volume integral is converted to the integral over a closed surface, disappears on the lateral surface but is different from zero on the bases. . The lateral surface of the cylinder tends to spatial infinity, where the charges are zero. Consequently, integral of lateral surface turns to zero. The result is an equality of two integrals over the three-dimensional hypersurfaces for different points in time. If we turn into the laboratory reference system, relatively to which insular system is moving inertially, then of the simultaneity hypersurfaces used in the insular system cease to be orthogonal to the time axis of the laboratory system. Because of the relativity of simultaneity, an unequal to one-half Reichenbach's parameter arises from it (6), the speed of light will be equal to the fundamental constant only on the average. The hypersurfaces of simultaneity in the four-dimensional Minkowski's space are invariant geometric objects. In the zone of relativity of simultaneity, any hypersurfaces can be chosen, while the numerical values of the integrals of the conserved quantities do not depend on the choice of the hypersurfaces of simultaneity.

3. Results

Thus, the non-standard Reichenbach's synchronization is necessary for the coincidence of the numerical values of the integrals of the conserved quantities in two different inertial frames of reference, thereby ensuring an integral covariant formulation of the laws of conservation of physical quantities.

The inclined axes of time in Minkowski's space have a physical meaning of time with non-standard clock synchronization according to Reichenbach.

References

[1] Michkewich N V 1969 (in Russian) Physical Fields in General Relativity (original Russian title: *Fizicheskie poliya v obshey teorii otnositelnosty* Moscow: Nauka)
[2] Feynman R 1987 (in Russian) The character of physical law (original Russian title: *Karakter fizicheskix zakonov* Moscow: Nauka)
[3] Einstein A 1905 *Ann. Phys.* **17** 891
[4] Ugarov V A 1977 (in Russian) Special Relativity (original Russian title: *Specialnaya Teoriya Otnositelnosty* Moscow: Nauka)
[5] Reichenbach H 1924 *Axiomatik der relativistischen Raum-Zeit-Lehre* Braunschweig: Verlag
[6] Grunbaum A 2010 *Found. Phys.* **40** 1285

Measurements of angular distribution and spectrum of transition radiation with a GridPix detector

N Belyaev[1], M L Cherry[2], K Desch[3], K Filippov[1], P Fusco[4,5], J Kaminski[3], S Konovalov[6], D Krasnopevtsev[1], F Loparco[4,5], M N Mazziotta[5], D Ponomarenko[1], C Rembser[7], A Romaniouk[1], A Savchenko[1], E Shulga[1], S Smirnov[1], Yu Smirnov[1], V Sosnovtsev[1], P Spinelli[4,5], M Strikhanov[1], P Teterin[1], V Tikhomirov[1,6], K Vorobev[1] and K Zhukov[6]

[1] National Research Nuclear University MEPhI (Moscow Engineering Physics Institute), Kashirskoe highway 31, Moscow, 115409, Russia
[2] Dept. of Physics & Astronomy, Louisiana State University, Baton Rouge, LA 70803 USA
[3] University of Bonn, Germany
[4] Dipartimento di Fisica "M. Merlin" dell'Università e del Politecnico di Bari, I-70126 Bari, Italy
[5] Istituto Nazionale di Fisica Nucleare, Sezione di Bari, I-70126 Bari, Italy
[6] P. N. Lebedev Physical Institute of the Russian Academy of Sciences, Leninsky prospect 53, Moscow, 119991, Russia
[7] CERN, the European Organization for Nuclear Research, CH-1211 Geneva 23, Switzerland

E-mail: Yury.Smirnov@cern.ch

Abstract. In recent years, developments of gaseous detectors based on a combination of electron multiplication gap in the gas and pixel readout chips as a part of the anode plane (GasPixel detectors) reached a level where they can offer unique opportunities for particle detection. Transition radiation (TR) detectors based on this technology can be one of the possible applications. In this work, measurements of energy spectra and angular distributions of transition radiation photons produced by particles with different gamma factors made with a GridPix detector prototype are presented. The observed results are compared with theoretical predictions.

1. Introduction

GasPixel detector is one of the new promising technologies, which combines the properties of silicon and gaseous detectors. It has been extensively developed over the last decades (see review in [1] and also references in [2]). The operating principle of these detectors is similar to that of Time Projection Chambers with a two-dimensional readout plane. In this work the GasPixel detector with a special design called GridPix ([3]) was used. A schematic view of the GridPix detector is shown in figure 1. A specially treated pixel electronics chip (TimePix-1, [2, 4]) is placed in the gas volume. The gas volume has a drift gap (the width of the gap depends on the exact detector requirements) and an electron multiplication region. Electrons produced by an ionizing particle drift towards the pixel plane and develop an avalanche in the multiplication

Figure 1. Schematic view of the GridPix detector (taken from [2]). The pixel chip is separated from a drift volume by a mesh. Electron amplification happens in the gap between the mesh and the pixel chip.

gap. The detectors are usually operated with an effective gas gain of $(2–5) \times 10^3$, which allows detection of single electrons with high efficiency. The chip has the possibility to choose an operation mode for each pixel: time or amplitude measurement. If time measurement mode is used, these detectors offer the possibility to reconstruct 3D track segments with a spacial accuracy of below 10 µm in a chip plane (see [1] and [2]). The GridPix detectors design (gas composition, drift gap, electronics) depends on the required functionality. Filled with Xe-based gas mixture, they may combine tracking and transition radiation detector properties (see [4]). In this case, the read-out electronics measures the signal amplitude in each pixel using Time-over-Threshold (ToT) information. The advantage of GridPix detectors as TRD devices is their capability to reconstruct a microscopic picture of ionization around a particle track and, hence, to maximize the separation between ionization and TR clusters. In a radiator medium, TR photons are generated at some angle to the particle direction, which depends on a particle's Lorentz factor. Using a GridPix detector, it is possible to identify photons not only by measuring ionization clusters along a track but also their distance from a track. This opens a possibility to improve particle identification properties of the detector. In this paper, spectra and angular distributions of TR photons measured by a GridPix detector are presented. They are compared with theoretical predictions.

2. Test beam setup

The experimental setup is shown schematically in figure 2. It contains a GridPix detector and TR radiator installed 2 meters upstream. In order to minimize TR photon absorption by air, a pipe filled with helium was placed between the radiator and the detector.

The GridPix test beam prototype was built on the basis of a Timepix chip (see details in [5]) with the pixel size of $55 \times 55\,\mu m^2$. The chip has a 256×256 pixel matrix providing $\approx 14 \times 14\,mm^2$ sensitive area. The detector is very similar to the one described in [6]. Only the cathode had been slightly modified and the readout electronics is based on the Scalable Readout System as described in [7]. The drift gap was chosen to be 3 cm and was filled with a mixture of Xe/CO_2 80%/20%. Electrical field in the drift gap was 900 V/cm. Charge deposited on each pixel was measured using Time-Over-Threshold method (ToT). A stack of 150 foils with a separation between foils of 2 mm was used as the TR radiator. Each radiator foil was made of 4 layers of polypropylene foils of 15.5 µm thick put on top of each other. The foils are held together by electrostatic forces without or with a very small (a few microns) air gap in between. Total thickness of such a radiator foil is about 62 µm. The tests were performed at the H8

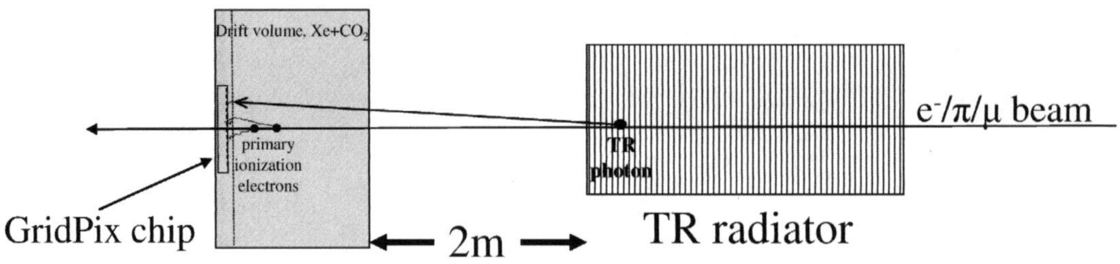

Figure 2. Schematic view of the experimental setup (not to scale).

SPS beam line at CERN with electrons and pions of 20 GeV ($\gamma = 3.9 \times 10^4$ and $\gamma = 1.4 \times 10^2$, respectively) and muons of 180 GeV ($\gamma = 1.7 \times 10^3$). The foils were oriented perpendicular to the beam and the detector plane had about 89° angle with respect to the beam line.

Figure 3 shows a typical event display with identified absorbed photons (black circles) and beam particle (red circle). Each point corresponds to one fired pixel and its color reflects the signal amplitude in this pixel. A group of adjacent fired pixels is referred to as a "cluster" throughout this paper, and it is an indication of a detected particle (beam particle, TR photon or a delta ray).

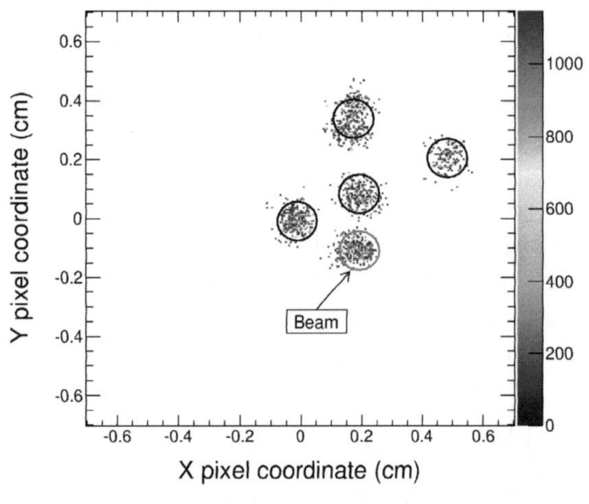

Figure 3. A typical event from the GridPix detector. Pixels fired by a beam particle are in the red circle.

3. Results and discussion

In order to identify clusters in each event as shown in figure 3, a cluster search algorithm has been developed. It assumes that clusters have a circular shape. Besides being able to identify position of the clusters, the algorithm provides cluster geometry parameters and deposited energy calculated as a sum of ToT of all pixels within a circle. This allows separating clusters from beam particle and those originating from TR photons. In order to translate the sum of ToT of all pixels belonging to a cluster into an energy value, a calibration using 5.95 keV Fe^{55} source was performed. An energy resolution $\frac{\sigma_E}{E} = 14\%$ was obtained for this photon energy.

Photoelectrons from energetic photons have a significant pathlength in the gas and not all the energy can be accounted when a cluster is identified. In order to understand how this effect may affect the energy measurement, dedicated simulations based on the program developed in [8, 9] were performed. They include all processes occurring at photon absorption, photoelectron thermalisation, diffusion of secondary electrons drifting towards the anode plane. ToT is considered to be proportional to a number of electrons collected to one pixel.

Simulations show that, indeed, the average photoelectron path is significant and for photon energy of 10 keV it is ≈0.12 mm and for 20 keV it is ≈0.5 mm. Tails are extended up to several millimeters. Diffusion of secondary electrons drifting towards the anode also significantly contributes to the size of the ionization cluster seen by Pixel chip. The transverse diffusion coefficient of $176\,\mu m/\sqrt{cm}$ was initially calculated using the simulation package Garfield v.7.39 ([10]). But after comparison of the experimental data for Fe^{55} calibration run with the corresponding simulations the coefficient was corrected and the value of $210\,\mu m/\sqrt{cm}$ gave the best agreement of simulation with the data.

Distributions of ToT sum in identified clusters from absorbed photons with energy values of 10, 15 and 20 keV obtained in simulations are shown in figure 4. One sees that the higher is

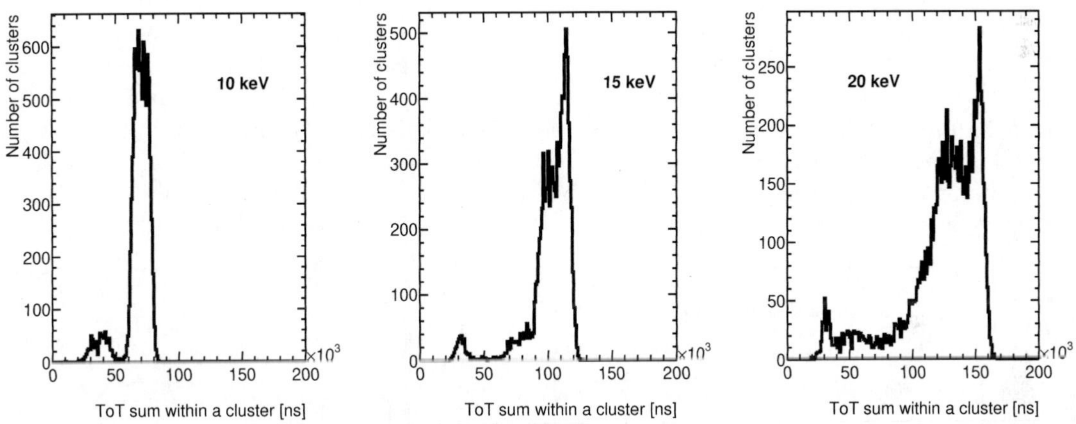

(a) (b) (c)

Figure 4. Simulated spectra of sums of pixels ToT for mono-energetic photons with energy values of 10 (a), 15 (b) and 20 keV (c). Simulation includes a detector response.

the photon energy, the wider is the distribution of measured ToT sum, because more pixels are not associated with the cluster. The initial simple algorithm for calculating the number of pixels in a cluster assumes its regular circular shape with a radius limited by 1.0 mm. Calculation of the cluster energy becomes less accurate as the cluster size grows and more pixels fall outside the assumed circular boundary. This unavoidably leads to a deformation of the measured TR spectrum in the energy range above ≈20 keV.

The measured distributions of the cluster multiplicity per event are shown in figure 5 for the case without radiator ((a), 20 GeV pions and electrons, 180 GeV muons) and with radiator ((b), 20 GeV electrons and 180 GeV muons).

Clusters not associated to the beam particle are referred to as “secondary” clusters in the text, whereas the other clusters are called the “primary” ones. In about 10% of events there are secondary clusters even without a radiator. They originate from delta-electrons and upstream bremsstrahlung photons (for electrons) absorbed in the detector. The number of events with

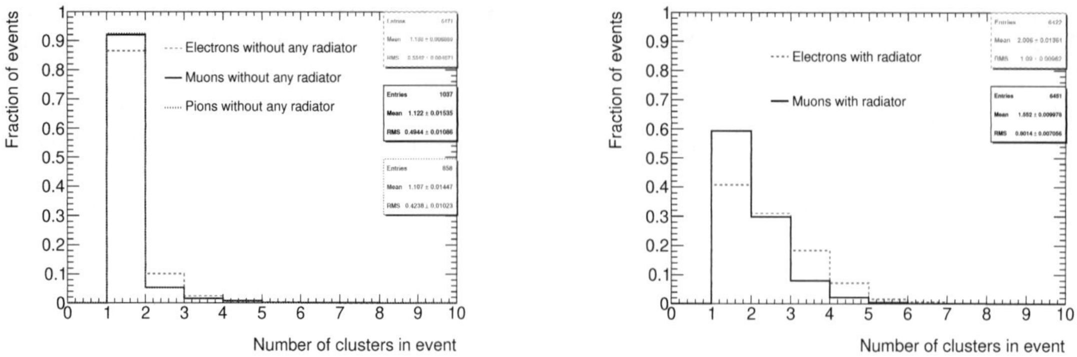

(a) (b)

Figure 5. Distributions of cluster multiplicity per event for different beam particles without (a) and with radiator (b).

clusters increases up to 60% for electrons and up to 40% for muons for the case with the radiator (plot (b) in figure 5). This corresponds to an average number of detected TR photons of 1.0 for electrons and 0.5 for muons. Not all of the clusters can be separated from the beam particle track. The typical size of a cluster is about 1 mm (see figure 3), which means that only clusters at least 1 mm apart from each other can be identified. This corresponds to an angle between a photon and a particle of about 0.5 mrad for the test beam geometry. A comparison of secondary cluster energy spectra for electrons of 20 GeV and 180 GeV muons is shown in figure 6 (a).

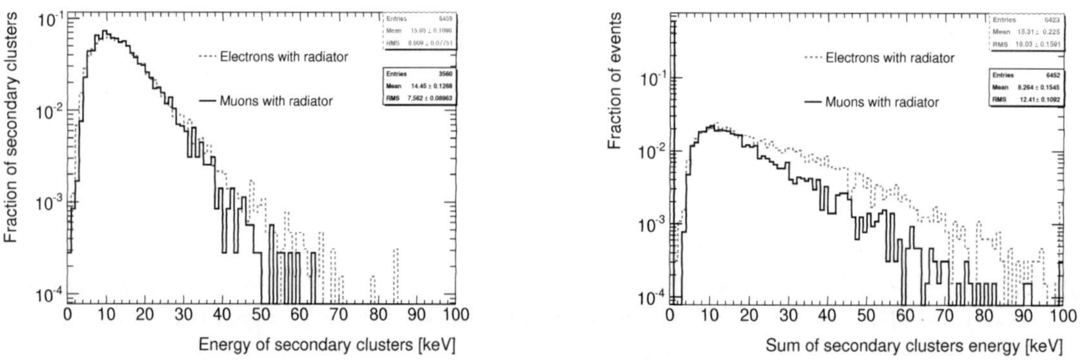

(a) (b)

Figure 6. (a): comparison of shapes of energy distributions of secondary clusters for electrons and muons. (b): distributions of sums of energies of secondary clusters in event for electrons and muons. Distributions are normalized to their own integrals. The rightmost bin on plot (b) includes overflown values.

The integral of each spectrum was normalized to 1 in this figure. One sees that in spite of the fact that the number of detected TR photons is almost two times less for muons, the shape of energy distribution is very close to that for electrons. The distributions of the sums of secondary clusters energies (total TR photons energy) per event are shown in figure 6 (b). These spectra are different because electrons produce more than one cluster more often.

The scatter plots of cluster energy vs. angle are presented in figure 7. Both distributions have the same number of beam particles.

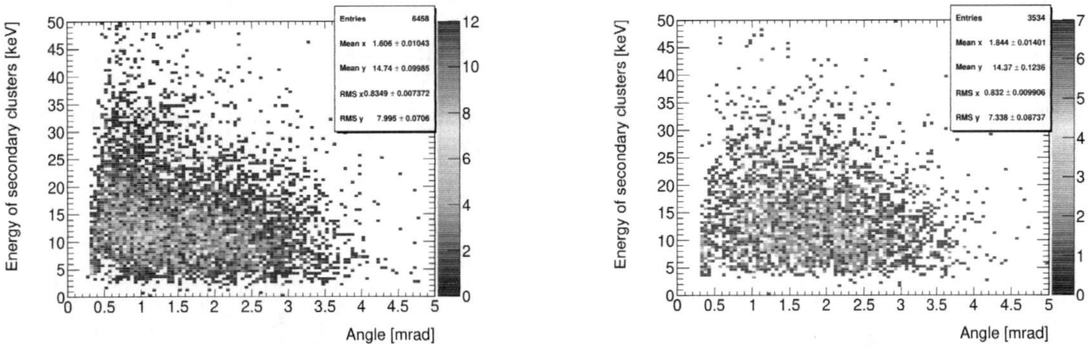

(a) (b)

Figure 7. Scatter plots of cluster energy vs. angle. Plot (a) is for electrons and plot (b) is for muons.

In general, the distributions look similar but one still observes some differences. In both cases TR is concentrated at angles below 4 mrad. Empty area below 0.3 mrad is defined by size of the cluster, which cannot be reconstructed too close to the primary particle cluster. This area increases with the photon energy because of a rise of the cluster size.

The projections of these distributions on the horizontal axis for the whole energy interval (a) and for energies below 15 keV (b) are shown in figure 8.

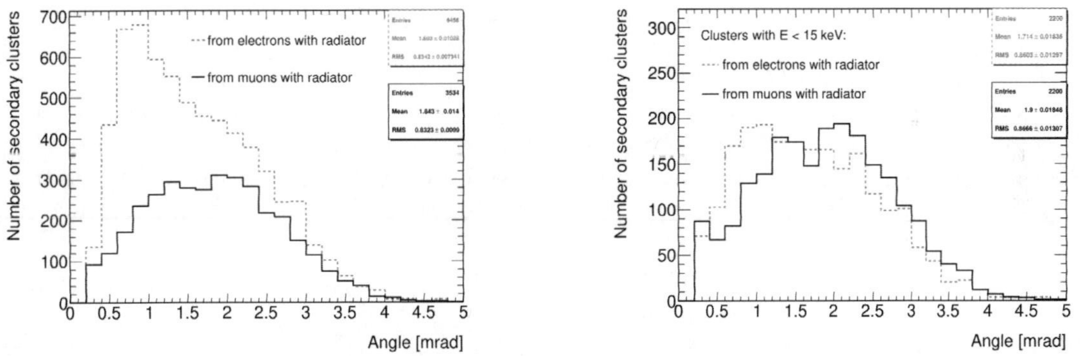

(a) (b)

Figure 8. Angular distribution of TR clusters for electrons and muons (plot (a) is for all energies, plot (b) is for energies below 15 keV).

TR photons originating from electrons have a tendency to concentrate at smaller angles. Some indication of a presence of two peaks in the muon distribution around 1 mrad and 2 mrad is also observed.

The TR is emitted along the particle at angles up to $\sqrt{\frac{1}{\gamma^2} + \frac{\omega_1^2}{\omega^2}}$, where $\omega_1 \approx 21$ eV is the plasma frequency of the CH_2 foils and ω is a TR photon energy. Angular distribution of photons at the last interference maximum is peaking up at the angle of $\sqrt{\frac{1}{\gamma^2} + \frac{\omega_2^2}{\omega^2}}$, where $\omega_2 \approx 0.7$ eV is the plasma frequency of air, which fills the space between the foils. For both 20 GeV electrons and 180 GeV muons one expects the maximum angle for TR photons in the energy range of 10–20 keV at ≈1 mrad. Less energetic photons should have wider distribution. In order to understand the observed behavior in more details, a dedicated calculation based on exact TR production formulas ([11]) was performed. The calculations also include TR absorption in the radiator and in the detector materials.

The results of these calculations are shown in figure 9 (plot (a) for electrons and plot (b) is for muons).

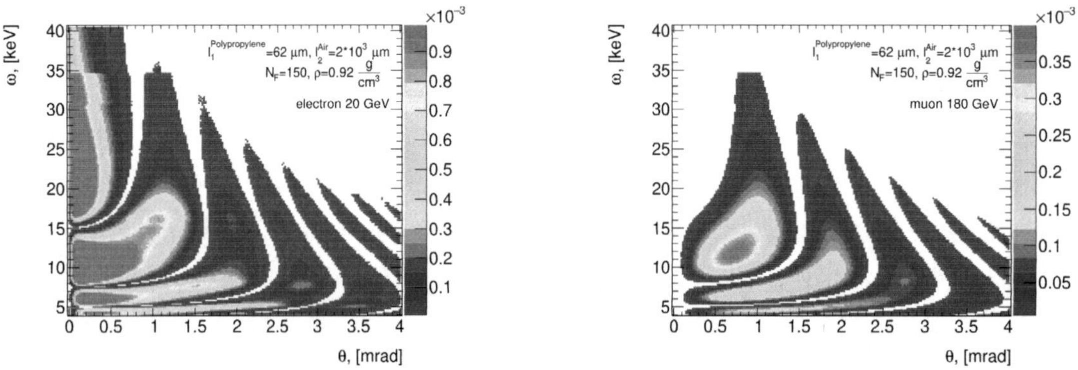

(a) (b)

Figure 9. Calculated scatter plots of cluster energy vs. angle: for electrons (a) and for muons (b).

One can clearly see that for electrons a significant part of TR is concentrated at angles below 0.5 mrad. As for the muons, only a small fraction of photons fall in this range. One also notices that the energy distribution of the TR clusters should have several peaks. The two most energetic ones are at ≈10 keV and ≈20 keV for electrons and ≈6 keV and ≈12 keV for muons. There is no significant difference between energy spectra observed in the data for these two sorts of particles. This can be explained by the fact that the most energetic photons produced by electrons are not reconstructed in the data because of minimal angle constraint. Limitations on the energy resolution for high energy photons also significantly deform spectrum. These two effects make measured spectra similar.

The projections of 2D distributions on horizontal axis are shown in figure 10.

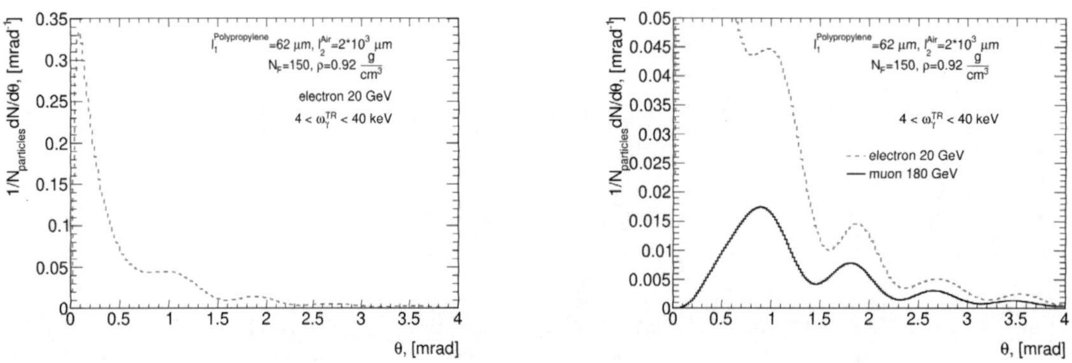

(a) (b)

Figure 10. Analytically calculated angular distributions of TR clusters for electrons (a) and for both electrons and muons (b).

Figure 10 (a) is for electrons and Figure 10 (b) is for muons and electrons. If we apply a requirement on a minimum reconstruction angle of 0.4 mrad, then about half of the TR photons are lost for electrons. The angular distributions for electrons and muons in the range above 0.5 mrad have quite a similar shape (plot (b) in figure 10), however, the number of TR photons is by about a factor of 2 smaller for muons, which is also observed in the data. The two strongest peaks in angular distributions in this range are clearly pronounced around 1 and 2 mrad. This is also seen in the data (see figure 8), however, the calculations predict a significantly stronger peak at $\approx$1 mrad than what is observed in the data.

4. Conclusions

Simultaneous measurement of angular and energy distributions of transition radiation photons were performed for the first time with the help of GridPix detector. No significant difference except in number of photons (by a factor of 1.8) between 20 GeV electrons ($\gamma = 3.9 \times 10^4$) and 180 GeV muons ($\gamma = 1.7 \times 10^3$) was observed. Angular distributions for electrons and muons are slightly different. Theoretical calculations of angular distributions predict a significantly different shape than the one observed in data. Calculations also predict a double-peak structure of the TR photons energy spectrum, which was not observed in data. The sources of all these discrepancies are the subject of future studies.

Acknowledgments

We gratefully acknowledge the financial support from Russian Science Foundation grant (project No. 16-12-10277).

References

[1] Olive K A *et al.* 2014 *Chin. Phys.* C **38** 090001
[2] Boldyrev A S *et al.* 2016 *Nucl. Instrum. Meth.* A **807** 47–55
[3] Kaminski J, Bilevych Y, Desch K, Krieger C and Lupberger M 2017 *Nucl. Instrum. Meth.* A **845** 233–5
[4] Hartjes F, Hessey N, Fransen M, Konovalov S, Koppert W, Morozov S, Romaniouk A, Tikhomirov V and Van der Graaf H 2013 *Nucl. Instrum. Meth.* A **706** 59–64
[5] Llopart X, Ballabriga R, Campbell M, Tlustos L and Wong W 2007 *Nucl. Instrum. Meth.* A **581** 485–94
[6] Krieger C, Kaminski J, Lupberger M and Desch K 2017 *Nucl. Instrum. Meth.* A **867** 101–7
[7] Lupberger M, Desch K and Kaminski J 2016 *Nucl. Instrum. Meth.* A **830** 75–81
[8] Azevedo C D R, Biagi S, Veenhof R, Correia P M, Silva A L M, Carramate L F N D and Veloso J F C A 2015 *Phys. Lett.* B **741** 272
[9] Biagi S 2017 Degrad URL `https://magboltz.web.cern.ch/magboltz/`
[10] Veenhof R 2010 Garfield – simulation of gaseous detectors URL `https://garfield.web.cern.ch/garfield`
[11] Cherry M L 1978 *Phys. Rev.* D **17** 2245

IOP Conf. Series: Journal of Physics: Conf. Series **934** (2017) 012050 doi:10.1088/1742-6596/934/1/012050

Optimising the Active Muon Shield for the SHiP Experiment at CERN

A Baranov[1,2,⋆], E Burnaev[3], D Derkach[1,⋆], A Filatov[1,⋆], N Klyuchnikov[3], O Lantwin[4,⋆], F Ratnikov[1,2,⋆], A Ustyuzhanin[1,2,⋆] and A Zaitsev[3]

[1] National Research University Higher School of Economics, Moscow, Russia
[2] Yandex School of Data Analysis, Moscow, Russia
[3] Skolkovo Institute of Science and Technology, Moscow, Russia
[4] Imperial College London, London, UK
[⋆] on behalf of the SHiP Collaboration

E-mail: `filatovartm@gmail.com`

Abstract. The SHiP experiment is designed to search for very weakly interacting particles beyond the Standard Model which are produced in a 400 GeV/c proton beam dump at the CERN SPS. The critical challenge for this experiment is to keep the Standard Model background level negligible. In the beam dump, around 10^{11} muons will be produced per second. The muon rate in the spectrometer has to be reduced by at least four orders of magnitude to avoid muon-induced backgrounds. It is demonstrated that new improved active muon shield may be used to magnetically deflect the muons out of the acceptance of the spectrometer.

1. Introduction

The Standard Model of elementary particles despite its enormous success, has several places for improvement [1]. In particular, the baryon asymmetry of the Universe [2] and a Dark Matter existence are yet to be explained. The discovery of heavy neutral leptons (HNLs), right handed partners of the Standard Model neutrinos, would solve both above mentioned riddles of Nature. The existence of such particles is strongly motivated by theory and were searched extensively previously. Cosmological constraints on the properties of HNLs now indicate that the majority of the interesting parameter space for such particles was beyond the reach of the previous searches at the PS191 [5], BEBC [7], CHARM [4], CCFR and NuTeV [6] experiments.
For this reasons, a new fixed-target experiment at the CERN SPS accelerator was proposed [8]. This experiment will use a 400 GeV proton beam on a fixed target to produce a large number of charm mesons. The HNLs from charm meson decays have a significant polar angle with respect to the beam direction, approximately 50 mrad on average. In order to maximise the geometric acceptance for a given transverse size of the detector, the detection volume must therefore be placed as close as possible to the target. The production of the charm mesons is accompanied by copious direct production of pions, kaons and short-lived light resonances. The subsequent decays of these particles would result in a large flux of muons and neutrinos. To minimise these decays, a combination of a target and a hadron absorber of a few metres length, both made of as dense a material as possible, is required. To reduce the detector occupancy and backgrounds induced by the residual muon flux, a muon shield is required downstream of the

IOP Conf. Series: Journal of Physics: Conf. Series **934** (2017) 012050 doi:10.1088/1742-6596/934/1/012050

hadron absorber. The experimental set-up must therefore balance the opposing requirements of locating the detector as close as possible to the target and of accommodating a sufficiently long muon shield upstream of the fiducial volume of the detector to reduce muon-induced backgrounds.

2. Bayesian Optimisation

The main goal of our research was to find a light and efficient magnetic shield. In order to achieve this, we applied a Bayesian optimisation algorithm. In this section we give an introduction to this method and motivation behind this approach.
Let $f(x)$ be a black-box function on $X \subset R^d$, which does not have an analytic expression and can be evaluated at a specific point only by conducting computationally intensive simulations, which in turn can yield noisy observations of the form $y_i = f(x_i) + \epsilon_i$, where ϵ_i is the random variable with zero mean and constant variance. Our ultimate goal is to find a point x_* where $f(x_*)$ is a true global optimum (henceforth without loss of generality we will consider a maximisation problem). In this problem setup, the general framework for optimum searching iteratively constructs the sample $D_N = \{(x_i, y_i)\}_{i=1}^{N}$, each iteration consists of three steps:

(i) Choose a new point x_i for simulation
(ii) Conduct simulations to obtain $y_i = f(x_i) + \epsilon_i$
(iii) If a stopping criterion is not met, add (x_i, y_i) to the sample D_N and proceed to iteration $i + 1$

Convergence to the exact solution of the problem may require infinite amount of iterations, however, under some theoretical assumptions any degree of its approximation can be obtained within the corresponding finite horizon.

Theorem 1

If $f(x)$ is Lipschitz-continuous with a Lipschitz continuity constant C, X is a d-dimensional unit hypercube, then in the noise-free case for any $\varepsilon > 0$ some point x_+ with near-optimal true value $f(x_+)$: $f(x_+) \geq \max_{x \in X} f(x) - \varepsilon$ is guaranteed to be found within $(C/2\varepsilon)^d$ iterations. (see [10])

Strong convergence guarantee is provided against the worst-case scenario, however the solution in this case will typically require an impractically large number of simulations for any reasonable accuracy of approximation. Since such scenarios are not plausible in practice and the number of simulations is very limited, one can come to the idea to relax guarantees in order to meet practical limitations, yet insure against average-cases. Bayesian approach to optimisation allows implementing this idea by providing specific choosing strategies (step i) in the framework for optimum search. In this approach, the objective function is approximated with *a surrogate model*, that is, a parametric function for which prior knowledge of parameters combined with likelihood of the observed sample defines a posterior distribution over the parametric family. Choice of new points to simulate is guided by *the principle of minimum expected risk* (its formalisation under the Bayesian framework has been provided in [9]) resulting in a secondary optimisation problem, which in turn depends not on the original function, but on its computationally cheaper surrogate-model.
The most popular choice for building surrogate models in the field of engineering design is based on taking *Gaussian processes* as a parametric family of priors along with assuming there is white noise in the observations $\epsilon_i = \mathcal{N}(0, \sigma^2)$. By definition [11, 22] Gaussian process is a collection of random variables, any finite number of which have a joint Gaussian distribution. Consequently it is completely specified by its mean function $m(x) = \mathbb{E}[f(x)]$ and the covariance function $k(x, x') = \mathbb{E}[(f(x) - m(x))(f(x') - m(x'))]$. The most important property of Gaussian processes

for Bayesian optimisation is that posterior takes also a form of Gaussian process, moreover its mean and covariance functions have analytic expressions:

- $\hat{f}_N(x) = \mathbb{E}[f(x)|D_N] = m(x) + \mathbf{k}(x)\mathbf{K}^{-1}(y_N - \mathbf{m})$,
- $\hat{\sigma}_N^2(x) = \mathbb{V}[f(x)|D_N] = k(x,x) - \mathbf{k}(x)\mathbf{K}^{-1}\mathbf{k}(x)^T$,

where $\mathbf{m} = \{m(x_i)\}_{i=1}^N$, $\mathbf{k}(x) = \{k(x,x_i) + \sigma^2\delta(x,x_i)\}_{i=1}^N$, $\mathbf{K} = \{k(x_i,x_j) + \sigma^2\delta(x_i,x_j)\}_{i,j=1}^N$
Several conditions on the family of priors that ensure convergence of the Bayesian optimisation to the optimum are specified in [12], they are satisfied by a Gaussian process with a constant mean function. Therefore, the convergence is guaranteed when the principle of minimum expected risk is used, as for instance an expected deviation from the global optimum:

$$x_{i+1} = \underset{x}{\text{argmin}}\ \mathbb{E}[||f(x) - f(x_*)|||D_i] \tag{2.1}$$

Expression 2.1 is computationally expensive since it requires estimation of $f(x_*)$, thus in present work we replace this criterion with its common approximation called *Expected Improvement* [13]:

$$x_{i+1} = \underset{x}{\text{argmin}}\ \mathbb{E}[\max\{0, f(x) - f(x_+)\}|D_i], \tag{2.2}$$

where $x_+ = \underset{x_i}{\text{argmax}}\ f(x_i)$. Expected Improvement has also been shown to converge under additional mild assumptions [14], but its main advantage is a closed-form expression, that doesn't require numerical integration:

$$\mathbb{E}[\max\{0, f(x) - f(x_+)\}|D_i] = \hat{\sigma}_N^2(x)Z\Phi(Z) + \hat{\sigma}_N^2(x)\phi(Z), \tag{2.3}$$

where $Z = \frac{\hat{f}_N(x) - f(x_+)}{\hat{\sigma}_N^2(x)}$, ϕ and Φ are PDF and CDF of the standard normal distribution respectively.

3. SHiP shield optimisation

The muon shield is a critical component of the SHiP experiment, which deflects the high flux of muons produced in the target, that would represent a very serious background for the particle searches, away from the detector. The shield consists of eight magnets and each magnet is parameterised by seven values: length, width, etc... Because the cost of the muon shield is significant we include the weight of the muon shield as a proxy for the cost. Therefore, our aim is to find the most efficient solution at a lowest cost possible. We apply Bayesian optimisation techniques to solve this problem. The optimisation is performed for already chosen material, thus the material balance is discussed elsewhere [15].
For the evaluation of the shield performance we use 18 million simulated events passed through the muon shield and measured at a scoring plane 64m downstream of the muon shield at the z of the first tracking station of the spectrometer. All other subsystems downstream of the muon shield are removed.
Proton fixed target collisions and inelastic muon interactions are simulated using PYTHIA8 [16] and PYTHIA6 [18] respectively.
Heavy flavour cascade production is also taken into account [19]. The SHiP detector response is simulated in the GEANT4 [20] framework. The simulation is done within the FAIRROOT framework [21].
The transverse (x, y) position of muons is obtained at the first tracking station. For muons with $|y| < 5$m, their x-position is converted into "importance weight" $\sigma_{\mu^\pm} = \sqrt{(260 \mp x_\mu/\text{cm})/560}$ for $-300 < x_\mu < 260$ and $-260 < x_\mu < 300$cm for positive and negative muons respectively, else $\sigma_{\mu^+} = 0$ [15].

IOP Conf. Series: Journal of Physics: Conf. Series **934** (2017) 012050 doi:10.1088/1742-6596/934/1/012050

Now we can introduce the complete loss function, that depends on the physical performance of the shield, its weight, and the length implicitly via the weight.

$$L(\Sigma, W) = (1 + \Sigma)(1 + \exp^{\frac{W - W_0}{W_0}}),$$

where $\Sigma = \sum_{\mu} \sigma_{\mu}$, W is a weight of configuration and W_0 is a weight of a baseline configuration [15] shown in Fig.1. The weight of the baseline is about 1900 tons and Σ for the baseline is approximately equal to 32. As can be seen, weight is penalised exponentially because we are not interested in the heavy regions, as we will not be able to construct such configurations due to budget and engineering constraints.

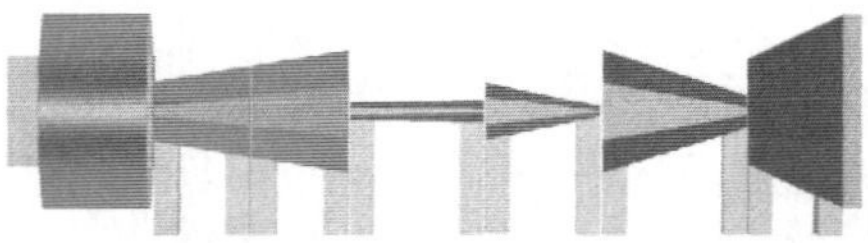

Figure 1: Baseline configuration of the SHiP muon shield, as described in [15].

The following problems are addressed during the optimisation:

(i) Optimisation is performed in 42-dimensional space

(ii) Computation of the Σ is time consuming.

(iii) Computation of the Σ is noisy due to limited statistics in Monte - Carlo simulations.

To ensure a more even coverage of the muon phase space, the number of muons is capped in bins of momentum and transverse momentum, and augmented through resampling available muons in bins with low occupancy. This also reduces the overall number of muons to be simulated to about 500 thousand, and helps us to decrease the time of computations by factor of 8 times in average. We also introduce noise as a prior knowledge into Gaussian Process: in this setup GP tries to estimate the noise of the computation and incorporate it into final variance. The points are computed in batches of 100 points to optimise the usage of the available computing resources. The process of optimisation is illustrated in Fig. 2.

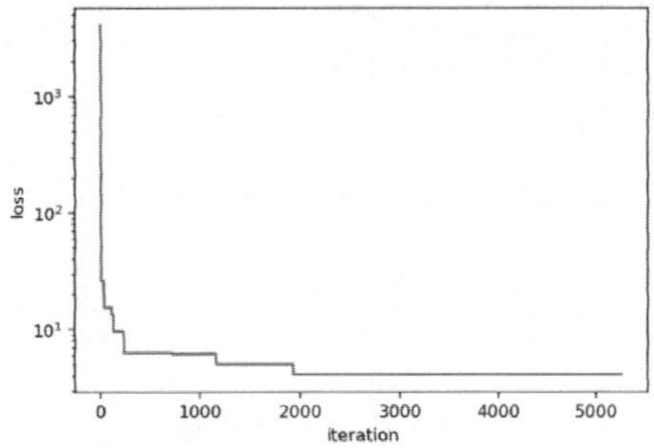

Figure 2: Evolution of the best found configuration with GP iterations.

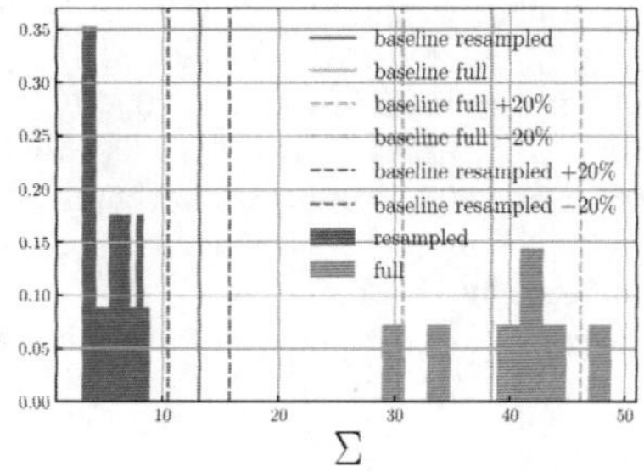

Figure 3: Distributions of rejection performance scores for resampled and full muon samples for the best found shield configuration, comparing with the baseline solution.

The optimisation procedure is stopped after 5000 iterations. The obtained configuration is found to be lighter by 25 % than the baseline while having about the same rejection capability as illustrated in Fig. 3. Fig. 4 illustrates the new found configuration.

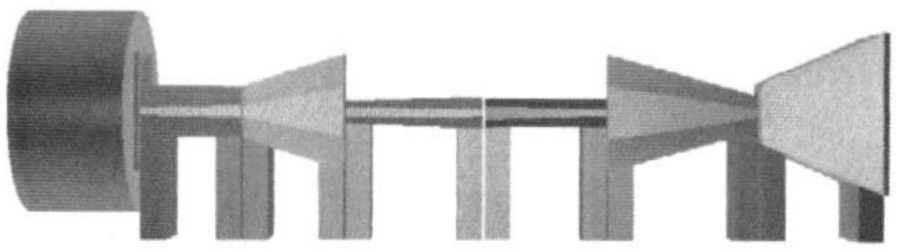

Figure 4: The best found muon shield configuration for the SHiP detector

4. Conclusion

Bayesian optimisation is a powerful tool which can be applied for the optimisation of non differentiable functions. Even though Gaussian processes can guarantee global optionality under ideal conditions, these are not satisfied under resouce restrictions constraints. However we have demonstrated that this method can be successfully applied even to complicated optimisation problems in physics.

References

[1] Ellis J *et. al* 2009 *Phys. Lett.* B **679** 369
[2] Asaka *et. al* 2005 *Phys. Lett.* B **620** 17
[3] Asaka *et. al* 2005 *Phys. Lett.* B **631** 151
[4] Bergsma F *et. al* 1986 *Phys. Lett.* B **166** 437
[5] Bernardi G *et. al* 1986 *Phys. Lett.* B **166** 479
[6] Vaitaitis A *et. al* 1999 *Phys. Rev. Lett.* **83** 4943
[7] Cooper-Sarkar *et. al* 1985 *Phys. Lett.* B **160** 207
[8] Bonivento W *et. al* 2013 *Preprint* 1310.1762
[9] Burnaev E and Panov M 2015 *Lecture Notes in Computer Science* vol 9047, ed A Gammerman, V Vovk and H Papadopoulos (Cham:Springer) pp 116–125
[10] Betrò B 1991 *J. of Glob. Opt.* **1** 1
[11] Rasmussen C E and Williams C K I 2006 *Gaussian Processes for Machine Learning* (University Press Group Limited)
[12] Mockus J 1994 *J. of Glob. Opt.* **4** 347
[13] Mockus J, Tiesis V and Zilinskas A 1978 *Towards Global Optimization* **2** (Elsevier Science Ltd.) pp 117–129
[14] Vazquez E and Bect J 2010 *J. of Stat. Plan. and Inf.* **140** 3088
[15] Akmete A *et al.* 2017 *JINST* **12** P05011
[16] Sjöstrand T, Mrenna S, and Skands P 2008 *Comput.Phys.Commun.* **178** 852
[17] Andreopoulos C *et al* 2010 *Nucl.Instrum.Meth.* **A614** 87
[18] Sjöstrand T, Mrenna S, and Skands P 2006 *JHEP* **05** 026
[19] Dijkstra H and Ruf T 2015 SHiP-NOTE-2015-009 (Geneva: CERN)
[20] Agostinelli S *et al.* 2003 *Nucl.Instrum.Meth.* **A506** 250
[21] Al-Turany M, Bertini D, Karabowicz R, Kresan D, Malzacher P, Stockmanns T and Uhlig F 2012 *J.Phys.Conf.Ser.* **396** 022001
[22] Burnaev E, Panov M and Zaytsev A 2016 *J. Commun. Technol. Electron.* **61** 661

Improving of RPC for Muon System of CMS experiment

E Voevodina[1,2] on behalf of the CMS Muon Group

[1] Physics Department, University âĂIJFederico IIâĂİ, via Cinthia 21, Napoli, 80126, Italy
[2] INFN Napoli,via Cinthia 21, Napoli, 80126, Italy

E-mail: elena.voevodina@cern.ch

Abstract. In the Endcap regions, CMS Muon spectrometer is using Cathode Strip Chambers (CSCs) as muon tracking and trigger detectors, also Resistive Plate Chambers (RPCs) serve as dedicated trigger detectors and improve the muon reconstruction by providing the excellent timing resolution for identification of muon particles. At the present, the four Endcap discs are not fully equipped: RPCs are covering only Endcap disks up to $|\eta| = 1.8$. During the Long Shutdown 3 (from 2023 to 2026) of the Large Hadron Collider, Endcap stations 3 and 4 will be further instrumented with new improved RPCs (iRPCs) that will be covering the pseudorapidly region $1.8 < |\eta| < 2.4$, increasing the redundancy in this region. Nowadays, the final design of RPC chambers has been developed and the RPC double gap prototypes with a gas thickness 1.4mm are being tested. The performance and stability of iRPCs at the future HL-LHC upgrades have been studied (detection efficiency, cluster size, rate capability) at the Gamma-Irradiation Facility (GIF++) at CERN, where high energy charged particle beams (mainly muons) are combined with gammas from a 14 TBq 137Cesium source which simulates the background expected at the CMS experiments. In this paper, the main results obtained during the test at GIF ++ are presented.

1. Introduction

In 2013, the program upgrade of the LHC, which was called High Luminosity of Large Hadron Collider (HL-LHC), was approved [1]. In HL-LHC era, the luminosity will raise from 1.5 - 2 up to 5×10^{34} cm^{-2} s^{-1}, i.e. the integrated luminosity will increase around threefold with respect to the original design values. For LHC upgrade, two long shutdown periods are scheduled to give the machine and the experiments the necessary time to anticipate these luminosity increases: Long Shutdown 2 (LS2) in 2019/2020 and Long Shutdown 3 (LS3) in 2023/2026 [1]. During these long shutdown periods the CMS Collaboration intends to hold up huge work to update and improve all systems in order to keep the high particle detection efficiency in view of the HL-LHC era.

The Muon System of CMS experiment will undergo a heavy upgrade. In particular, Endcaps region will be extended in both sides, where new muon detectors will be installed: Gas Electron Multiplier and improved Resistive Plate Chambers (iRPCs) covering the pseudorapidity from 1.8 to 2.4 as shown in figure 1 [2]. At present, pseudorapidity region of present Muon System of CMS is covered by Cathode Strip Chambers only and lacks of redundant coverage despite the fact that it is a challenging region for muons in terms of backgrounds and momentum resolution. In order to maintain good efficiency for the muon trigger in this region additional new iRPCs

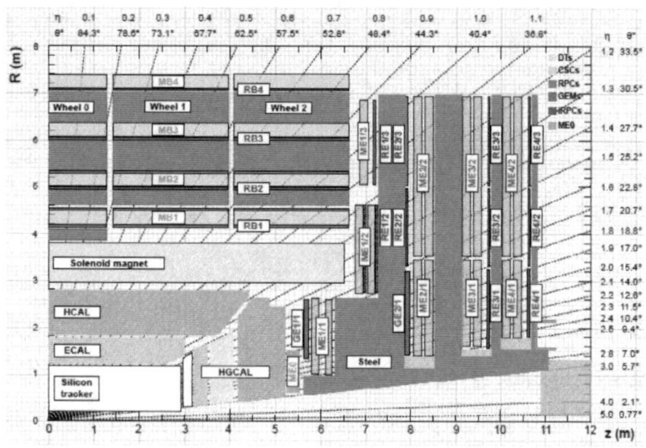

Figure 1. A quadrant of the CMS experiment, where improved RPCs are labeled with RE3/1 and RE4/1.

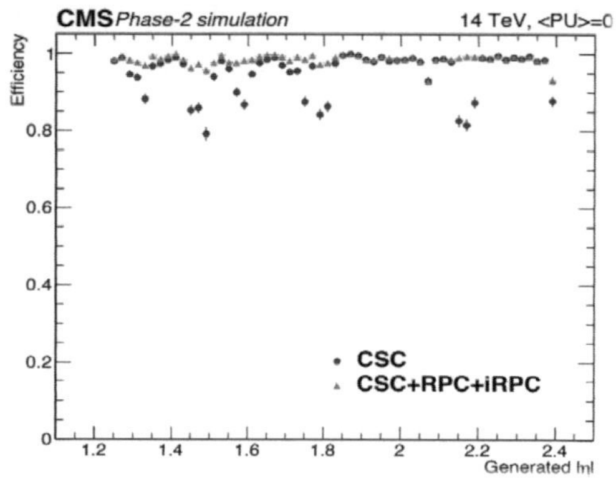

Figure 2. Simulated efficiency of the trigger primitive stubs at the level of 3-/4- station with and without the addition of the RPCs.

are planned to be installed. These iRPCs will have a good timing resolution and a special resolution with few centimeters to mitigate background effects and to increase the redundancy of the system as shown in figure 2.

The first large size iRPC prototype for high eta region was built at the Korea Detector Laboratory (KODEL) and was tested at the Gamma Irradiation Facility (GIF++) at CERN. The gap size as well as the electrode thickness is equal to 1.4mm, allowing to decrease the operating high voltage. The loss in gas gain will then be compensated by the higher signal amplification of improved front-end electronics. The lower deposited charge per crossing particle will also reduce the integrated charge and slow down the aging.

2. Testing of the large size iRPC's prototype at GIF++ CERN

iRPCs will be installed in the most forward region of the Muon system, they will be located very close to the LHC beam pipe, thus, both the chambers and readout electronics will be subjected to high radiation. It is estimated that in these regions the total expected background rate will be the order of 0.6 kHz/cm^2. In this harsh radiation environment, new RPC stations must have a stable and high performance. Such a value of rate is equivalent to an integrated charge of approximately 1 C/cm^2 during the lifetime of the detectors assuming a mean charge deposition $< q >$ per avalanche of 20 pC and a safety factor 2 [3]. In order to study their performance and stability, improved RPCs have been tested at GIF++.

The GIF++ was specially designed in collaboration of the $CERN's$ Engineering and Physics Department to test the future detectors for HL-LHC program. GIF ++ consists of a 14 TBq 137 Cs gamma source (gamma with an energy of 662 keV corresponding to the typical energy of the neutron âĂŞ induced background radiation in the CMS experiment) with two sets of adjustable filters to change the intensity and a high-energy muon beam with momentum up to 100 GeV/c from the secondary SPS beam line H4 in EHN1 [4]. RPC Collaboration of the CMS experiment is actively using the GIF++ to investigate efficiency, cluster size, multiplicity and so on of improved detectors under different background conditions.

In May 2017 during the test beam at SPS, a large size prototype of iRPC from KODEL, having a trapezoidal shape (larger base 92 cm, lower base 63 cm and height 166.3 cm) was tested at GIF++ CERN. The prototype was equipped with a strip panel with 2 cm wide strips, 96 in total. To read the charge deposited on the strips, the prototype has been equipped with

IOP Conf. Series: Journal of Physics: Conf. Series **934** (2017) 012051 doi:10.1088/1742-6596/934/1/012051

front-end electronics, using commercial amplifier chips and a threshold ranging in between 1 mV ~ 160~180 fC and 0.3 mV ~ 50 fC which was tested in KODEL. The front-end electronics threshold has been fixed of 300 μV. The prototype iRPC chamber was placed at a distance about 2m form ^{137}Cs source. Several High Voltage (HV) scans have been performed using different radiation conditions (with and without radiation source), to define the optimal operating voltage, which is referred to as the working point (WP).

3. Determination of efficiency and cluster size for iRPC's prototype

During the test beam that took place in May 2017 at GIF ++, a beam trigger was used to determine of muon detection efficiency. The beam trigger consists of three scintillators, the first with a size of 45 cm by 45 cm and located upstream outside of the bunker, the second of the same size located downstream, outside of the bunker, and the last one a small scintillator of size 15 cm by 15 cm located in front of chamber (see the figure 3).

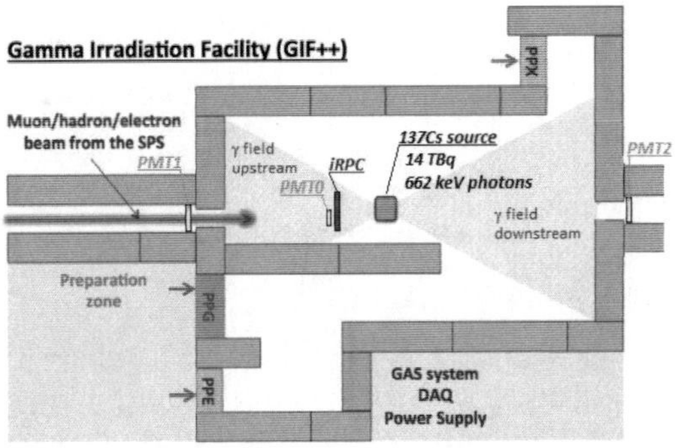

Figure 3. Principal scheme of muon test beam configuration at GIF++, where iRPC is indicated with blue box and the trigger system consisting in a series of three scintillators (scintillators PMT0, PMT1 and PMT2) is indicated with a yellow box.

In this case, efficiency (E) for different background radiation levels is calculated using the appropriate formula:

$$E = \frac{N_{quadruple}}{N_{trigger}},$$

$whereN_{quadruple}$ corresponds to the number of muons, which were detected by the chamber and scintillators and $N_{trigger}$ is the number of events, which were detected by scintillators only. The efficiency of the RPC detectors strongly depends not only on the applied high voltage, but on the environment. Using the environmental control station inside GIF ++ bunker, the temperature and pressure were determined. Thus, due to temperature and pressure variations, the applied high voltage (HV_{app}) needs to be corrected to obtain the effective voltage (HV_{eff}). The correction is performed according to the formula below [5]:

$$HV_{app} = \beta HV_{eff} = HV_{eff}[(1-\alpha) + \alpha\frac{P}{P_0}\frac{T_0}{T}],$$

where $\alpha = 0.8$ and $T_0 = 293K$, $P_0 = 990mbar$ are the reference temperature and pressure. T and P are measured during the test using the environmental control station. Usually for each RPC chamber a HV scan is performed in order to determine the optimal working point. The HV scan allows to measure the efficiency as a function of the effective voltage. This dependency can be fitted using the sigmoidal curve described by the subsequent formula[6]:

$$<E>= \frac{E_{max}}{1+e^{-\lambda(HV_eff-HV_{50})}},$$

where E_{max} is the maximum efficiency reached by chambers at $HV \longrightarrow \infty$ and λ is proportional to the slope of sigmoid at inflection. The HV_{50} used in the fitting sigmoid function is defined as the high voltage at which every roll reaches 50% of the efficiency plateau. The WP is defined as $WP = Knee + 120V$ for every RPC chamber[6]. The Knee is the high voltage point on the sigmoid where the efficiency is 95% of the efficiency plateau, shown in figure 4. After the determination of the optimal WP at which the iRPC is stable and efficient, the cluster size is calculated. A consecutive set of strips, each collecting an induced charge defines a cluster. The number of the strips in the cluster gives the Cluster size, it was indicated blue dots in figure 4.

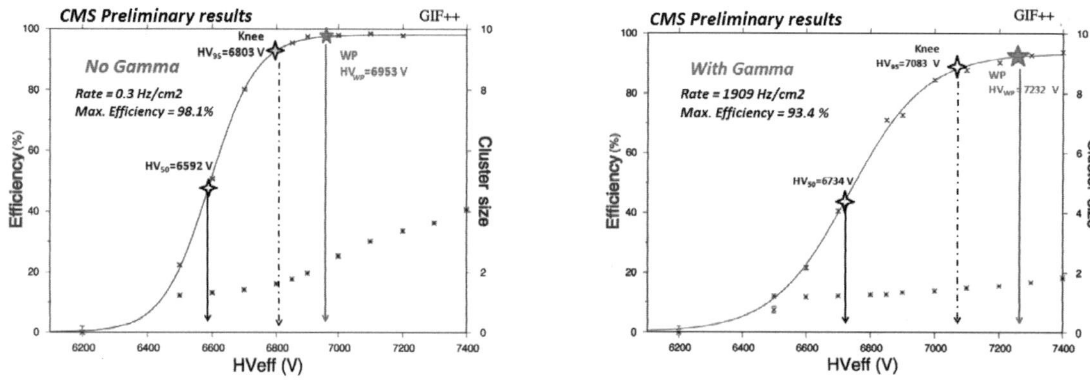

(a) (b)

Figure 4. Efficiency and average cluster size of large size double gas gap iRPC prototype with thickness of 1.4mm as a function of the effective voltage, tested without gamma background (a) and with gamma background rate of 1.91 kHz/cm^2 (b). The data were measured at the fixed threshold of 300 μV.

Conclusions

During the third long shutdown of LHC 2023/2026, the RPC collaboration will build and install 36 improved RPC chambers with thinner gas gaps and new electronics which will cover the region of $1.8 < |\eta| < 2.4$ in Endcap. Nowadays RPC collaboration is actively working on developing RPC chambers for this highly irradiated region. Using the unique capabilities of the Gamma Irradiation Facility at CERN, the performance of our detectors can be studied with a muon beam under different background conditions. In this paper, the first results of the testing of the large size iRPCs prototype were presented. This detector was successfully tested in a high radiation environment at GIF++. The chamber has stable working efficiency of 93.4% (at gamma rate= $1.91kHz/cm^2$) and the average cluster size is persistently below 2, which is in good agreement with the expectations [7] and the performance of all the present RPCs from the CMS Muon System. Work is ongoing and a second prototype iRPC has already been constructed to investigate of new low-noise front-end electronics.

Acknowledgments

Special thanks go to Kyong Sei Lee and Sung Keun Park that provided a strong basis and the material needed to construct this article. I would like to give a really big thank for help all my colleagues from the RPC collaboration and the CMS muon system.

References

[1] Schmidt B 2016 *J. Phys.: Conf. Ser* **706** 022002

[2] CMS Collaboration 2015 Technical Proposal for the Phase-II Upgrade of the CMS Detector CERN-LHCC-2015-010
[3] Fagot A *et al.* 2016 *JINST* **11** C09017
[4] Pfeiffer D, Gorinec G, Reithler H *et al.* 2017 *arXiv:*1611.00299v3
[5] Abbrescia M *et al.* 1995 *Nucl. Instr. Meth.* **359** 603
[6] Thyssen F *et al.* 2013 *J. Instr.* **7** C01104
[7] CMS collaboration The Muon Project, CMS Technical Design Report CERN-LHCC-97-32

Commissioning of the first chambers of the CMS GE1/1 muon station

Martina Ressegotti[1,2] on behalf of the CMS Muon Group

[1] Unversity of Pavia, via Bassi 6, Pavia, IT
[2] INFN Pavia, via Bassi 6, Pavia, IT

E-mail: martina.ressegotti@cern.ch

Abstract. The upgrades of the LHC planned in the next years will increase the instantaneous luminosity up to $5 \times 10^{34}\ cm^{-2}s^{-1}$ after Long Shutdown 3, a value about five times higher than the nominal one for which the CMS experiment was designed. The resulting larger rate of interactions will produce a higher pileup environment that will challenge the trigger system of the CMS experiment in its original configuration, in particular in the endcap region. As part of the upgrade program of the CMS muon endcaps, additional muon detectors based on Gas Electron Multiplier (GEM) technology will be installed, in order to be able to sustain a physics program during high-luminosity operation without performance losses. The installation of the GE1/1 station is scheduled for Long Shutdown 2 in 2019–2020; already a demonstrator composed of five superchambers has been installed during the *Extended Year-End Technical Stop* at the beginning of 2017. Its goal is to test the system's operational conditions and also to demonstrate the integration of the GE1/1 chambers into the CMS online system. The status of the installation and commissioning of the GE1/1 demonstrator is presented.

1. Introduction

The CMS muon system [1] employs several detection technologies: Drift Tubes (DTs) and Cathode Strip Chambers (CSCs) are present in the endcaps in the region $1.0 < |\eta| < 2.4$ to provide precise position measurements and trigger, while Resistive Plate Chambers (RPCs) are installed up to $|\eta| < 1.8$ both in the barrel and in the endcaps to provide redundant trigger and coarse position measurement. In addition, among other future upgrades the installation of a new station, the GE1/1 station, based on Gas Electron Multiplier (GEM) technology [2] is scheduled in 2019–2020 in the region $1.6 < |\eta| < 2.2$.

1.1. Triple-GEM detectors

The fundamental element of a GEM detector is the GEM foil, a 50 μm thick polymer foil coated with copper on each side and covered with biconical holes arranged in a regular pattern. A triple-GEM detector is composed of a stack of three GEM foils, completed at its ends by a drift cathode and a readout plane, as shown in figure 1. It is operated by applying an appropriate potential to the seven conductor surfaces, so that the electric field between the foils causes the electrons to drift toward the next GEM foil and the field inside the holes causes electron avalanche multiplication. The signal produced on the readout plane is collected by an appropriate electronic.

It has been measured that triple-GEM detectors have a high rate capability up to 100 MHz/cm^2 and a muon detection efficiency of about 97% or more, hence they are suitable for the operation in the environment of the forward muon region.[3]

The voltage can be applied to each detector's stage either using one single high voltage channel per triple-GEM detector and distributing it with a divider, or by supplying the seven high voltage channels independently from each other. The naming convention used for the seven channels is the following: the surface of the first GEM foil facing the drift cathode is named *G1top* (*GEM 1 top*), the one facing the readout plane is named *G1bot* (*GEM 1 bottom*). Similarly the surfaces of the other GEM foils are named *G2top*, *G2bot*, *G3top*, *G3bot*.

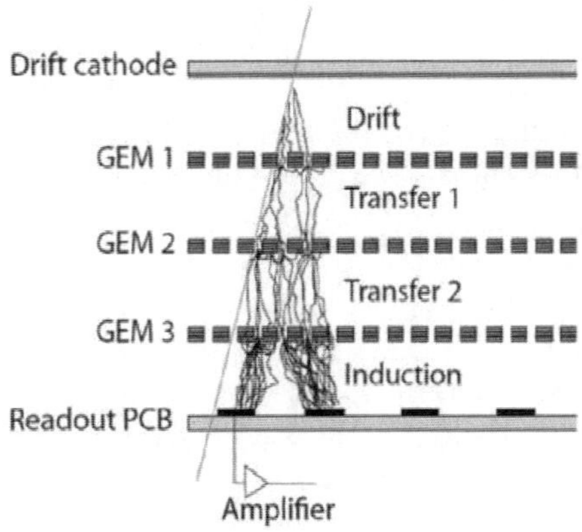

Figure 1: Schematic representation of a triple-GEM detector. From the top down, the drift cathode, the three GEM foils and the readout plane are encountered. The sorrounding volume is filled with an $Ar : CO_2$ gas mixture. The voltage is applied to the drift cathode and to the top and bottom surfaces of each GEM foil. In the figure the detection of a particle and the subsequent production of charge is also represented.

1.2. The GE1/1 station

The GE1/1 station consists of 36 chambers per endcap, called *Gemini*, each one spanning 10°. Each Gemini will be composed of a stack of two triple-GEM detectors, called *Layers*.

Its installation is motivated by the increase in luminosity, up to about $5 \times 10^{34}\ cm^{-2}s^{-1}$, that the Large Hadron Collider (LHC) has scheduled in the next years. As a consequence, the background rate in the $1.6 < |\eta| < 2.2$ region is expected to reach about 1000 Hz/cm^2, so that achieving an acceptable L1 trigger rate for muons with transverse momentum $p_T < 25\ GeV$ will not be possible with the current muon system configuration without increasing the threshold on muon momentum.

The GE1/1 station will allow keeping a trigger rate smaller than 5 kHz without increasing such threshold, by adding redundancy in front of the existing ME1/1 station and working combined with CSCs to measure the muon bending angle in the magnetic field.[3]

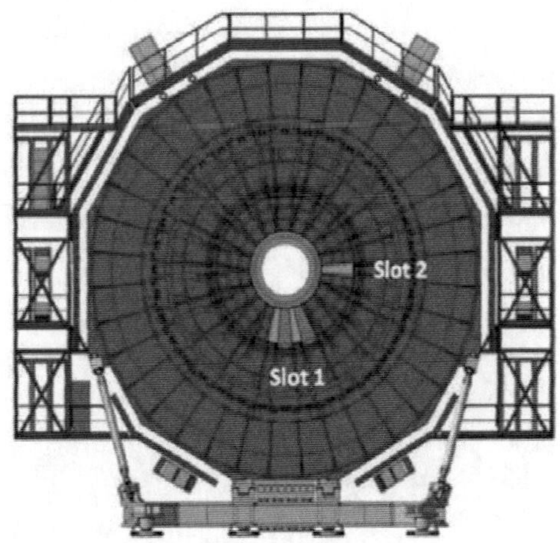

Figure 2: Outline of the negative endcap of the CMS experiment showing the positions where the five Gemini chambers of the Slice Test have been installed. Four of them are placed almost vertically in the portion indicated as *Slot 1*, one of them is located horizontally in the *Slot 2*.

2. The GE1/1 Slice Test

Even if the full installation of the GE1/1 station is scheduled in 2019–2020, five Gemini chambers have already been installed in the CMS experiment at the beginning of 2017 as shown in figure 2. They consist of the GE1/1 *Slice Test*, whose goal is to prove the system's operational conditions and to demonstrate its integration into the CMS online system.

3. The High Voltage system

The high voltage is supplied in two different ways: four Gemini chambers are supplied with a CAEN A1526N module (one HV channel per Layer), and a ceramic divider is used to distribute the voltage to each detector's stage; one Gemini chamber is supplied with a CAEN A1515TG module (one HV module per Gemini chamber) providing seven independent HV channels per Layer. The latter method will be used for the production chambers for the full GE1/1 system.

The stability of the high voltage provided with both methods has been evaluated in the first months of operation, both with and without collisions. Time intervals of 7 and 12 hours during collisions have been taken into account; time intervals of 7 and 10 days without collisions have been considered. In all cases an overall stability $< 1\%$ or better has been observed. As an example, figures 3 and 4 show the monitored voltage during collisions for a Layer supplied with a CAEN A1526N and a CAEN A1515TG module respectively.

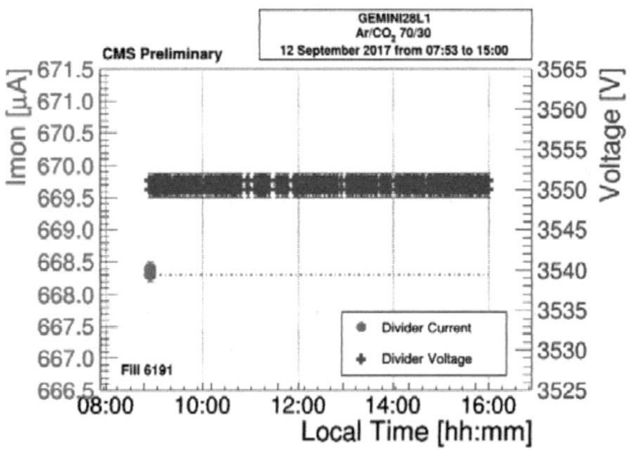

Figure 3: Measured voltage (crosses) and current (dots) through the divider for one Layer supplied with a CAEN A1526N module, in a time period of about 10 hours with collisions. Each marker corresponds to a value that has been archived into the database. Values are archived only if a change is observed, so time intervals without markers indicate that the value has remained unchanged during that time interval. The dashed line is added to complete such time intervals.

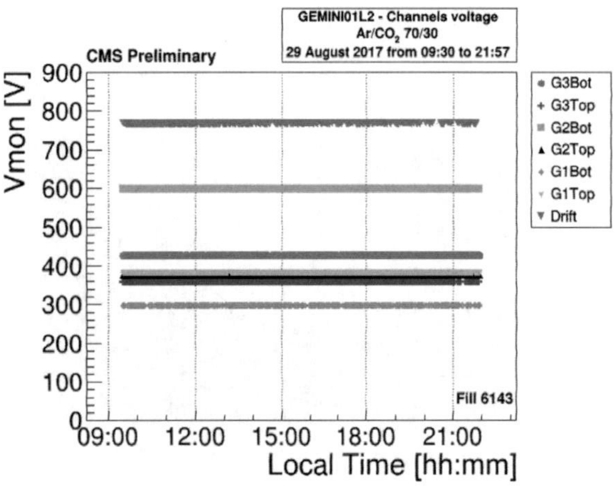

Figure 4: Measured voltage during 12 hours time interval with collisions, for one Layer supplied with a CAEN A1515TG module. Each curve corresponds to a voltage applied in a different part of the detector. They are named *Drift*, *G1top*, *G1bot*, *G2top*, *G2bot*, *G3top*, *G3bot* following the naming convention explained in figure 1.1.

4. The Readout and Low Voltage system

The readout system of the slice test is based on VFAT2 chips and OptoHybrids V2B (OH). For the production chambers VFAT3 chips will be used instead of VFAT2. In addition 8 optical fibers for data flow and control are used per Layer.[4, 5] VFAT2 chips are powered with about 3.3 V, while OHs need two low voltage channels of about 4 V and 1.7 V. As a consequence three low voltage channels are necessary per detector Layer.

As an example, figure 5 shows the monitored voltage and current of one of the LV channels powering the VFAT2 chips. Two different modes are clearly visible: when the VFAT2 chips are powered but in *sleep mode* the current is about 2 A, when they are in *running mode* the current is higher, up to about 6.5 A. Figure 5 shows an overall stability over 10 days.

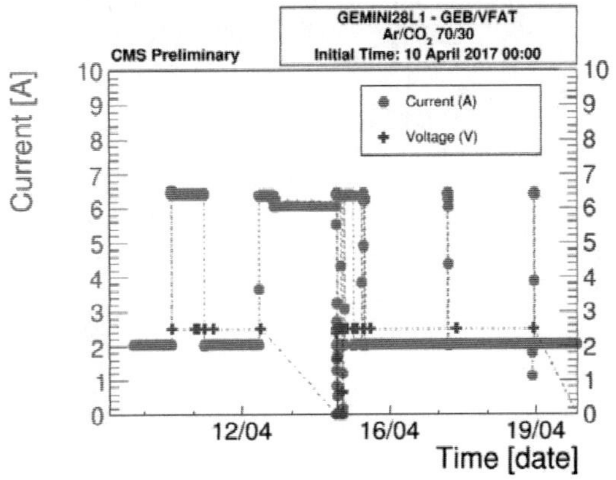

Figure 5: Measured voltage (crosses) and current (dots) supplying the VFAT2 chips of one Layer. Each marker corresponds to a value that has been archived into the database. Values are archived only if a change is observed, so time intervals without markers indicate that the value has remained unchanged during that time interval. Initially the VFATs were in *sleep mode* and moved to the *running mode* on April 11 at about 12 pm. On April 14 both curves assume zero values because the LV has been off for a short time.

5. System Calibration and Timing Performance

Several steps are performed to calibrate the electronics.

Threshold scans scan the number of measured hits per channel as a function of applied threshold.

S-curves scan the response of the channels to an injected pulse calibrated to a given charge at a given threshold. It indicates at which amplitude of the calibration pulse a signal becomes visible, i.e. a conversion between the threshold and the charge, to evaluate the equivalent noise charge of the system. As the threshold value of each strip can be adjusted using programmable registers, this scan allows the perform such a correction in order to level out the response of the channels. Figure 6 shows an example of S-curves performed on a VFAT before and after trimming the threshold.

Latency scans scan the ratio of events with detected hits over the total number of events, per different latency values. The latency is the time difference between the time of arrival of a L1Accept (L1A) and the time at which the related event was stored.

After such calibration, it has been possible to detect the first real muon data. Figure 7 shows the arrival time of some events attributed to the detection of a cosmic ray muon.

Summary

The installation of the GE1/1 station, based on GEM technology, in the CMS muon system has been scheduled in order to allow to maintain an acceptable trigger rate after the LHC upgrades. A Slice Test composed of five Gemini chambers has started at the beginning of 2017 and is currently under commissioning. The operation and stability of the high voltage and low voltage systems have been verified, the electronics have been successfully calibrated. It has been possible to successfully detect first muon data, coming from cosmic ray muons. Several other aspects being part of the commissioning of the slice test have not been covered here: for example the gas system, cooling system, cable routing and other necessary services have been installed and are working properly. In addition the production of GE1/1 chambers for the full installation of the GE1/1 station is currently in full swing.

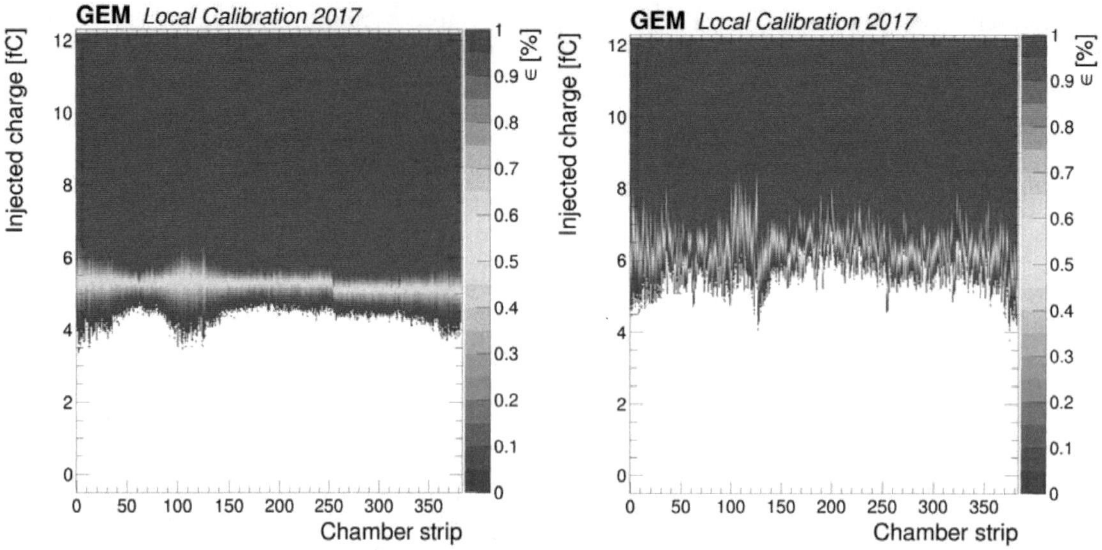

Figure 6: S-curves performed on two VFATs reading one sector [3] of a Layer, before and after adjusting the thresholds of each channel. Before trimming the channels display a dispersion of the 50% of hit-per-pulse ratio, indicating that the effective threshold is not constant across the chips. After trimming the channels display a reduced dispersion of the 50% of hit-per-pulse ratio around the average one.

Acknowledgments

We gratefully acknowledge support from FRS-FNRS (Belgium), FWO-Flanders (Belgium), BSF-MES (Bulgaria), BMBF (Germany), DAE (India), DST (India), INFN (Italy), NRF (Korea), QNRF (Qatar), and DOE (USA).

References

[1] CMS collaboration 2008 The CMS experiment at the CERN LHC JINST 3 S08004
[2] Sauli F 1997 GEM: A new concept for electron amplification in gas detectors *Nucl. Instrum. Meth.* **A** 386(1997)531
[3] Colaleo A et al. 2016 CMS Technical Design Report for the Muon Endcap GEM Upgrade CERN-LHCC-2015-012 CMS-TDR-013
[4] Aspell P 2007 VFAT2: A front-end system on chip providing fast trigger information, digitized data storage and formatting for the charge sensitive readout of multi-channel silicon and gas particle detectors *Topical Workshop on Electronics for Particle Physics (TWEPP07)*, Prague, Czech Republic, 3-7 September 2007 p 292
[5] Lenzi T 2017 A micro–TCA based data acquisition system for the Triple-GEM detectors for the upgrade of the CMS forward muon spectrometer *Journal of Intrumentation* JINST **12** C01058

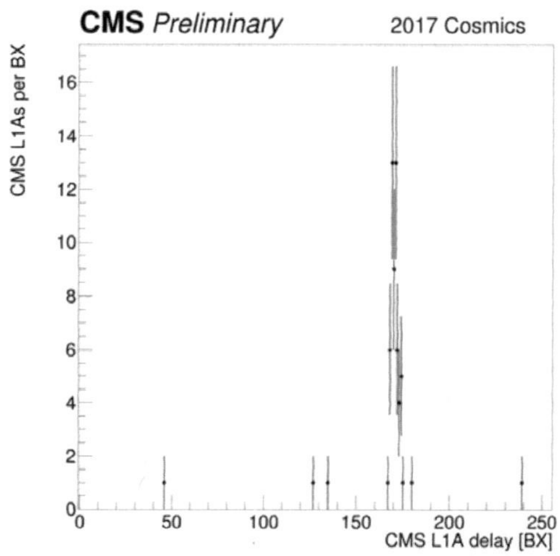

Figure 7: Number of measured hits, integrated over all VFATs, as a function of the delay from a L1A for cosmic ray muon data. The observed delay (175 BX) corresponds to the expected one for cosmics ray muon data.

Test beam studies of possibilities to separate particles with gamma factors above 10^3 with straw based Transition Radiation Detector

N Belyaev[1], M L Cherry[2], S A Doronin[1,3], K Filippov[1], P Fusco[4,5], S Konovalov[3], D Krasnopevtsev[1], V Kramarenko[6], F Loparco[4,5], M N Mazziotta[5], D Ponomarenko[1], D Pyatiizbyantseva[1], R Radomskii[1,3], C Rembser[7], A Romaniouk[1], A Savchenko[1], E Shulga[1], S Smirnov[1], Yu Smirnov[1], V Sosnovtsev[1], P Spinelli[4,5], P Teterin[1], V Tikhomirov[1,3], K Vorobev[1] and K Zhukov[3]

[1] National Research Nuclear University MEPhI (Moscow Engineering Physics Institute), Kashirskoe highway 31, Moscow, 115409, Russia
[2] Dept. of Physics & Astronomy, Louisiana State University, Baton Rouge, LA 70803 USA
[3] P. N. Lebedev Physical Institute of the Russian Academy of Sciences, Leninsky prospect 53, Moscow, 119991, Russia
[4] Dipartimento di Fisica "M. Merlin" dell' Università e del Politecnico di Bari, I-70126 Bari, Italy
[5] Istituto Nazionale di Fisica Nucleare, Sezione di Bari, I-70126 Bari, Italy
[6] Skobeltsyn Institute of Nuclear Physics Lomonosov Moscow State University, Moscow, 119991, Russia
[7] CERN, the European Organization for Nuclear Research, CH-1211 Geneva 23, Switzerland

E-mail: `Vladimir.Tikhomirov@cern.ch`

Abstract. Measurements of hadron production in the TeV energy range are one of the tasks of the future studies at the Large Hadron Collider (LHC). The main goal of these experiments is a study of the fundamental QCD processes at this energy range, which is very important not only for probing of the Standard Model but also for ultrahigh-energy cosmic particle physics. One of the key elements of these experiments measurements are hadron identification. The only detector technology which has a potential ability to separate hadrons in this energy range is Transition Radiation Detector (TRD) technology. TRD prototype based on straw proportional chambers combined with a specially assembled radiator has been tested at the CERN SPS accelerator beam. The test beam results and comparison with detailed Monte Carlo simulations are presented here.

1. Introduction

Study of hadron production at the Large Hadron Collider (LHC) at small angles with respect to the beam is being actively discussed now (see, for example, [1]). Apart from a better understanding of the fundamental QCD processes, the study of high energy particle production in the forward direction is an extremely important topic for cosmic ray physics. Such measurements could remove uncertainties in physics models explaining particle production with energies up to 10^{17} eV in the Universe (the problem of so-called knee of the cosmic ray spectrum at these energies). The proposed experiment at the LHC is aimed at measurement of the secondary particle composition (protons, kaons, pions, muons and electrons) in the rapidity range of $\sim 5 < |\ \eta\ | < 9$ produced in proton collisions. The energy range of the particles extends from 1 TeV to 6 TeV that corresponds to Lorentz factor value from $\approx 1 \times 10^3$ to $\approx 4 \times 10^4$ for protons, kaons and pions. The only particle identification technique able to effectively separate hadrons with these γ-factors is based on the properties of the X-ray transition radiation (TR) production. Transition radiation detectors (TRD) have been used for accelerator experiments and cosmic ray experiments on the ground, at balloon altitudes, and in space (see reviews, [2, 3, 4]). Most of the transition radiation detectors are designed to separate electrons and pions and they use the threshold effect of the TR production. In these detectors the TR yield starts to be significant at γ-factor of $\sim 5 \times 10^2$ and saturates already at $\sim 2 \times 10^3$. There exist cosmic ray experiments and developments [5] with TRDs for large γ-factor range. They use the fact that TR production is proportional to Z^2 and are aimed to separate nuclei with charges $Z > 1$. Separation technique for single-charged particles with γ-factors up to $\approx 4 \times 10^4$ does not exist yet.

The TRD operation range is defined by the radiator and the detector properties: material, thickness and number of radiator foils, the gap between the foils, detector material and its thickness. For efficient particle separation it is important to exploit all features of the TR. The probability density of TR exiting from the radiators is a complex function of the particle's γ-factor, the radiator parameters, photon emission angle and photon energy. Integrating over the angle, one obtains the TR spectrum with many maxima. TR production at each maximum has its own γ-factor dependence. This feature can be used to develop a single detector with responses to a few γ-factor regions, which could significantly enhance its performance.

In order to study this possibility, a dedicated setups based on straw proportional tube arrays were built and tested at the CERN SPS accelerator [6, 7]. The straw layers are interleaved with blocks of radiator material. Dedicated Monte Carlo (MC) simulations were performed and compared with the experimental data. The results of these measurements with a Mylar radiator and comparison with MC predictions are presented in this paper.

2. Test beam setup

The experimental set-up and schematic view of the TRD prototype are shown in Figure 1. Beam particles, triggered by $8 \times 20\ mm^2$ scintillators, cross 22 layers of thin-walled proportional chambers (straws) which detect ionization losses and TR photons. The straws have 4 mm diameter and are spaced by 5 mm within one layer. Four layers of straws are grouped together and the gaps between the groups are used to install the TR radiator blocks. The layers within one group are shifted with respect to each other in the vertical direction to minimize fluctuations of the active gas thickness crossed by the particles. The straws are made from a special conductive polyimide (Kapton) film of 70 μm thickness and are filled with gas mixture of 71.8% Xe, 25.6% CO_2 and 2.6% O_2. The anode wire has a diameter of 30 μm. Similar straw chambers are used in the Transition Radiation Tracker detector [8] of the ATLAS experiment [9] at the LHC. The gas gain is about $2.5 \cdot 10^4$ and is controlled with an accuracy of about 1.5%. An energy calibration using an Fe^{55} source was done for each straw to convert QDC channel counts to energy. In order to separate signals from noise, only energy depositions above 300 eV were considered. Each radiator block contains 30 Mylar foils of 50 μm thick spaced by 3 mm. The first radiator, installed in front of the first straw layer, contains 55 foils.

A few types of the beams were used for tests: 20 GeV mixed beam containing electrons and pions as well as muon beams with energies of 120, 180 or 300 GeV. These beams allows to cover the range of γ-factors from 140 to 3.9×10^4. Tests with and without radiators were performed at all beam energies.

Figure 2 shows the energy spectra obtained with a 20 GeV electron beam with and without radiators. The spectra are presented in two forms: differential – an energy registered in straws,

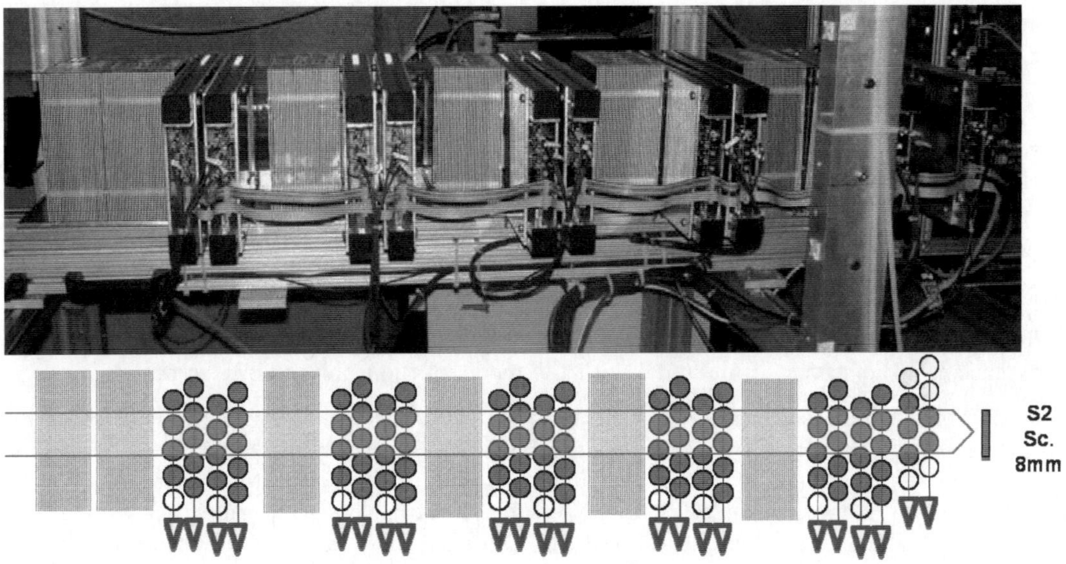

Figure 1. Top: side view of the TRD straw layers interchanged with radiator blocks in the H8 beam line at the CERN SPS. Bottom: schematic view of the straw TRD.

and integral – probability to exceed some threshold of registered energy. For the configuration with radiators, energy depositions above ~6 keV are defined mainly by absorbed TR photons. At an energy above 6 keV, one distinguishes two peak structures – one peak occupies the energy range of 6-14 keV and another one – above 14 keV.

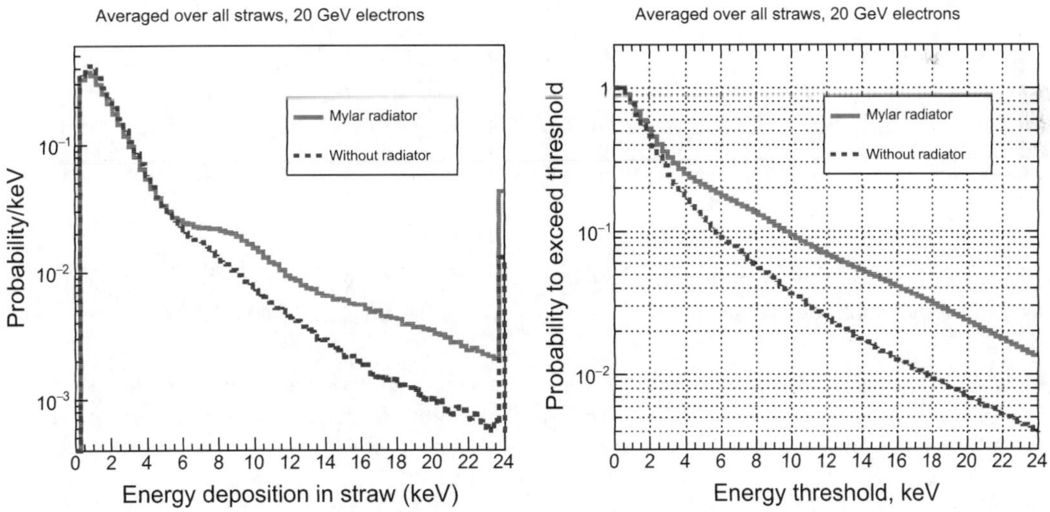

Figure 2. Experimental spectra of the energy registered in the straws by a 20 GeV electron beam with and without radiators. Left: differential spectra, right: integral. The rightmost bin on the differential spectrum includes the overflows. The spectra are averaged over all TRD straws.

IOP Conf. Series: Journal of Physics: Conf. Series **934** (2017) 012053 doi:10.1088/1742-6596/934/1/012053

3. Monte Carlo simulations and comparison with data

In order to understand in detail the measured spectra, dedicated Monte Carlo simulations were performed. The MC software is based on the TRT simulation code [10] and includes detailed descriptions of the detector geometry and the materials (radiator blocks and straw tubes, including straw walls and anode wires).

The simulation also takes into account:

- An energy loss in the straw gas of a charged particle according to the Photo-Absorption Ionization (PAI) model [11].
- Production and absorption of transition radiation photons.
- Response of the straws to the energy deposited by ionizing particles and photons.
- Convolution of the straw output signal with the model of the front-end electronics.

Some additional factors, such as space charge effect, photo electron path in the active gas and TR generation on the straw walls, were also included in the Monte Carlo model.

A comparison of the simulation results and data was done for all particle energies. As an example, spectra averaged over all straw layers for 20 GeV electron beam are shown in Figure 3. A good agreement between data and MC can be seen. The two peak structures in the same energy ranges as in data are clearly seen. The fact that they are more pronounced in MC than in the data might be explained by the simplified model of the photoelectron path in the gas volume used in MC.

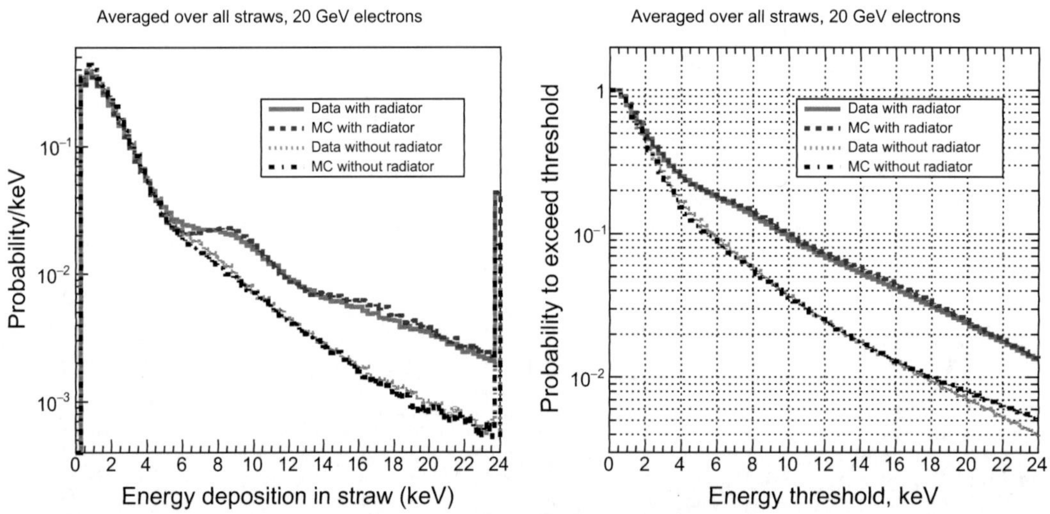

Figure 3. Comparison of experimental and simulated spectra of energy registered in the straws, averaged over all straw layers. Left: differential spectra, right: integral. The rightmost bin on the left plot includes overflow values. Results are shown for 20 GeV electron beam, and using Mylar radiators.

The comparison of data and MC for probabilities to exceed a certain energy deposition threshold as a function of straw layer number is shown in Figure 4. As the TR photons generated in radiators are absorbed in the active gas, it is not surprising that the probability to detect TR photon drops with increasing straw layer number with respect to radiator. On the other hand, high energy TR photons have a relatively low absorption efficiency and can be accumulated along the beam. This TR accumulation effect is clearly observed in the data and MC. In general, the agreement between data and MC is very good but some difference still exists. This difference can be explained by the beam structure (beam profile) which is not exactly known. In MC the beam profile is considered to be flat, with a width of 8 mm. For data, the 8 mm scintillation counter defines the beam limits, however the beam profile within the 8 mm still has some structure.

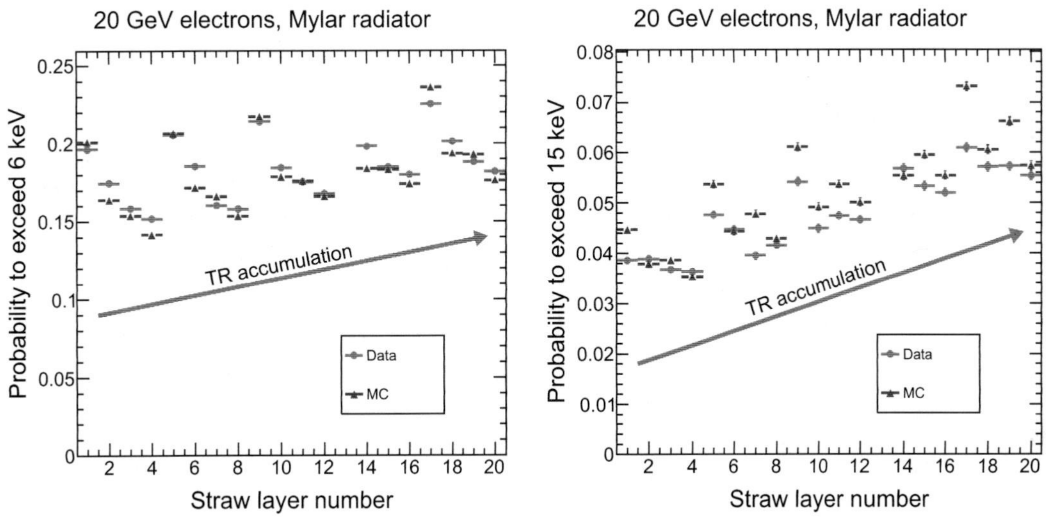

Figure 4. The measured and simulated probabilities to exceed an energy threshold depend on straw layer number along the beam. Left: for 6 keV threshold, right: for 15 keV threshold. Results are shown for a 20 GeV electron beam, TRD with Mylar radiators.

4. Lorentz factor dependence

As it was mentioned in section 1, different parts of the TR spectrum have different γ-factor dependencies. Figure 5 shows the averaged probability to have an energy deposition in one straw layer in different energy intervals as a function of γ-factor. Both, data and MC simulation results are presented.

This figure clearly shows a different γ-factor response for the soft TR energy range (6-14 keV) and for the TR energy range above 14 keV. The soft TR range is better suitable for particle separation with γ-factors from $\sim 1\times10^3$ to $\sim 8\times10^3$. The high energy range allows to move the TR production threshold to $\sim 3\times10^3$ and it can be effectively used for particle separation with close mass values in the γ-factor range from $\sim 3\times10^3$ to $\sim 2\times10^4$.

In general, a very good data/MC agreement is obtained. These studies confirm that two energy intervals have significantly different gamma dependencies, which allows to obtain responses on two γ-factor regions in a single detector.

5. Conclusions

A transition radiation detector with a specially designed radiator was tested in CERN SPS beam line with different particle types and energies. Predictions by a dedicated Monte Carlo simulations program, which was developed to describe the particular TRD, show good agreement with the data obtained. The possibility to use different energy intervals to obtain a different detector response as a function of the particle's γ-factor was demonstrated. These studies give

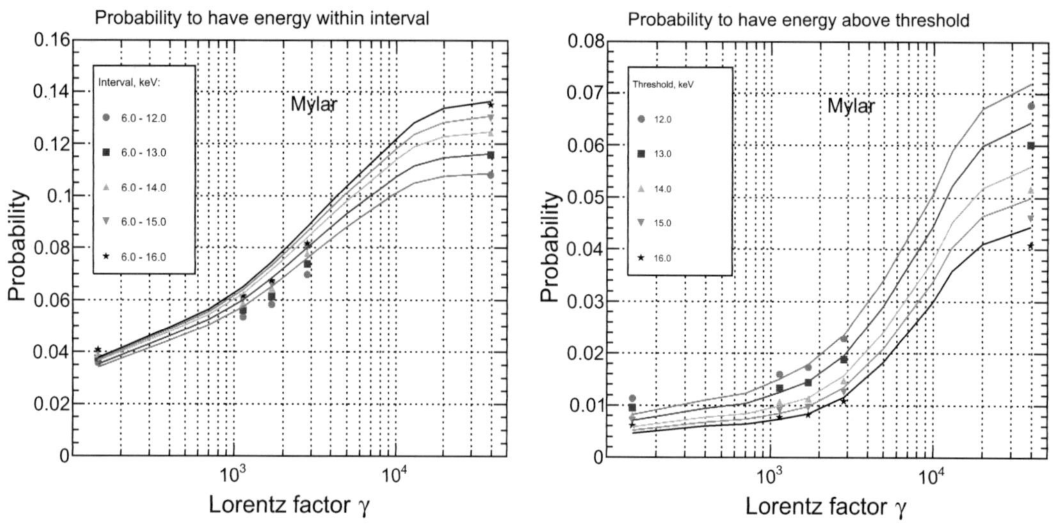

Figure 5. Averaged probabilities per straw layer to have an energy deposition in the low energy interval (left) and above a high energy threshold (right) as a function of the particle's γ-factor. Data (markers) and MC (points, connected by lines) are presented.

a solid basis for an optimization of the detector configuration which would allow to separate particles with γ-factors in the range of $\sim 10^3$-4×10^4.

Acknowledgments

We gratefully acknowledge the financial support from Russian Science Foundation grant (project No. 16-12-10277).

References

[1] `https://indico.cern.ch/event/379705/contributions/1806467`
`https://indico.cern.ch/event/563277/contributions/2488034`
[2] Dolgoshein B 1993 *Nucl. Instr. and Meth.* A **326** 434-469
[3] Andronic A, Wessels J P 2012 *Nucl. Instr. and Meth.* A **666** 130-174
[4] Patrignani C *et al.* 2016 (Particle Data Group) *Chin. Phys.* C **40** 100001, chapter "Transition radiation detectors"
[5] Cherry M L 2013 Measuring the Lorentz factors of energetic particles with transition radiation. *Nucl. Instr. and Meth.* A **710** 39-42
[6] Celebi E. *et al.* 2017 *Journal of Physics Conference Series* **798** 012172
[7] Tikhomirov V.O. *et al.* 2017 *Journal of Physics Conference Series* **798** 012183
[8] Abat E *et al.* 2008 The ATLAS Transition Radiation Tracker (TRT) proportional drift tube: design and performance *JINST* **3** P02013
[9] Aad G *et al.* 2008 The ATLAS Collaboration, The ATLAS Experiment at the CERN Large Hadron Collider. *JINST* **3** S08003
[10] Nevski P 2004 *Nucl. Instr. and Meth.* A **522** 116
[11] Grishin V M, Ermilova V K and Kotelnikov S K 1991 *Nucl. Instr. and Meth.* A **307** 273

IOP Conf. Series: Journal of Physics: Conf. Series **934** (2017) 012054 doi:10.1088/1742-6596/934/1/012054

Study of the influence of ADC sampling rate on the efficiency of neutron-gamma discrimination by the pulse shape

A S Chepurnov[1*], O I Gavrilenko[2], M A Kirsanov[3+], S G Klimanov[3], A S Kubankin[4,5]

[1] Skobeltsyn Institute of Nuclear Physics, Lomonosov Moscow State University, Leninskie gory, GSP-1, Moscow, 119991, Russia
[2] Technofinancetrade JSC, Krasnopresnenskaya nab. 12, off 712, Moscow, 123610, Russia
[3] National Research Nuclear University MEPhI (Moscow Engineering Physics Institute), Kashirskoe highway 31, Moscow, 115409, Russia
[4] Laboratory of Radiation Physics, Belgorod National Research University, Koroleva St. 2a, Belgorod, 308034, Russia
[5] Lebedev Physical Institute, Leninskiy Prospekt 53, Moscow, 119333, Russia

E-mail: aschepurnov@yandex.ru*, makirsanov@mephi.ru+

Abstract. The influence of a sampling rate of ADC on the efficiency of the pulse shape discrimination procedure (PSDP) developed for gamma-neutron discrimination was studied. Pu-Be neutron source and two types of digitizers (CAEN DT5730 and CAEN DT5743) were used. Both digitizers together with application software allow to store sequences of waveforms from a scintillation detector. The functional features of the CAEN DT5730 and CAEN DT5743 are described, and experimental characteristics of their operation are compared. Experimental values of an efficiency of neutron/gamma signal discrimination using two ADCs with different sampling frequencies are presented.

1. Introduction

There are several applications which require a parallel detection of fast neutrons and gamma radiation: 1) a monitoring of neutron and gamma background in low-background underground experiments (neutrino and Dark Matter detectors) [1], 2) a measurement of a fast neutron yield from neutron generators [2, 3], 3) a monitoring of a spent nuclear fuel , 4) an environmental monitoring on nuclear power plants.

Scintillation detectors with organic crystals, plastic and liquid scintillators are used for such a technique of a radiation detection. The shape of the output pulses from such detectors depends on the particle type. Fast ADC are used to digitize the output pulse and, than, different procedures of digital processing are used to discriminate the type of the particle. We studied and present in this paper how differs an efficiency of the same PSDP from the type of ADC with a different sampling rate and an accuracy.

2. Experimental setup

The experimental setup consists of the Pu-Be neutron source, a scintillation detector, ADCs and a personal computer with an application software for digital pulse shapes discrimination. The

IOP Conf. Series: Journal of Physics: Conf. Series **934** (2017) 012054 doi:10.1088/1742-6596/934/1/012054

scintillation detector consists from an organic p-terphenyl crystal of 25x25 mm cylindrical form. Hamamatsu R6094 photomultiplier registers scintillation flashes. Output of the PMT's high voltage divider is connected directly to the digitizer input. The distance between the Pu-Be neutron source and the detector was 100 mm. USB interface embedded in to the ADCs was used to communicate with the personal computer.

We used two Desktop Waveform Digitizers alternately. One was CAEN DT5730 (ADC#1) with the following technical characteristics: 8 channels, 14 bit, 500 MS/s, bandwidth 250 MHz. The second digitizer was CAEN DT5743 (ADC#2) with the following technical characteristics: 8 channels, 12 bit, 3.2 GS/s, bandwidth 500 MHz) [4]. Both digitizers have 50 Ohm input impedance. Both digitizers together with appropriate software allow to store in to files input waveforms together with a time stamp of each sampling point relatively external or internally generated trigger. The feature when the samples are precisely time stamped is very important for such experiential tasks as neutron/gamma background monitoring near accelerators [5, 6] and a measurement of pulsed particle fluxes [7, 8]. ^{137}Cs and ^{60}Co radioactive gamma sources have been used to calibrate the scintillation detector.

3. The influence of ADC sampling rate on the efficiency of neutron-gamma discrimination

The digitizers can store different number of waveform's samples from the detector. ADC#2 stores a fixed number of the samples per event which is equal to 1024, while the number of the points per event for the ADC#1 is programmable. We used 500 samples from ADC#1 per event for the analysis. The number of samples per a pulsed waveform of a certain amplitude for the ADC#2 was six times more than for the waveform from ADC#1. This difference could influence the efficiency of gamma-neutron discrimination when these two digitizers are used.

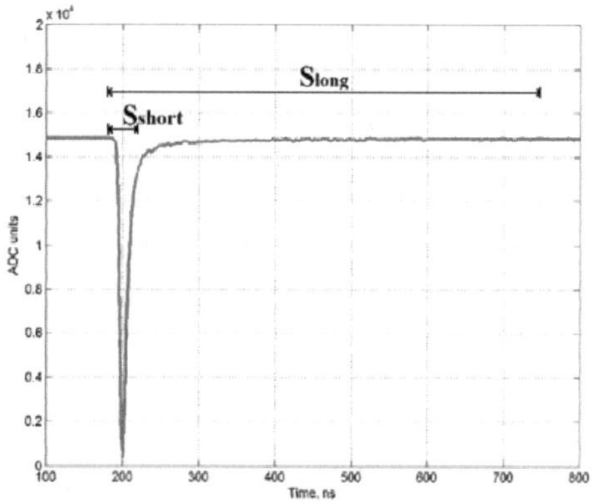

Figure 1. A waveform with indication of the short and the long gates.

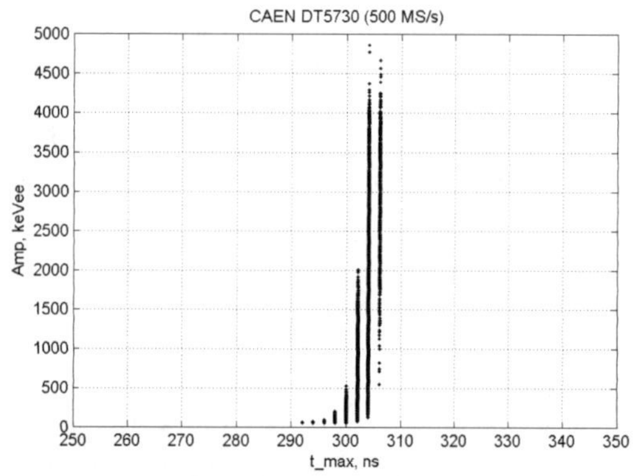

Figure 2. A distribution of a maximum position of pulses within a waveform when ADC#1 is used.

The application software was developed to discriminate gamma and neutron events according to the shape of the waveform. We calculate the area within the "short" (Sshort) and the "long" (Slong) gates separately (Figure 1). The area corresponds to the charge at the output of the detector. The length and the time position of the gates are optimized. A distribution of a maximum position of pulses within a waveform relatively the waveform start when ADC#1 is used is shown on Figure 2 and looks fixed.

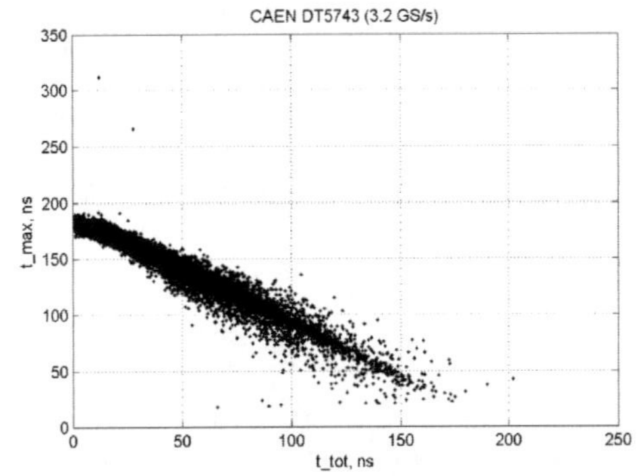

Figure 3. The distribution of the position of the maximum of the pulses within the waveform for ADC # 2.

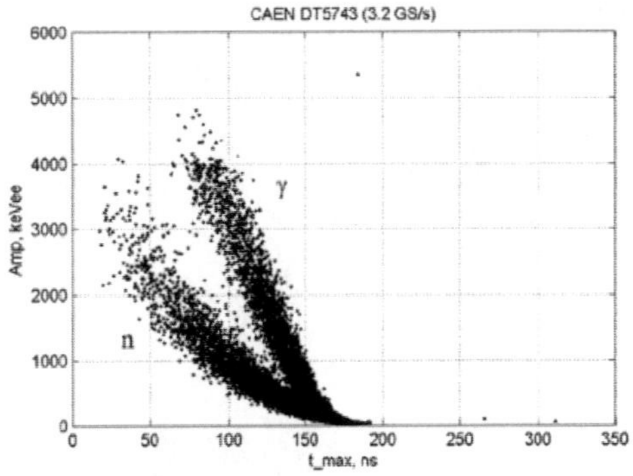

Figure 4. The dependence of pulse amplitude on the maximum position for ADC # 2 is used.

A position of the maximum of the pulses for the ADC#2 is distributed in the wide range and depends on the amplitude of the signal. It is shown on the Figure 3. The dependence of the moment of the maximum of the signal on the amplitude makes it possible to carry out a rough separation of the signals from neutrons and gamma (Figure 4).

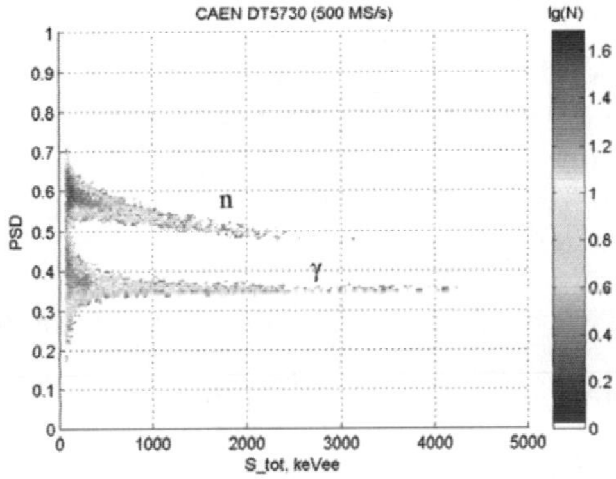

Figure 5. The dependence of the PSD parameter on the total pulse area when using ADC # 1.

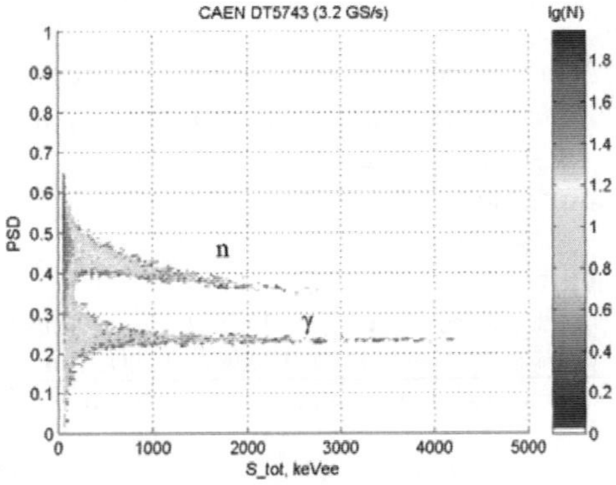

Figure 6. The dependence of the PSD parameter on the total pulse area when using ADC # 2.

PSDP as a procedure could be characterized by PSD parameter which describes numerically the discrimination process - PSD = $(S_{long} - S_{short}) / S_{long}$. Figure 5 and Figure 6 show the parametrized result of PSD procedure operation for gammas and neutrons when ADC#1 and ADC#2 are used.

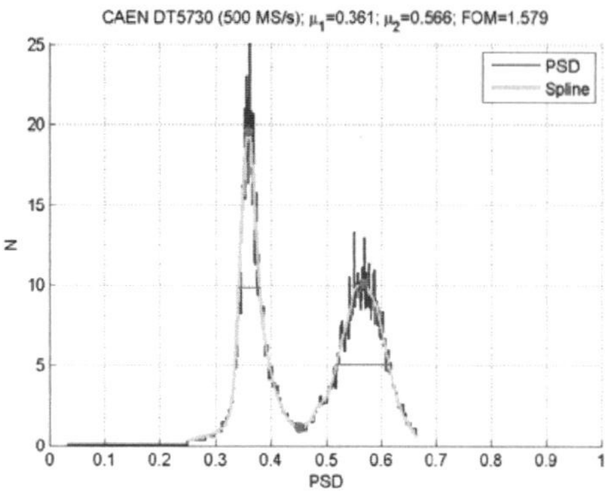

Figure 7. Discrimination of peaks from neutrons and gamma quanta using ADC # 1.

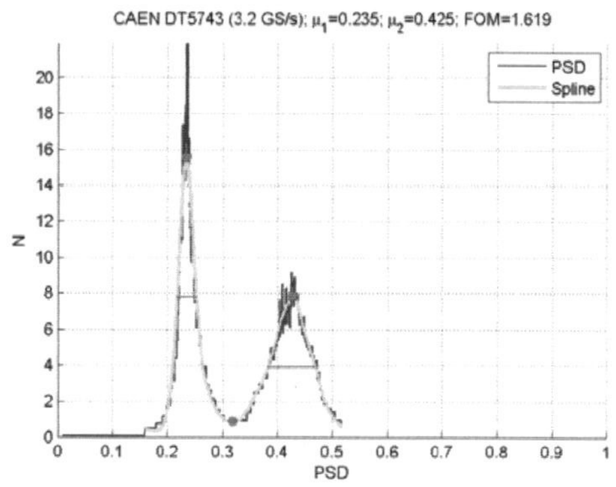

Figure 8. Discrimination of peaks from neutrons and gamma quanta using ADC # 2.

PSD parameter histograms together with corresponding spline approximations are represented on the Figure 7 and Figure 8. The quantitative criterion for the efficiency of PSDP could be defined as the value of FOM (Figure of Merit) when $FOM = (max_n - max_\gamma) / (FWHM_n + FWHM_\gamma)$. When ADC #1 is used the criteria FOM = 1.58, whereas when ADC #2 is used the criteria FOM =1.62.

4. Conclusion

Our experimental results show that the efficiency of the PSDP from neutrons and gamma rays is almost identical for both digitizers operating with 500Ms/s and 3.2Ms/s sampling rates. Thus, an increase in the sampling frequency of the ADC does not significantly increase the efficiency of pulse shape discrimination when detecting neutrons and gamma quanta using a scintillation detector with an organic crystal. Currently, a scintillation detector with a p-terphenyl crystal is used to measure the neutron yield in a new fast neutron generator being developed using carbon nanotubes.

Acknowledgments

The work is performed with financial support of the Ministry of Science and Education of Russia, the Project no. 14.578.21.0192 (RFMEFI57816X0192).

The authors are grateful for the equipment and software provided for the experiments by CAEN S.p.A. and TFT JSC.

References

[1] Du Q *et al.* 2017 *Preprint* arXiv:1706.06831 [physics.ins-det]
[2] Chepurnov A S *et al.* 2016 *J. Phys.: Conf. Series* **675** 032031
[3] Chepurnov A S *et al.* 2017 *J. Phys.: Conf. Series* **798** 012119
[4] *http://www.caen.it/csite/Function.jsp?parent=14&idfun=1*
[5] Kirsanov M A 2017 *J. Phys.: Conf. Series* **798** 012212
[6] Kirsanov M A 2017 *J. Phys.: Conf. Series* **798** 012214
[7] Kirsanov M A, Obodovski I M and Gangrski Yu P 1993 *Nucl. Instrum. Meth.* A **327** 48
[8] Kirsanov M A 2016 *J. Phys.: Conf. Series* **675** 042014

Light output distribution in scintillator strips with wave length shifting fibers of DANSS spectrometer

N Pogorelov[1], I Aleksev[1,2,3], D Kalinkin[1,3], I Makhichkulyan[1,2], V Nesterov[1], V Rusinov[1], A Starostin[1,2], D Svirida[1,2], E Tarkovsky[1]

[1] Alikhanov Institute for Theoretical and Experimental Physics, B. Cheremushkinskaya 25, Moscow, 117218, Russia

[2] National Research Nuclear University MEPhI (Moscow Engineering Physics Institute), Kashirskoe highway 31, Moscow, 115409, Russia

[3] Moscow Institute of Physics and Technology, 9 Institutskiy per., Dolgoprudny, Moscow Region, 141701, Russia

E-mail: n.pogorelov94@gmail.com

Abstract. DANSS is a highly segmented plastic scintillator detector, which uses scintillator strips with a Gd-loaded reflective cover to detect reactor antineutrino by inverse beta-decay. Light is collected with wave length shifting fibers (3 per strip) placed in grooves. Therefore the distribution of light output could be significantly non-uniform. Transverse profile of light output was studied in ITEP at a test bench consisting of proportional chambers and scintillator strips. Tracks of cosmic particles, which crossed chambers, were reconstructed with high accuracy, whereby transverse profiles of light output were built with step 1 mm for six scintillator strips. This result is important for calibration of DANSS and the method could be useful in constructing similar detectors.

The main goal of DANSS[1] (Detector of Anti-Neutrino based on Solid Scintillator) is search for sterile neutrino. The detector is placed 10 meters below the core of nuclear reactor WWER-1000 at Kalinin Power Plant. Sensitive area of the detector is a cube of polystyrene-based solid scintillator with size 1x1x1 m^3, covered by passive shielding and muon veto. Reactor antineutrino is detected by inverse beta-decay. Energy outputs of positron annihilation and neutron capture are both registered.

Spectrometer consists of 2500 polystyrene-based strips with gadolinium-containing cover, arranged in modules. Each strip has size 1x4x100 cm^3. Each strip contains 3 wavelength shifting fibers, placed in grooves. Two side fibers are read out by PMT, the central fiber is read out by SiPM. Relative dependence of the signal in SiPM(PMT) of the position of energy release is very important for Monte-Carlo model of the detector and data analysis. Light in scintillator is divided between different fibers, so transverse profile of light collection must be taken into account. In order to address this problem, we constructed a test bench consisting of proportional chambers (PC) and scintillator strips. General outline of the test bench is given in fig. 1.

Distance between the top and the bottom chambers is 252 mm. Sensitive area of the chambers is 20x20 cm2[2], which covers only 20% of strip length, so the measurements were done in three positions with center of the sensitive area in 20, 50 and 80 cm from the SiPM's. PC`s with 1 mm pitch help us to find crossing of cosmic particle with Inner Strips. Coordinate resolution of system is about 0.3 mm for transverse plane and 0.33 mm for longitudinal plane with track efficiency not less than 98%. IS's are similar to strips, used in DANSS, but with all wave length shifting fibers connected to SiPMs. Calibration to photoelectrons (ph.e.) was done individually for each SiPM, using its noise. Additional **T**rigger **S**trips (TS) cover gaps between inner strips for better trigger uniformity. Trigger is produced when summary signal in TS's and IS's is above 3 MeV.

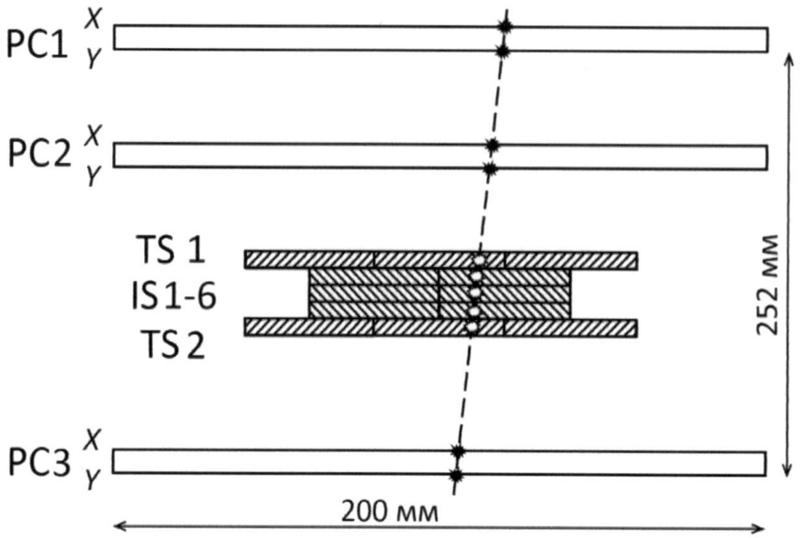

Figure 1: General outline of the test bench. PC1-3 – proportional chambers, TS1-2 – trigger strips, IS1-6 – inner strips.

Transverse profile of light output was built in 1 mm steps based on the length-normalized spectrum (in ph.e./cm). A hit was included in spectrum only if there are at least 5 p.e., track was found in both projections ($\chi^2 \leq 6$), slope of the track in the transverse plane is less than 0.1. The later condition limits spread of light produced to neighbor slices. For each spectrum, we found a median – the vertical line, which bisects square below histogram. The error is defined as the root-mean-square deviation divided by the root of the number of events. Dependence of the median on the coordinate is profile of the light output.

Typical profiles of light output are given in figure 2. Locations of grooves could be easily found as little deeps of energy response in the center of strip and in 10 mm in both directions from it. A slight asymmetry in the profile of the central photodetector may indicate inaccuracy in the manufacture of lateral grooves or a difference in their reflective properties due to uneven laying of fibers or gel filling. The difference in the magnitude of the light output of the left and right fibers can be further explained by the efficiency of the optical contact between the fiber and the photodetector. Decrease of the measured light output at the edges of the strip is connected with muon tracks leaving the strip and is discussed later. All features of the profiles are the same for all three positions of the chambers.

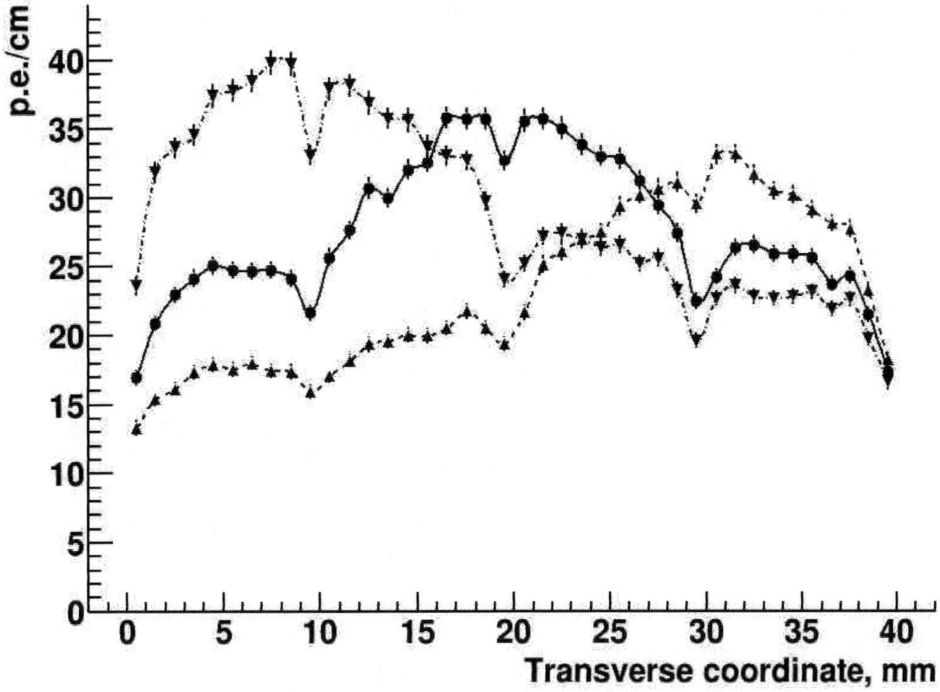

Figure 2: Example of the profile in three channels of one strip

Inner strips were arranged in three layers to resolve a question about insensitive area at their edges, which is due to gadolinium-containing cover and impossibility to brink strips closer. Predicted width of this area is 0.5±0.05 mm. Real width is estimated from energy loss at the edge. We built graph of total energy in pair of channels, adjacent to the gap, for all three strip layers for two detector positions. An example of such a graph is given in figure 3. Measured sizes of gaps are given in table 1 and are in good agreement with expected.

Table 1: Comparison of measured and predictable size of gap

	Size of insensitive area at strip center (50 cm position), mm	Size of insensitive area at close to SiPM edge (20 cm position), mm	Predicted size of insensitive area (gap + cover)
Top layer	0.66±0.08	0.54±0.14	0.50±0.05
Middle layer	0.48±0.08	0.50±0.11	0.50±0.05
Bottom layer	0.62±0.08	0.52±0.17	0.50±0.05

Impact of the non-uniformity to energy resolution could be estimated by average dependence of the signal from the coordinate in the strip in relative values. Normalization was done without 2 points at both edges of the strip. Because in DANSS light from side fibers goes to one PMT, for our estimation we add up signals from these fibers. The Result of this procedure is given in figure 4a. Projection of all normalized profiles on the vertical axis (figure 4b) gives us a quantitative estimate of

IOP Conf. Series: Journal of Physics: Conf. Series **934** (2017) 012055 doi:10.1088/1742-6596/934/1/012055

the light output distribution for both types of photodetectors. Root-mean-square is about 14.2% for SiPM and about 6.7% for PMT.

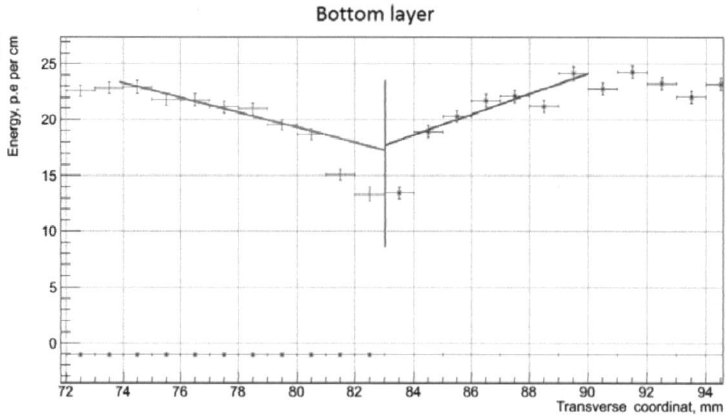

Figure 3: Example of total energy in pair of channels, adjacent to the gap

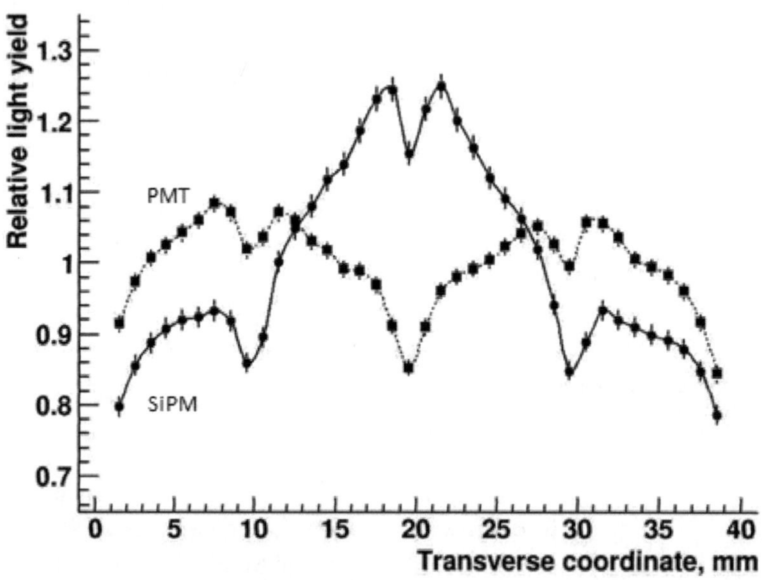

Figure 4a: Impact of the non-uniformity to energy resolution for SiPM and PMT

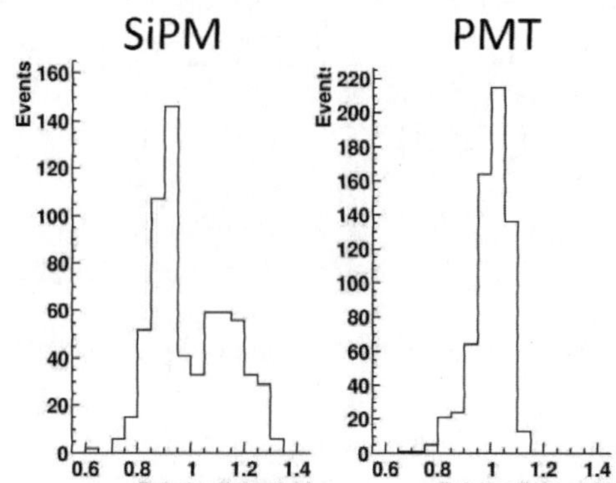

Figure 4b: Light output distribution for SiPM (left) and PMT (right)

In the analysis of data from DANSS we sum response of both types of the detectors with the same weights. This procedure results in the total non-uniformity of about 7.8% (figure 5a,b). Thereby non-uniformity of light yield causes small but accountable contribution to the energy resolution of the DANSS detector. Though the measurements were completed with a particular strip design the results could be used for construction of other detectors with scintillation plates

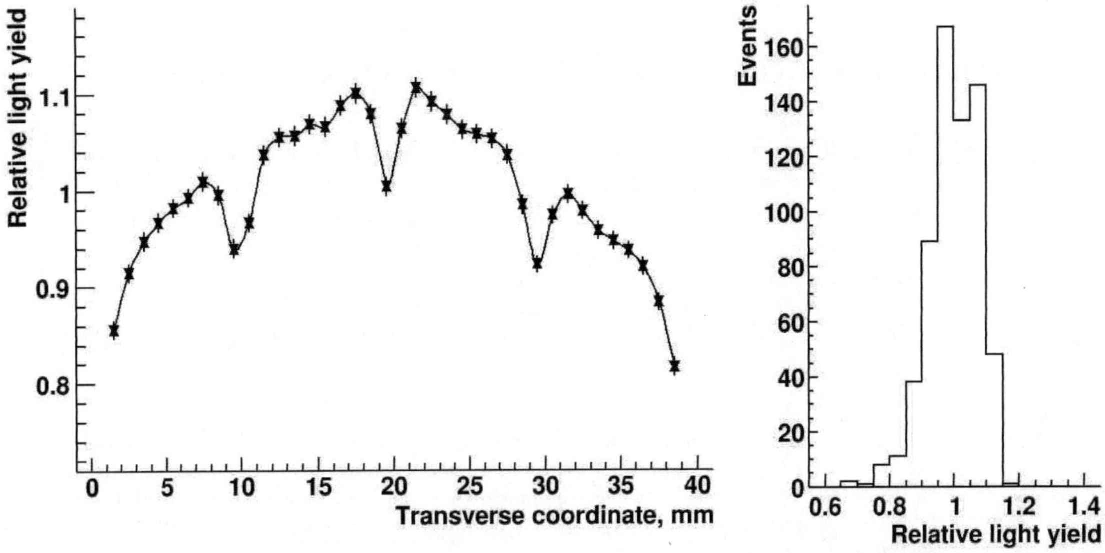

a) b)

Figure 5: a) Impact of the non-uniformity on the respond of one strip in DANSS-spectrometer, b) Light output distribution for one strip in DANSS-spectrometer

The test bench construction became possible thanks to the support of State Corporation Rosatom in the framework of state contracts № H.4x.44.90.13.1119 and № H.4x.44.9Б.16.1006. Data taking and analysis is supported by Russian Science Foundation grant 17-12-01145.

References

[1] Alekseev I, Belov V, Brudanin V, Danilov M, Egorov V, Filosofov D, Fomina M, Hons Z, Kazartsev S, Kobyakin A et al. 2016 DANSS: Detector of the reactor AntiNeutrino based on Solid Scintillator *JINST* **11** no.11, P11011

[2] Alekseev I G, Borduzhin I G, Budkovsky P E et al. 2014 *Instrum Exp Tech* **57**: 535.

A Silicon Detector Based Beta-spectrometer

I E Alekseev[1], S V Bakhlanov[2], N V Bazlov[2,3], E A Chmel[2], A V Derbin[2], I S Drachnev[2], I M Kotina[2], V N Muratova[2], N V Pilipenko[2], D A Semenov[2], E V Unzhakov[2], V K Yeremin[4].

[1] V. G. Khlopin Radium Institute, St. Petersburg, Russia
[2] Petersburg Nuclear Physics Institute, NRC "Kurchatov Institute", Gatchina, Russia
[3] St. Petersburg State University, St. Petersburg, Russia
[4] Ioffe Physical-Technical Institute of the Russian Academy of Sciences, St. Petersburg, Russia

E-mail: derbin_av@pnpi.nrcki.ru

Abstract. Here we present the specifications of the newly developed beta-spectrometer based on thick full absorption Si(Li) detector. The spectrometer was designed for precision measurements of various beta-spectra, namely for the beta-spectrum shape study of ^{144}Pr, which is considered to be one of the most promising anti-neutrino source for sterile neutrino searches.

1. Introduction

Precision measurements of beta-spectra have always been and are still playing an important role in several fundamental physical problems predominantly in neutrino physics. It was the continuous shape of beta-decay spectrum, that has driven W. Pauli towards the neutrino hypothesis. The initial beta-spectrum measurements for the purpose of determining the neutrino mass were carried out by G. Hanna and B. Pontecorvo [1] via gas counter that registered electrons produced by decays of tritium added into the detector volume.

Magnetic and electrostatic spectrometers possess the superior energy resolution, but at the same time such devices appear to be very complex and large-scale installments. Since the electron free path at 3 MeV (which is, basically, the maximum beta-transition energy for long-living isotopes) does not exceed 2 g/cm^3, solid state scintillation and ionization detectors were effectively employed for detection of electrons [2, 3]. The main drawback of solid state scintillators is their relatively poor energy resolution, which stands at approximately 10% at 1 MeV. In case of semiconductor detectors there is a significant probability of back-scattering from the detector surface that depends on the detector material. The most widespread silicon-based semiconductors have the backscattering probability of the order of 10% for 100 keV electrons at normal incidence [4]. In case of electron energies above 1 MeV and high Z detector materials, it also becomes important to take the bremsstrahlung into account.

The considered Si(Li) spectrometer was developed for precision measurement of ^{144}Ce-^{144}Pr beta-spectrum in order to determine the antineutrino energy spectrum. The ^{144}Ce-^{144}Pr antineutrino source will be used for experimental sterile neutrino searches by Borexino SOX collaboration [5].

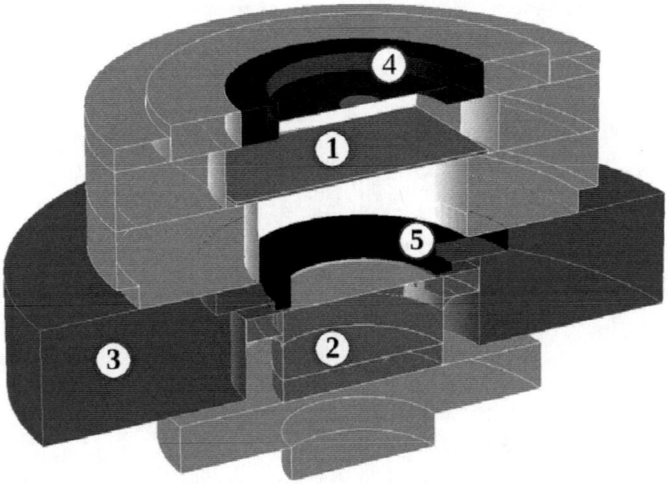

Figure 1. The layout of beta-spectrometer: 1 – Transmission detector (thickness 0.3 mm), 2 – Full absorption detector (thickness 6.5 mm), 3 – Al-Mg alloy assembly ring, 4 – source (Cerium acetate on mylar substrate), 5 – tungsten collimator ring.

2. Experimental setup

The photo of the experimental setup based on full absorption Si(Li) detector is shown in figure 1. The sensitive region of the Si(Li) detector fabricated at NRC KI PNPI has a diameter about 15 mm and a thickness of 6.5 mm. These dimensions ensure effective absorption of electrons with energies up to 3 MeV. The detector was equipped with an tungsten collimator with a diameter of 14 mm and a thickness of 2 mm. The negative bias voltage of 1 kV was applied directly to the gold coating of the detector. The energy resolution determined for 59.6 keV gamma-line of ^{241}Am turned out to be FWHM = 900 eV.

The entire setup was placed inside the vacuum cryostat and cooled down to the liquid nitrogen temperature. The detector was equipped with charge-sensitive preamplifier with resistive feedback and cooled field-effect transistor. As noted above, the spectrometer was designed with intent of measuring the beta-spectrum of ^{144}Pr in the energy range of (0 – 3) MeV. In order to perform measurements in such a broad dynamic range, the Si(Li) detector was equipped with two separate spectrometric channels, each with spectrometric amplifier and 14-bit analogue-to-digital converter set up as a standalone module. Channel settings were adjusted to register events within (0.01 – 0.5) MeV and (0.05 – 6.0) MeV energy intervals. The choice of the upper energy limit at 6 MeV was conditioned by our intention to monitor the possible alpha-activity of the sample under investigation. A dedicated DAQ control software allows one to acquire and record two 16000-channel spectra from the Si(Li) detector.

3. Results

In order to determine the main characteristics of the spectrometer we used a ^{207}Bi source, providing gamma-rays, X-rays and conversion and Auger electrons. The ^{207}Bi spectrum, measured with the Si(Li) detector, is shown in figure 2(a) for the intervals (0.01 – 2.0) MeV and (450 – 580) keV, respectively. The ^{207}Bi source with an activity of 10^4 Bq was placed inside the vacuum cryostat at a distance of 14 mm from the Si(Li) detector surface. Three of the most intense ^{207}Bi gamma-lines have energies of 569.7 keV, 1063.7 keV and 1770.2 keV and are emitted with probabilities of 0.977, 0.745 and 0.069 per single ^{207}Bi decay, respectively. The corresponding peaks of conversion electrons form K-, L- and M-shells are clearly visible

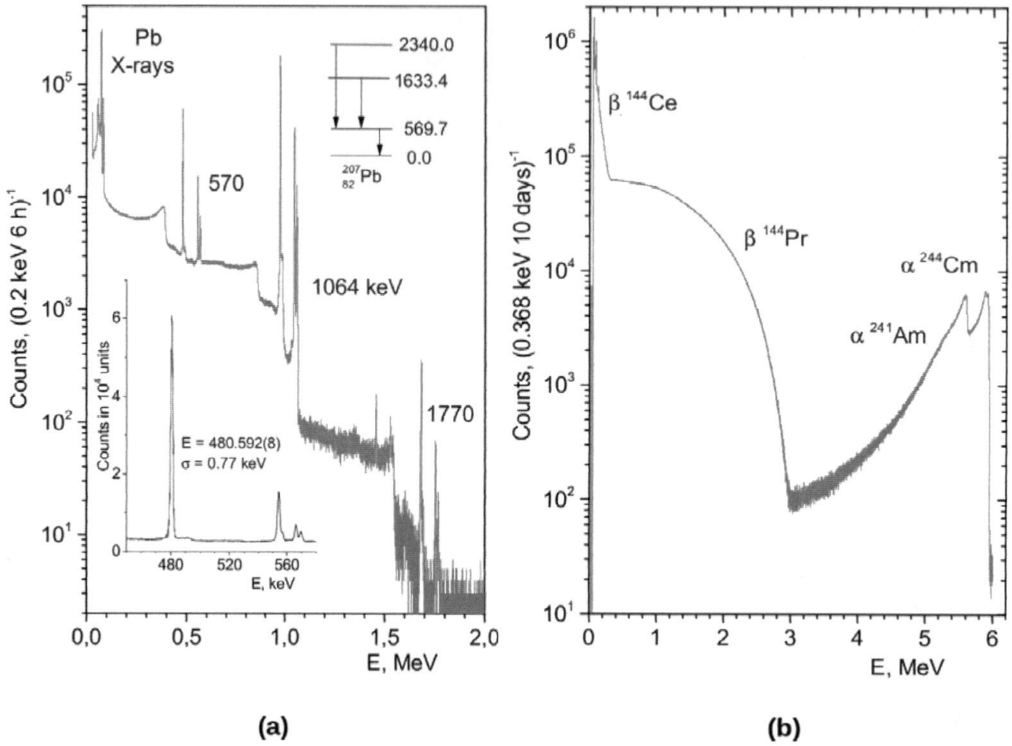

Figure 2. (a) The spectrum of ^{207}Bi source measured with the Si(Li) detector in energy range of (0.01 – 2.0) MeV. The inset shows the electron peaks corresponding to internal conversion from K-, L-, M-shells of the 570 keV nuclear level. (b) The spectrum of ^{144}Ce-^{144}Pr measured with the Si(Li) detector.

in the spectrum in figure 2(a). The electron energy resolution determined via 480 keV line is FWHM = 1.8 keV.

The low-energy part of the spectrum was used to evaluate the thickness of non-sensitive layer on the surface of Si(Li) detector. This area contains a set of peaks corresponding to Pb X-rays from K- and L-series and Auger electrons. The observed position of 56.94 keV Auger peak ($e_{KL_1L_2}$) appeared to be 56.22 keV. Inclusive of the golden coating thickness of (500 A°), the measured 59 keV electron energy loss of 720 eV corresponds to 4700 A° of non-sensitive layer.

The energy spectrum of the ^{144}Ce-^{144}Pr source, measured with the Si(Li) detector during 10 days, is shown in figure 2(b). The total spectrum contains gamma- and electron peaks. The spectrum contains characteristic X-ray lines of Np at 13.9 keV, 17.8 keV and 20.8 keV, produced by trace amounts of alpha-decaying ^{241}Am contained in the ^{144}Ce source. The peaks at 59.6 keV, 80.1 keV and 133.5 keV are related to the gamma-transitions of ^{237}Np and ^{144}Pr nuclei. The most intense peaks at 35.8 keV and 41 keV correspond to K_{α_1,α_2}- and K_β-lines of Pr X-rays. The spectrum of electrons from beta-decay of ^{144}Ce$-^{144}$Pr consists of also conversion and Auger electrons from 133.5 keV gamma-transition. After the main measurements were completed the thin silicon detector was mounted and γ- and X-rays activity was measured by Si(Li)-detector in anti-coincidence with thin Si-detector. Inclusion of this spectrum into the analysis allows us to account for overall gamma- and X-ray contribution.

The differential energy spectrum of electrons in $\beta-$decay is described as:

$$N(W)dW \sim pW(W - W_0)^2 \cdot F(Z,W) \cdot H(W) \cdot \cdot L_0(Z,W) \cdot C_{A,V}(Z,W) \cdot \cdot S(Z,W) \cdot G(Z,W) \cdot B(W), \tag{1}$$

where W and p are total energy and momentum of electron and $F(W, Z)$ is Fermi function. Additionally, the following correction factors have to be taken into account: finite size of the nucleus correction for electromagnetic $L_0(Z, W)$ and weak $C_{A,V}(Z, W)$ interaction; screening corrections of the nuclear charge by electrons $S(Z, W)$; radiative corrections $G(Z, W)$ and weak magnetism correction $B(W)$. The antineutrino spectrum can be calculated after the parameters of shape factor $H(W)$ will be extracted from the analysis of experimental spectrum.

4. Conclusion

The beta-spectrometer with of 6.5 mm thick full absorption Si(Li)-detector has been developed. The spectrometer can be used for precision measurements of the beta-spectrum shapes of various radioactive nuclei, in particular to measure the beta-spectra of ^{144}Pr, which is the most promising antineutrino source for searching for neutrino oscillations to a sterile state.

5. Acknowledgments

This work was supported by the Russian Science Foundation (grant 17-12-01009).

References

[1] Hanna G C and Pontecorvo B 1949 The β-spectrum of H^3 *Phys. Rev.* **75** 983
[2] Simpson J J 1985 Evidence of heavy-neutrino emission in beta decay *Phys. Rev. Lett.* **54** 1891
[3] Derbin A V, Egorov A I, Muratova V N *et al.* 1993 Search for 17 kev neutrinos in ^{63}Ni beta-decay *JETP Lett.* **58** 3
[4] Derbin A V , Egorov A I, Bakhlanov S V and Muratova V N 1997 Measurement of the ^{45}Ca β spectrum in search of deviations from the theoretical shape *JETP Lett.* **66** 81
[5] Bellini G *et al.* 2013 SOX: Short distance neutrino oscillations with Borexino *JHEP* **2013(8)** 38

IOP Conf. Series: Journal of Physics: Conf. Series **934** (2017) 012057 doi:10.1088/1742-6596/934/1/012057

Digital pulse shape discrimination between fast neutrons and gamma rays with para-terphenyl scintillator

A S Chepurnov[1*], M A Kirsanov[2+], A A Klenin[3], S G Klimanov[2], A S Kubankin[3,4]

[1] Skobeltsyn Institute of Nuclear Physics, Lomonosov Moscow State University, Leninskie gory, GSP-1, Moscow, 119991, Russia
[2] National Research Nuclear University MEPhI (Moscow Engineering Physics Institute), Kashirskoe highway 31, Moscow, 115409, Russia
[3] Laboratory of Radiation Physics, Belgorod National Research University, Koroleva St. 2a, Belgorod, 308034, Russia
[4] Lebedev Physical Institute, Leninskiy Prospekt 53, Moscow, 119333, Russia

E-mail: aschepurnov@yandex.ru[*], makirsanov@mephi.ru[+]

Abstract. In the presented work, we investigated several digital methods of a discrimination signals from fast neutrons and gamma quanta. The experimental setup consists of a Pu-Be neutron source, a scintillation detector with an organic para-terphenyl monocrystal, and a digitizer (CAEN DT5730, 500 MS/s). Mixed waveform sequences were stored and then separated by pulse shape. Four methods were used for signals separation. Comparison of the traditional and the new methods of Figure of Merit (FOM) calculation is given. FOM = 1.5 was obtained in our setup for the minimum threshold value. A scintillation detector with a para-terphenyl crystal was used to measure neutron yield in the neutron generator with carbon nanotubes.

1. Introduction

A registration of fast neutrons flux against the background of gamma radiation is an important task both for the case of low particle intensity (measurement of the neutron and gamma background in neutrino and dark matter particle detectors [1], environmental monitoring) and for the case of high mixed fluxes of neutrons and gamma rays (neutron flux from neutron generators [2, 3] and particle accelerators, monitoring of spent nuclear fuel). To detect fast neutrons, scintillation detectors with organic scintillators are widely used. In such scintillators, two components of de-excitement are present: the fast and the slow. Intensity of the slow component in such scintillators depends on the type of the detected particle. A modern technique of digital acquisition with the following digital processing of the output pulses from a scintillation detector allows to discriminate signals from gamma quanta and neutrons.

2. Experimental setup

The experimental setup includes a Pu-Be neutron source, a scintillation detector with an organic crystal, CAEN DT5730 (8 channels, 14 bit, 500 MS/s, bandwidth 250 MHz) digitizer and personal computer. The scintillation detector contains a para-terphenyl single crystal (formula $C_{18}H_{14}$, density 1.23 g / cm^3) of a cylindrical shape 25x25mm. The Hamamatsu R6094 photoelectric multiplier (PMT) registers scintillation photons from a para-terphenyl crystal. Signals from the PMT are fed to the

IOP Conf. Series: Journal of Physics: Conf. Series **934** (2017) 012057 doi:10.1088/1742-6596/934/1/012057

analog input of the digitizer [4]. The impedance of the analog inputs of the digitizer is 50 Ohm. The digitizer can acquire a different number of samples per event. We stored 500 samples per event. Waveform signals together with the time stamp per each sample were recorded in a data file via USB interface. The time stamp which is fixed relatively the external or internal trigger is important feature which is critical while the following tasks are performed: 1) a monitoring of neutron and gamma background near accelerators and pulsed neutron generators [5, 6]; a registration of gamma spectra in intervals between irradiation pulses [7, 8]. The distance from the Pu-Be neutron source to the scintillation detector was varied between 10 to 22 cm. An energy calibration of the scintillation detector was done by ^{137}Cs and ^{60}Co gamma sources.

3. The efficiency of neutron-gamma discrimination by the pulse shape

The intensity of the slow component of the light flash from the para-terphenyl scintillator is larger when the fast neutrons is detected in comparison with the case of gamma quanta detection [9, 10]. This phenomena allows to discriminate the signals from neutrons and gamma quanta by analysis of the pulse shapes.

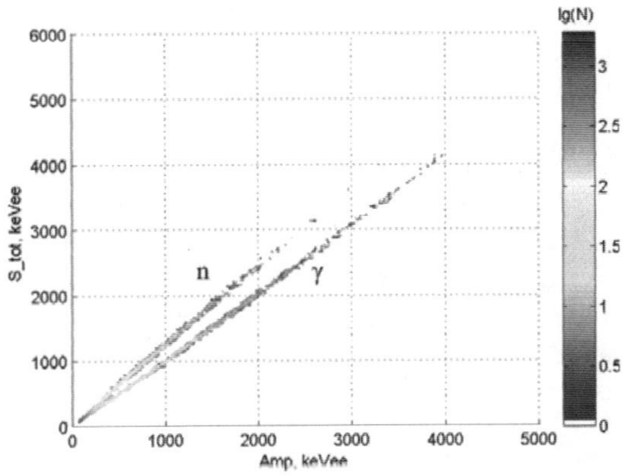

Figure 1. Dependence of the pulse area from the pulse amplitude.

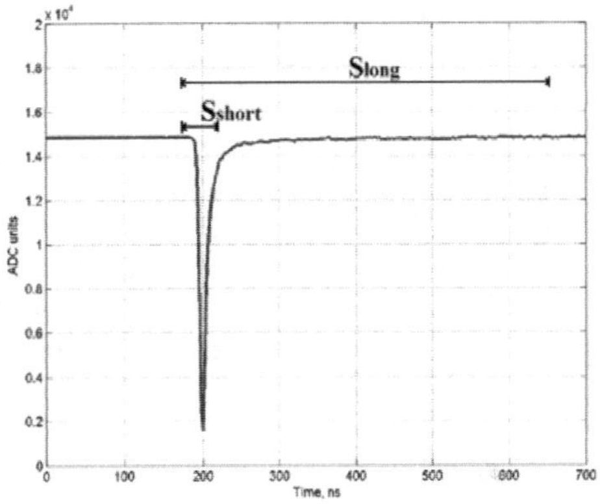

Figure 2. Waveform with indication of the short and long gates.

Several digital methods for fast neutrons and gamma quanta pulse shape discrimination have been used. Original application software was developed for this purpose. The most simple method is based on the analysis of the dependence of the area of pulsed signals on their amplitudes (Figure 1). As can be seen from Figure 1, the signals form two separate branches. The disadvantage of this method appears in the poor discrimination at low energies. Other methods use a calculation of the areas in short S_{short} and long S_{long} gates for each pulsed signal (Figure 2). The length and time position of both gates are optimized. The dependence of S_{short} from S_{long} allows to separate neutrons from gamma quanta (Figure 3). This method shows the same problem of poor separation of signals at low energies.

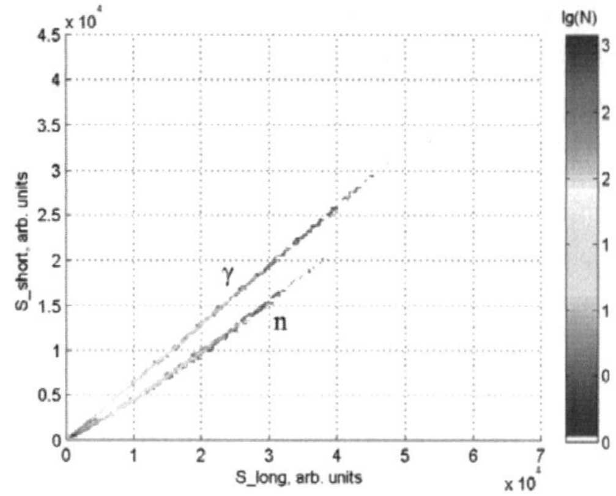

Figure 3. The dependence of S_{short} from S_{long}.

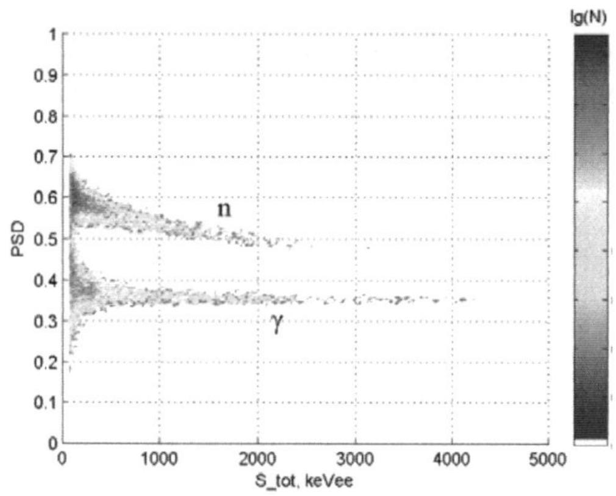

Figure 4. The dependence of the PSD parameter on the total signal area.

The best signal discrimination was achieved by using the following criteria - PSD (Pulse Shape Discrimination) = (S_{long} - S_{short}) / S_{long}. Figure 4 shows the dependence of the PSD parameter on the total area of the signals. It can be seen that this method makes it possible to separate the signals, including for low energies of neutrons and gamma quanta.

The quantitative criterion for the efficiency of PSD procedure is the value of Figure of Merit (FOM). The value of FOM = (max_n – max_γ) / ($FWHM_n$ + $FWHM_\gamma$). The FOM value is traditionally estimated by fitting peaks on the PSD histogram by the sum of two Gaussian distributions. However, such an approximation could be incorrect rather often. We use for this task spline approximation of the peaks (Figure 5). Figure 5 shows the discrimination of neutrons and gamma quanta at different threshold levels: "no threshold", 100 keVee, 200 keVee, and 300 keVee. The efficiency of the gamma-neutron PSD increases from FOM = 1.5 (without a threshold) to FOM = 1.8 (threshold = 300 keVee).

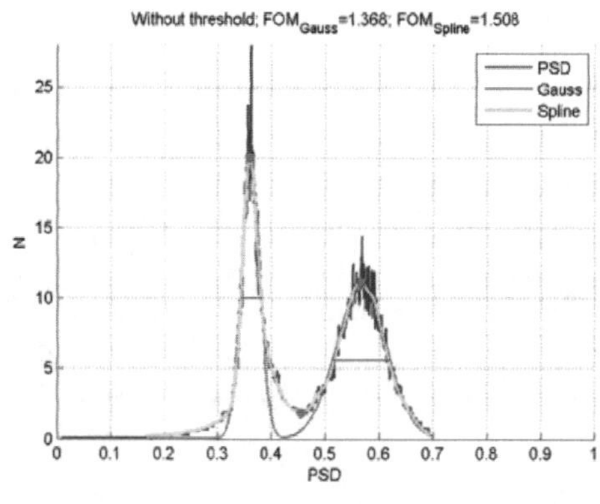

Figure 5a. Neutrons and gamma rays discrimination, without a threshold.

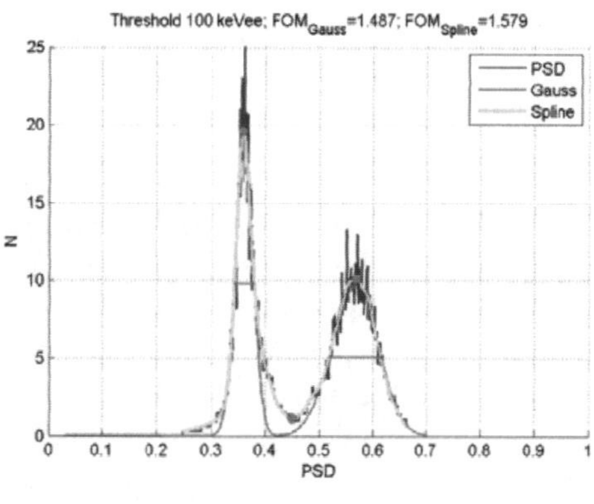

Figure 5b. Neutrons and gamma rays discrimination, threshold = 100 keVee.

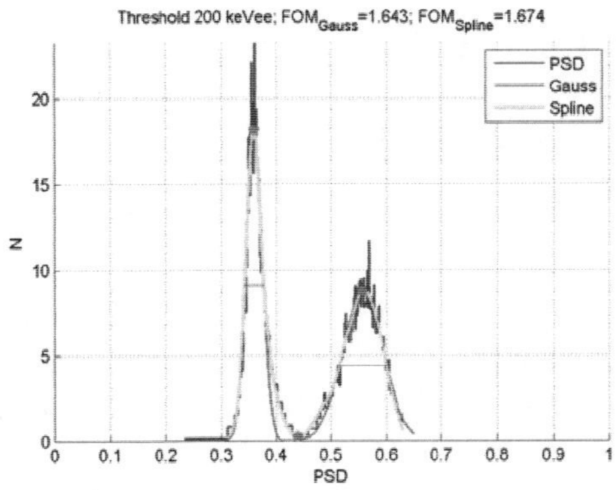

Figure 5c. Neutrons and gamma rays discrimination, threshold = 200 keVee.

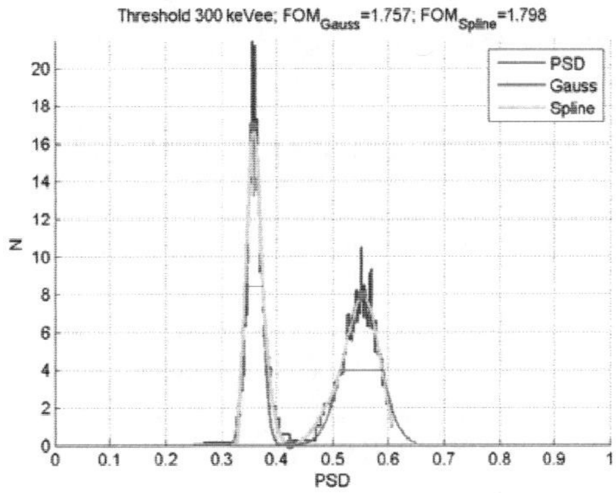

Figure 5d. Neutrons and gamma rays discrimination, threshold = 300 keVee.

4. Conclusion

A scintillation detector with a para-terphenyl crystal is a good instrument to solve the problem of fast neutrons detecting against a background of gamma radiation. Digital methods of pulse shape analysis make it possible to achieve an effective discrimination of the signals from neutrons and gamma quants. The value of FOM = 1.5 was obtained in our setup for the minimum threshold, while FOM ≈ 1 for the known BC-501A liquid scintillator. The scintillation detector with the p-terphenyl crystal was used by us to measure the neutron yield in the newly developing neutron generator based on carbon nanotubes.

Acknowledgments

The work was financially supported by the grant from the Russian Science Foundation (project №16-19-10535).

References

[1] Du Q *et al.* 2017 *Preprint* arXiv:1706.06831 [physics.ins-det]
[2] Chepurnov A S *et al.* 2016 *J. Phys.: Conf. Series* **675** 032031
[3] Chepurnov A S *et al.* 2017 *J. Phys.: Conf. Series* **798** 012119
[4] *http://www.caen.it/csite/CaenProd.jsp?parent=14&idmod=757*
[5] Kirsanov M A 2017 *J. Phys.: Conf. Series* **798** 012212
[6] Kirsanov M A 2017 *J. Phys.: Conf. Series* **798** 012214
[7] Kirsanov M A, Obodovski I M and Gangrski Yu P 1993 *Nucl. Instrum. Meth.* A **327** 48
[8] Kirsanov M A 2016 *J. Phys.: Conf. Series* **675** 042014
[9] Angelone M *et al.* 2013 *Preprint* arXiv:1305.0442 [physics.ins-det]
[10] Sardet A *et al.* 2015 *Nucl. Instrum.Meth.* **A 792** 74

IOP Conf. Series: Journal of Physics: Conf. Series **934** (2017) 012058 doi:10.1088/1742-6596/934/1/012058

Total reaction cross sections and neutron-removal cross sections of neutron-rich light nuclei measured by the COMBAS fragment-separator

B M Hue[1,3], T Isataev[1], B Erdemchimeg[1,2], A G Artukh[1], D Aznabaev[1], S Davaa[2], S A Klygin[1], G A Kononenko[1], G Khuukhenkhuu[2], K Kuterbekov[4], S M Lukyanov[1], T I Mikhailova[1], V A Maslov[1], K Mendibaev[1,4], Yu M Sereda[1], Yu E Penionzhkevich[1], A N Vorontsov[1]

[1]Joint Institute for Nuclear Research, 141980, Dubna, Russia
[2]National University of Mongolia, NRC, Ulaanbaatar, Mongolia
[3]Institute of Physics, Hanoi, Vietnam
[4]Eurasian Gumilev University, Astana
E-mail: erdem@jinr.ru

Abstract. Preliminary results of measurements of the total reaction cross sections σ_R and neutron removal cross section σ_{-xn} for weakly bound ^{6}He, ^{8}Li, ^{9}Be and ^{10}Be nuclei at energy range (20-35) A MeV with ^{28}Si target is presented. The secondary beams of light nuclei were produced by bombardment of the ^{22}Ne (35 A MeV) primary beam on Be target and separated by COMBAS fragment-separator. In dispersive focal plane a horizontal slit defined the momentum acceptance as 1% and a wedge degrader of 200 μm Al was installed. The Bρ of the second section of the fragment-separator was adjusted for measurements in energy range (20-35) A MeV. Two-neutron removal cross sections for ^{6}He and ^{10}Be and one –neutron removal cross sections ^{8}Li and ^{9}Be were measured.

1. Introduction

Radioactive nuclear beam (RNB), or also called radioactive ion beam (RIB) technology was intensively developed since the middle of 80s (see for review Ref.[1]). It has become possible to determine the nuclear size for unstable nuclei, mainly based on the interaction and reaction cross sections for intermediate energies (<100 A MeV). Measurements of reaction cross section (σ_R) started at GANIL [2]. There are two major methods to measure σ_R at intermediate energies: associated γ -ray detection method [3]; and the transmission method using Bρ–ΔE–TOF identification or ΔE–E identification [4]. Our experiment has taken place using the COMBAS fragment-separator with experimental set-up similar to one used in [5]. The experiment was aimed an attempt to measure the neutron-removal cross section of several light exotic isotopes. The secondary beams of the stable and short-lived light nuclei ^{6,8}He, $^{7\text{-}9}$Li, $^{9\text{-}12}$Be were produced by bombardment of the ^{22}Ne (35/A MeV) primary beam on the Be (89 mg/cm^2) target and separated by COMBAS fragment-separator [6]. The secondary products were detected by a telescope consisting of four Si detectors (ΔE) with 300, 300, 300 and 500 μm thicknesses, respectively, and E-detector CsI(Tl) with 15mm thickness. In the present paper, preliminary results of measurements of total cross section and the n-removal cross sections σ_{xn} for weakly bound ^{6}He, ^{8}Li, ^{9}Be and ^{10}Be nuclei at energy range (20–35)/A MeV are presented and compared with all other available previous experimental data.

2. The experimental procedure

Our σ_R measurements require magnetic separation analysis and particle identification (PID) to obtain a narrow monoenergetic projectile beam which is stopped in one or more Si or CsI detectors. A target

(89 mg/cm^2) target of ^{9}Be placed at the entrance focus of the COMBAS separator was irradiated separately with ^{22}Ne (35 AMeV) beam accelerated at the U-400M cyclotron of FLNR, JINR. The secondary beams were then transmitted through the COMBAS magnetic analyzing system [7]. In dispersive focal plane a horizontal slit defined the momentum acceptance as 1% and a wedge degrader of 200 μm Al was installed. Figure 1 (not to scale) shows our detector system. The first Si detector ΔE_1 (300 μm) was set into a cylinder chamber to obtain ΔE information. Behind ΔE_1, there is a rectangular parallelepiped box containing ΔE_0 and two detectors with 16 horizontal X strips and 16 vertical Y strips in near side of (300 μm) ΔE_2 and (300 μm) ΔE_3, respectively. These detectors are designed for both measuring spectra of the secondary beams and determining their space distribution. And the silicon (500 μm) ΔE_4 detector was also installed inside the box to obtain ΔE information. In order to measure remain energy of particles, a thick CsI(Tl) detector is used and placed at the end of the telescope.

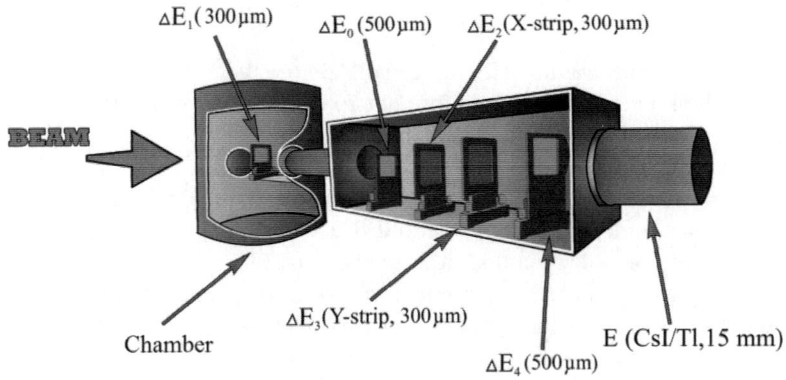

Figure 1. Detectors system (not to scale) used in this experiment.

3.Data analysis and results

Each detector was calibrated to take into account the known energy losses of both the projectiles of interest. All analyzed events were required to pass tight energy gates, appropriate to the projectile of interest, for both position sensitive detectors and the ΔE counter. We obtained total kinetic energy spectra are shown for ^{6}He, ^{8}Li, ^{9}Be and ^{10}Be in figure 2. The off-scale peaks are produced by nonreacting projectiles. For the two-neutron-halo nuclei as ^{6}He, 2n removal is one of the strongest reaction channels [8]. Both the magnitude and energy dependence of ^{6}He (a) demonstrate a prominent two-neutron-removal peak above the continuum caused by core reactions. On average, ^{6}He breaks up two neutrons then carry off two-sixth of the remaining kinetic energy.

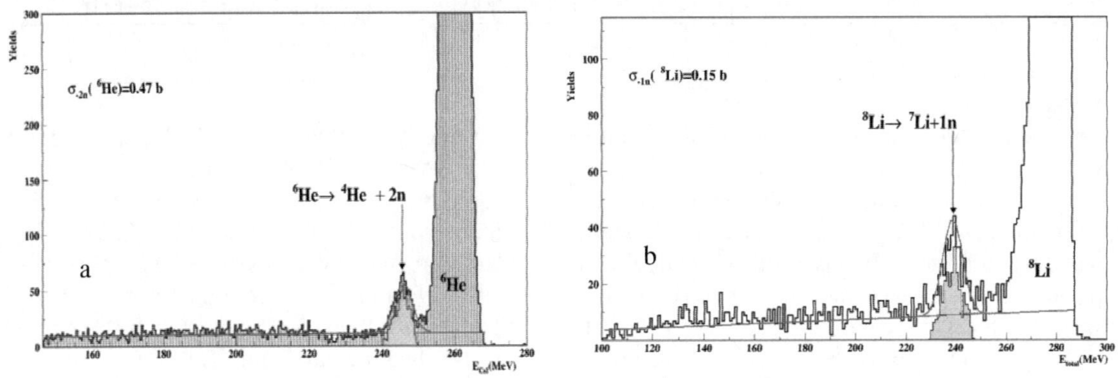

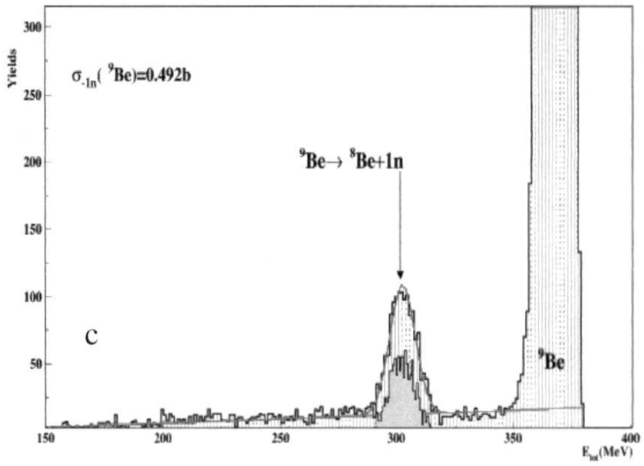

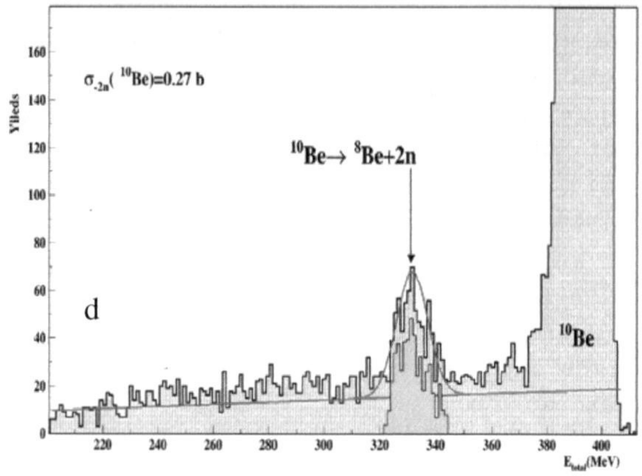

Figure 2. Total kinetic energy spectra for ^{6}He (a), ^{8}Li (b), ^{9}Be (c) and ^{10}Be (d) projectiles in Si telescope and CsI(Tl) detector, with structure due to 2n-removal from ^{6}He, ^{10}Be and 1n removal from ^{8}Li and ^{9}Be indicated.

A similar effect for ^{10}Be (d) is attributed to the weak binding of two last neutrons. As for ^{9}Be, due to its Borromean structure, the breakup of which may occurred directly to two alpha particles and a neutron, or via one of two unstable intermediate nuclei: ^{8}Be and ^{5}He. We determined ^{8}Be nucleus and one neutron removal from the breakup of ^{9}Be (c) in this experiment. That fact is very intrigue because ^{8}Be is unbound nucleus and we are going to clarify this effect in further data analyze. In figure 2 (b) the similar peak in energy spectra for ^{8}Li was identified as corresponding to ^{7}Li. This also indicates that ^{7}Li is the core of ^{8}Li nucleus.

We determined σ_R by transmission method. This procedure is described clearly in Ref[9].Yields in the n-removal peaks for ^{6}He, ^{8}Li, ^{9}Be and ^{10}Be were obtained by fitting and background subtraction. We report measurements of σ_R and σ_{-2n} for ^{6}He and ^{10}Be, and σ_{-1n} for ^{8}Li and ^{9}Be in the table 1 with comparisons of data from [8,10].

Table 1. Measured cross sections on Si for total, 2n and 1n removal from ^{6}He, ^{10}Be, ^{8}Li and ^{9}Be

Type	Energy (MeV/nucleon)	σ_{-xn} (b) [this work]	σ_{-xn} (b)	σ_R (b)	σ_R (b) [this work]
^{6}He, σ_{-2n}	23.7	0.68	0.47±0.04 [8]	1.59±0.06 [8]	1.53±0.15
^{8}Li, σ_{-1n}	29.4	0.15		1.58±0.05 [8]	1.65±0.15
^{9}Be, σ_{-1n}	33.4	0.49	0.085±0.006 [10]	1.63±0.05 [10]	1.63±0.16
^{10}Be, σ_{-2n}	32.6	0.23	0.046±0.004 [10]	1.57±0.05 [8]	1.73±0.16

4. Conclusion

The particle identification for secondary beam was performed by employing the ΔE−E method for a various isotopes ranging from He to Be. The ΔE and E were measured by the Si and CsI(Tl) detector telescope, respectively. Preliminary results on the neutron-removal cross sections of some neutron-rich light isotopes (^{6,8}He, ^{7,8}Li and ^{9,10}Be) were obtained. Data analys is still in progress. The obtained data was compared to previous results [8,10]. We found good agreement with the previous experiment results of other laboratory. The next experiment on total reaction cross section, n-removal cross section and elastic scattering measurements of other light neutron-rich nuclei are planned by using this technique and COMBAS fragment separator. Both 2*n* and 4*n* removal from ^{8}He were observed. Subtraction analysis of the data indicates that ^{4}He is a good core within ^{6}He and ^{8}He, as is ^{8}Be within ^{10}Be and ^{9}Be.

IOP Conf. Series: Journal of Physics: Conf. Series **934** (2017) 012058 doi:10.1088/1742-6596/934/1/012058

References

[1] Otten E 1989 *Treatise on Heavy-Ion Science* **8** pp. 517-638

[2] Saint-Laurent MG, Anne R, Bazin D, Guillemaud-Mueller D, Jahnke U, Jin Gen Ming, Mueller AC, Bruandet J F, Glasser F, Kox S, Liatard E, Tsan Ung Chan, Costa G J, Heitz C, El-Masri Y, Hanappe F, Bimbot R, Arnold E and Neugart R 1989 *Z. Phys.* A **332** pp. 457-465

[3] Bruandet J F 1986 *J. Phys. Colloques* **47** C4-125-C4-139

[4] Villari A C C, Mittig W, Plagnol E, Schutz Y, Lewitowicz M, Bianchi L, Fernandez B, Gastebois J, Gillibert A, Stephan C, Tassan-Got L, Audi G, Wenlong Zhan, Cunsolo A, Foti A, Belezyorov A, Lukyanov S and Penionzhkevich Y 1991 *Phys.Lett* B **268** pp. 345-350

[5] Warner R E, Kelley J H, Zecher P, Becchetti F D, Brown J A, Carpenter C L, Galonsky A, Kruse J, Muthukrishnan A, Nadasen A, Ronningen R M, Schwandt P, Sherrill B M, Wang J, and Winfield J S 1995 *Phys. Rev.* C **52** pp. R1166-1170

[6] Artukh AG, Sereda Yu M, Klygin S A, Kononenko G A, Teterev Yu G, Vorontzov A N, Kaminski G, Erdemchimeg B, Ostashko V V, Pavlenko Yu N, Litovchenko P G, Kovtun V E, Koshchiy E I, Foshchan A G and Kislukha D A 2011 *INSTRUM EXP TECH* **54** pp. 668–681

[7] Artukh A G, Gridnev G F, Grushezki M, Koscielniak F, Semchenkov A G, Semchenkova O V, Sereda Yu M , Shchepunov V A, Szmider J, Teterev Yu G, Bondarenko P G, Rubinskaya L A, Severgin Yu P,Myasnikov Yu A, Rozhdestvenski B Vand Konstantinov A Yu 1999 *Nucl. Instrum. Methods Phys.* A **426** pp. 605-617

[8] Warner R E, Patty R A, Voyles P M, Nadasen A, Becchetti F D, Brown J A, Esbensen H, Galonsky A, Kolata J J, Kruse J, Lee M Y, Ronningen R M, Schwandt P, Schwarzenberg J, Sherrill B M, Subotic K, Wang J and Zecher P 1996 *Phys. Rev.* C **54** pp. 1700-1708

[9] Erdemchimeg B, Artukh A G , Davaa S , Klygin S A, Kononenko G A, Khuukhenkhuu G, Lukyanov S M, Mendibaev K, Mikhailova T I , Sereda Yu M, Penionzhkevich Yu E and Vorontzov A N 2017 *Proceedings of the Int. Symp. on Exotic EXON-2016* pp. 41-45

[10] Warner R E, McKinnon M H, Needleman J S, Shaner N C, Becchetti F D, Roberts D A, Galonsky A, Ronningen R M, Steiner M, Brown J A, Kolata J J, Nadasen A, and Subotic K 2001 *Phys. Rev.* C **64** 044611

IOP Conf. Series: Journal of Physics: Conf. Series **934** (2017) 012059 doi:10.1088/1742-6596/934/1/012059

Resonance phenomena in the φ^8 kinks scattering

E Belendryasova[1] **and V A Gani**[1,2]

[1]National Research Nuclear University MEPhI (Moscow Engineering Physics Institute), 115409 Moscow, Russia

[2]National Research Center Kurchatov Institute, Institute for Theoretical and Experimental Physics, 117218 Moscow, Russia

E-mail: vagani@mephi.ru

Abstract. We study the scattering of the φ^8 kinks with power-law asymptotics. We found two critical values of the initial velocity, $v_{\rm cr}^{(1)}$ and $v_{\rm cr}^{(2)}$, which separate different regimes of the kink-antikink collision. At the initial velocities $v_{\rm in} < v_{\rm cr}^{(1)}$ kinks can not collide due to repulsive force between them. At $v_{\rm in} > v_{\rm cr}^{(2)}$ the kinks escape to spatial infinities after one collision. In the range $v_{\rm cr}^{(1)} \le v_{\rm in} \le v_{\rm cr}^{(2)}$ we observed kinks capture and formation of their bound state. Besides that, at these initial velocities we found resonance phenomena — escape windows.

1. Introduction

Scattering of one-dimensional topological defects (kinks) is very interesting and fast developing area. Many field theoretical models in (1+1)-dimensional space-time have kink-like solutions. The most famous are models with polynomial potentials, such as φ^4, φ^6 [1, 2], φ^8 [3]–[6], and ones with non-polynomial potentials — modified sine-Gordon, double sine-Gordon [7, 8], sinh-deformed φ^4 [9, 10], etc., [11, 12]. Note that models with two or more fields, which have topologically non-trivial solutions, are also actively studied recently [13]–[21].

The (1+1)-dimensional field-theoretical models are very important for numerous applications. For example, models with polynomial potentials are often used in the description of domain walls, phase transitions in condensed matter and in the early Universe, etc. Models with non-polynomial potentials, such as sine-Gordon, modified sine-Gordon, and double sine-Gordon, are used to describe disoriented chiral condensate [22, 23], some processes in ferro-magnets, Josephson contacts, in some nematic liquid crystals, and so on. One of the classical examples is that the dynamics of planar domain walls can be reduced to the kink-(anti)kink interaction [24]. Topologically non-trivial field configurations can be responsible for formation of the primordial black holes in the early Universe [25, 26].

In this short letter we present our first results of numerical simulation of the kink-antikink collisions withing the φ^8 model. Kinks of this model have one power-law and one exponential asymptotics. In our numerical experiments the kink and antikink are turned to each other by power-law tails. This leads to long-range interaction between the kinks, which substantially affect their scattering. Here we address only phenomenology of the kink-antikink scattering. A detailed study of mechanisms, which lead to these phenomena, will be a subject of a separate publication.

2. The φ^8 model

We consider a field-theoretical model with a real scalar field $\varphi(x,t)$ in two-dimensional space-time with the Lagrangian

$$\mathcal{L} = \frac{1}{2}\left(\frac{\partial\varphi}{\partial t}\right)^2 - \frac{1}{2}\left(\frac{\partial\varphi}{\partial x}\right)^2 - V(\varphi), \tag{1}$$

where potential has the form

$$V(\varphi) = \varphi^4(1-\varphi^2)^2. \tag{2}$$

The equation of motion for the field φ is partial differential equation of second order:

$$\frac{\partial^2\varphi}{\partial t^2} - \frac{\partial^2\varphi}{\partial x^2} + \frac{dV}{d\varphi} = 0. \tag{3}$$

The static kinks can be obtained from the first order ordinary differential equation

$$\frac{d\varphi}{dx} = \pm\sqrt{2V(\varphi)} \tag{4}$$

The solution with $\varphi(-\infty) < \varphi(+\infty)$ is called 'kink', while 'antikink' stands for the solution with $\varphi(-\infty) > \varphi(+\infty)$. Nevertheless, in many cases below we call both of them simply 'kinks'.

In a model of φ^8 the integration of equation (4) leads to implicitly defined kinks:

$$\pm 2\sqrt{2}\,x = -\frac{2}{\varphi} + \ln\frac{1+\varphi}{1-\varphi}, \tag{5}$$

see figure 1 The kinks (5) belong to the topological sectors $(-1,0)$ and $(0,1)$, while antikinks —

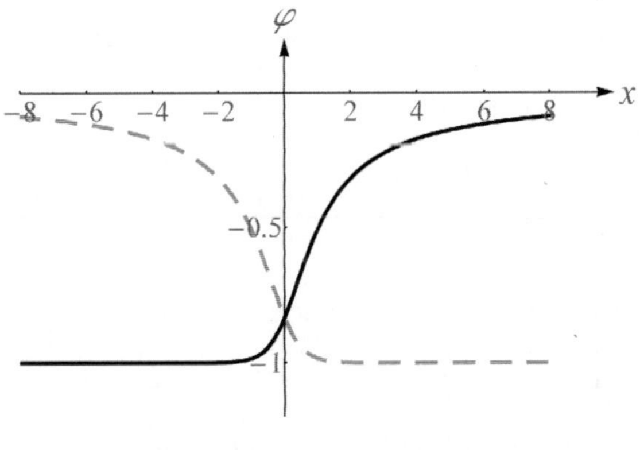

Figure 1. Kink $(-1,0)$ and antikink $(0,-1)$ of the φ^8 model.

to the sectors $(1,0)$ and $(0,-1)$. Asymptotics of all kinks are the following:

$$\varphi_{(-1,0)}(x) \approx -1 + \frac{2}{e^2}\,e^{2\sqrt{2}\,x}, \quad \text{at } x\to-\infty, \tag{6}$$

$$\varphi_{(-1,0)}(x) \approx -\frac{1}{\sqrt{2}\,x}, \quad \text{at } x\to+\infty, \tag{7}$$

$$\varphi_{(0,1)}(x) \approx -\frac{1}{\sqrt{2}\,x}, \quad \text{at } x\to-\infty, \tag{8}$$

$$\varphi_{(0,1)}(x) \approx 1 - \frac{2}{e^2}\,e^{-2\sqrt{2}\,x}, \quad \text{at } x\to+\infty. \tag{9}$$

As one can see from (6)–(9), each kink has one exponential and one power-law asymptotics.

3. Resonance phenomena in the kinks scattering in the sector $(-1, 0, -1)$

We study collisions of kinks with power-law tails, i.e. in our numerical simulation the two colliding kinks are faced to each other by algebraic tails. Let us take the initial condition in the form of the kink $\varphi_{(-1,0)}$ and the antikink $\varphi_{(0,-1)}$, which are initially separated by distance 2ξ and are moving towards each other:

$$\varphi(x) = \varphi_{(-1,0)}\left(\frac{x+\xi}{\sqrt{1-v_{\rm in}^2}}\right) + \varphi_{(0,-1)}\left(\frac{x-\xi}{\sqrt{1-v_{\rm in}^2}}\right), \tag{10}$$

see figure 2. The initial velocity of each kink is $v_{\rm in}$. In our numerical simulation we used the initial kink-antikink distance $2\xi = 30$.

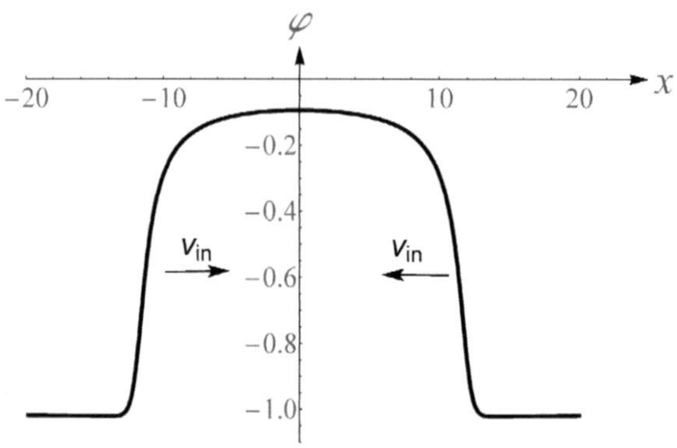

Figure 2. The initial condition of the type of (9).

We solved the second order partial differential equation (3) using the explicit finite-difference scheme,

$$\frac{\partial^2\varphi}{\partial t^2} = \frac{\varphi_j^{k+1} - 2\varphi_j^k + \varphi_j^{k-1}}{\delta t^2}, \quad \frac{\partial^2\varphi}{\partial x^2} = \frac{\varphi_{j+1}^k - 2\varphi_j^k + \varphi_{j-1}^k}{\delta x^2}, \tag{11}$$

on a grid with steps $\delta t = 0.008$, $\delta x = 0.01$.

Depending on the initial velocity, the kink-antikink collision looks very different. We have found two critical values of the initial velocity, $v_{\rm cr}^{(1)} \approx 0.08067$ and $v_{\rm cr}^{(2)} \approx 0.14791$, which separate different regimes of the kinks collisions.

(i) At $v_{\rm in} < v_{\rm cr}^{(1)}$ the kink and the antikink do not collide because they can not overcome mutual repulsion. The kinks approach, stop, and then move away from each other, see figure 3(a).

(ii) At $v_{\rm in} > v_{\rm cr}^{(2)}$ the kinks collide and escape to infinities, this scenario is shown in figure 3(b).

(iii) At $v_{\rm cr}^{(1)} \le v_{\rm in} \le v_{\rm cr}^{(2)}$ we observed
  (a) kinks capture and formation of their bound state (a bion), see figure 4(a);
  (b) two-bounce escape windows, see figure 4(b);
  (c) multi-bounce escape windows — three-and four-bounce escape windows.

The escape windows appear due to resonant energy exchange between translational mode (kinetic energy) and vibrational mode(s) of the colliding kinks. For more detailed discussion of this phenomenon see, e.g., [1, 3].

An attentive reader can ask, how can it be that kinks repel, and at the same time they can form a bound state? At the moment we do not have definite answer. This situation is a challenge for us, and we will study the forces between the φ^8 kinks with power-law tails in the near future.

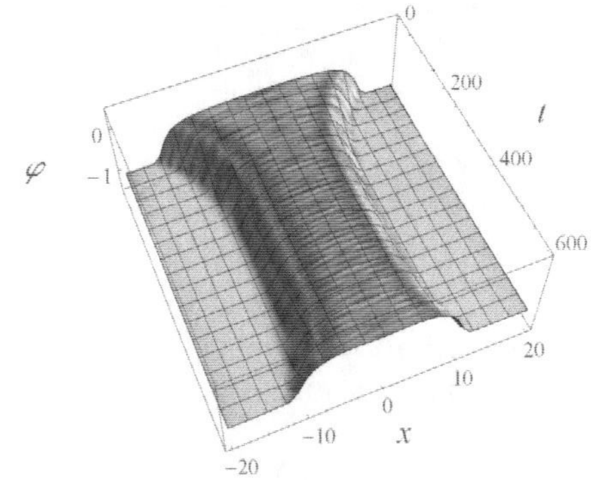

(a) Scattering at velocity $v_{\rm in} = 0.0806 < v_{\rm cr}^{(1)}$

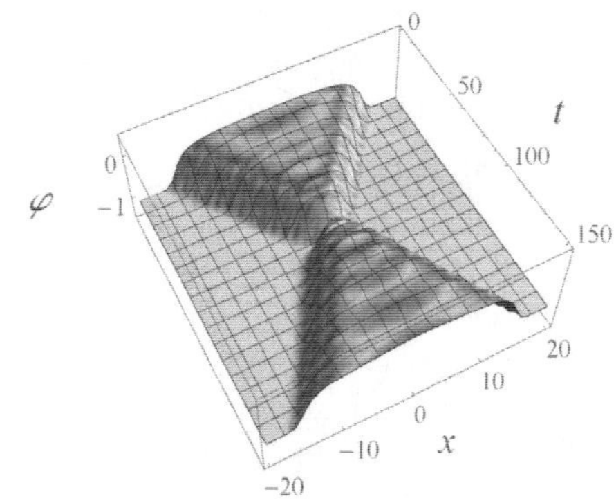

(b) Scattering at velocity $v_{\rm in} = 0.2000 > v_{\rm cr}^{(2)}$

Figure 3. Space-time picture of the kinks scattering at $v_{\rm in} < v_{\rm cr}^{(1)}$ and $v_{\rm in} > v_{\rm cr}^{(2)}$.

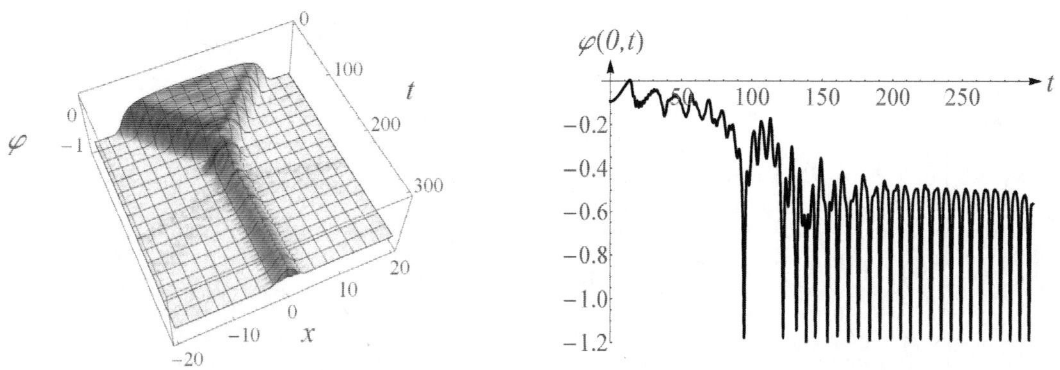

(a) Bion formation, $v_{\rm in} = 0.14510$

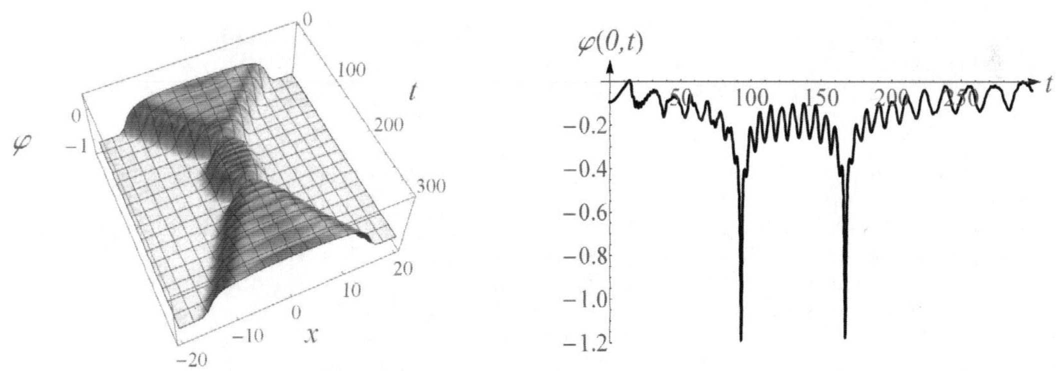

(b) Two-bounce window, $v_{\rm in} = 0.14724$

Figure 4. Space-time picture of the kinks scattering at $v_{\rm cr}^{(1)} \leq v_{\rm in} \leq v_{\rm cr}^{(2)}$.

4. Conclusion

We have investigated numerically the scattering of the φ^8 kinks with power-law asymptotics. Such kinks are highly interactive in the sense that the force between them falls off algebraic with distance [5]. This, in turn, leads to long-range interaction between kinks — they 'feel' each other at much bigger distances than kinks with exponential tails.

The results of our numerical simulation are very interesting. On the one hand, we observed repulsive forces between kinks. On the other hand, we have found resonant structures — escape windows, which are typical to models with attractive kink-antikink interaction. Hence, this our study opens wide prospects for future work.

- First, it is interesting to investigate the force between kink and antikink as a function of kink-antikink distance. This force seems to be repulsive at large distances, and it could be attractive at small distances.
- Second, kinks of the φ^8 model with the potential (2) have no vibrational mode, which could accumulate energy and could be responsible for the appearance of the escape windows. If so, where is the kinetic energy stored between the first and the second kinks collisions within the two-bounce escape window?

These and other issues will be the subject of a future study.

5. Acknowledgements

This work was performed using resources of the NRNU MEPhI high-performance computing center. The research was supported by the MEPhI Academic Excellence Project under contract № 02.a03.21.0005, 27.08.2013.

References

[1] Gani V A, Kudryavtsev A E and Lizunova M A 2014 *Phys. Rev.* **D 89** 125009
[2] Moradi Marjaneh A, Gani V A, Saadatmand D, Dmitriev S V and Javidan K 2017 *JHEP* **07** 028
[3] Gani V A, Lensky V and Lizunova M A 2015 *JHEP* **08** 147
[4] Gani V A, Lensky V, Lizunova M A and Mrozovskaya E V 2016 *J. Phys.: Conf. Ser.* **675** 012019
[5] Radomskiy R V, Mrozovskaya E V, Gani V A and Christov I C 2017 *J. Phys.: Conf. Ser.* **798** 012087
[6] Belendryasova E and Gani V A 2017 *Scattering of the φ^8 kinks with power-law asymptotics*, *Preprint* arXiv:1708.00403
[7] Gani V A and Kudryavtsev A E 1999 *Phys. Rev.* **E 60** 3305
[8] Gani V A et al 2017 *Scattering of the double sine-Gordon kinks*, *Preprint* arXiv:1711.01918
[9] Bazeia D, Belendryasova E and Gani V A 2017 *Scattering of kinks of the sinh-deformed φ^4 model*, *Preprint* arXiv:1710.04993
[10] Bazeia D, Belendryasova E and Gani V A 2017 *Scattering of kinks in a non-polynomial model*, *Preprint* arXiv:1711.07788
[11] Bazeia D, Losano L and Malbouisson J M C 2002 *Phys. Rev.* **D 66** 101701
[12] Bazeia D, González León M A, Losano L and Mateos Guilarte J 2006 *Phys. Rev.* **D 73** (2006) 105008
[13] Gani V A, Ksenzov V G and Kudryavtsev A E 2010 *Phys. Atom. Nucl.* **73** 1889 [*Yad. Fiz.* **73** 1940]
[14] Gani V A, Ksenzov V G and Kudryavtsev A E 2011 *Phys. Atom. Nucl.* **74** 771 [*Yad. Fiz.* **74** 797]
[15] Bazeia D and Mohammadi A 2017 *Eur. Phys. J.* **C 77** 203
[16] Lensky V A, Gani V A and Kudryavtsev A E 2001 *Sov. Phys. JETP* **93** 677 [*Zh. Eksp. Teor. Fiz.* **120** 778]
[17] Bazeia D, Freire W, Losano L and Ribeiro R F 2002 *Mod. Phys. Lett.* **A 17** 1945
[18] Gani V A, Konyukhova N B, Kurochkin S V and Lensky V A 2004 *Comput. Math. Math. Phys.* **44** 1968 [*Zh. Vychisl. Mat. Mat. Fiz.* **44** 2069] [(*Preprint* arXiv:0710.2975)]
[19] Bazeia D, Lobão Jr A S, Losano L and Menezes R 2014 *Eur. Phys. J.* **C 74** 2755
[20] Gani V A, Lizunova M A and Radomskiy R V 2016 *JHEP* **04** 043
[21] Gani V A, Lizunova M A and Radomskiy R V 2016 *J. Phys.: Conf. Ser.* **675** 012020
[22] Gani V A et al. 1999 *Phys. Atom. Nucl.* **62** 895 [*Yad. Fiz.* **62** 956] [(*Preprint* hep-ph/9712526)]
[23] Belova T I, Gani V A and Kudyavtsev A E 2001 *Phys. Atom. Nucl.* **64** 140 [*Yad. Fiz.* **64** 143] [(*Preprint* hep-ph/0003308)]
[24] Gani V A and Kudryavtsev A E 2001 *Phys. Atom. Nucl.* **64** 2043 [*Yad. Fiz.* **64** 2130]
[25] Gani V A, Kirillov A A and Rubin S G 2017 *Classical transitions with the topological number changing in the early Universe*, *Preprint* arXiv:1704.03688
[26] Gani V A, Kirillov A A and Rubin S G 2017 *Transitions between topologically non-trivial configurations,Preprint* arXiv:1711.07700